DYNAMICS of SURFACTANT SELF-ASSEMBLIES

Micelles, Microemulsions, Vesicles, and Lyotropic Phases

SURFACTANT SCIENCE SERIES

ADDITIONAL VOLUMES IN PREPARATION

DYNAMICS of SURFACTANT SELF-ASSEMBLIES

Micelles, Microemulsions, Vesicles, and Lyotropic Phases

edited by
Raoul Zana

CRC Press
Taylor & Francis Group
Boca Raton London New York

CRC Press is an imprint of the
Taylor & Francis Group, an **informa** business
A TAYLOR & FRANCIS BOOK

CRC Press
Taylor & Francis Group
6000 Broken Sound Parkway NW, Suite 300
Boca Raton, FL 33487-2742

First issued in paperback 2019

ISBN-13: 978-0-8247-5822-6 (hbk)
ISBN-13: 978-0-367-39312-0 (pbk)

Library of Congress Cataloging-in-Publication Data

Dynamics of surfactant self-assemblies : micelles, microemulsions, vesicles, and lyotropic phases / edited by Raoul Zana.
    p. cm. — (Surfactant science series ; v. 125)
Includes biographical references and index.
ISBN 0-8247-5822-6 (alk. paper)
  1. Surface active agents. 2. Self-assembly (Chemistry) 3. Micelles. 4. Aggregation (Chemistry) I. Zana, Raoul, 1937- II. Series.

QD75.2D96 2005
541'.33—dc22   2004058263

**Visit the Taylor & Francis Web site at
http://www.taylorandfrancis.com**

**and the CRC Press Web site at
http://www.crcpress.com**

Dedication

To the late E.A.G. Aniansson, for his teachings

and

To Doris, Caroline, and Jennifer, for their patience

Contents

PREFACE

Surfactants belong to the class of compounds referred to as amphiphiles. Conventional surfactants are made up of two moieties having antagonistic properties: a hydrophobic moiety, generally an alkyl or alkyl-aryl chain, and a polar head group that can be ionic, nonionic, or zwitterionic. In aqueous solutions, surfactants self-associate in order to reduce contacts between hydrophobic moieties and water and maximize water–water contacts. This self-association is cooperative and starts at a certain concentration, the so-called *critical micellization concentration* (cmc). The cmc is the most important characteristic of a surfactant. The surfactant self-association gives rise to *micelles*, which are surfactant aggregates where the alkyl chains are in contact with one another, forming an *oily core*, the surface of which is coated by the polar head groups. This structure of micelles largely prevents contact between water and alkyl chains.

Close to the cmc the micelles are spherical and their aggregation number — that is, the number of surfactants per micelle — ranges from, say, 20 to 100 for conventional surfactants. Micelles grow and change shape as the surfactant concentration or the ionic strength is increased.[1,2] Also, micelles are capable of solubilizing in their oily core compounds that are sparingly soluble in water, giving rise to *solubilized systems.*[2] Surfactants can also pack in the form of *vesicles,* which are closed bilayers of surfactants of varied shape, spherical or tubular, for instance.[3] In the presence of cosurfactant (mainly alcohols) and oils, one may obtain one phase, low viscous, thermodynamically stable systems called *microemulsions* that can have different structures.[2] In concentrated aqueous solutions, surfactants give rise to *lyotropic liquid* crystals or *mesophases*, which can have a variety of structures, lamellar, cubic, and hexagonal, for instance.[4] Last, surfactants can adsorb on, and desorb from, interfaces (air-solution or solid-solution). On solid–solution interfaces, adsorbed surfactants can form aggregates of well-defined shape: spherical, cylindrical, disklike, or meshlike.[5]

The above refers to surfactants. However, it also holds for amphiphilic block copolymers. These compounds are structurally related to surfactants, but their hydrophilic and hydrophobic moieties are of polymeric nature (*macromolecular surfactants*). They can form micelles, microemulsions, vesicles, and lyotropic mesophases, just like surfactants.[6] Amphiphilic block copolymers are currently attracting much interest in view of their many potential and actual uses.[6]

Micelles are not frozen objects. They are in equilibrium with free surfactants. In a micellar solution, surfactants are constantly exchanged between micelles and surrounding (intermicellar) solution. This implies processes of entry (or incorporation or association) of surfactants into micelles. Conversely, surfactants can exit (or dissociate) from micelles. The entry/exit processes are usually referred to as *exchange processes.*[7] Owing to these processes a given surfactant resides in a micelle a finite time, which is the *surfactant residence time.* Likewise, one can define the residence time of a solubilizate in a micelle.

The aggregation number of a given micelle fluctuates as a result of the exchange processes. Some of these fluctuations can result in the complete dissociation of micelles into molecularly dispersed surfactants. Conversely, free surfactants can self-associate and form micelles. Since micelles constantly form/break down, they have a *finite lifetime.*[7,8]

The dynamics of micellar equilibria — that is, of surfactant exchange and micelle formation/breakdown processes — have been investigated a great deal. Indeed, in addition to providing a better knowledge of micellar systems, a good understanding of the dynamics of micelles is required for the interpretation of experimental results obtained in other areas of surfactant science. The most important areas are surfactant diffusion in solutions; the interaction between surfactant assemblies; solubilization in, and emulsification, wetting, and foaming by, micellar solutions; rheology of surfactant solutions; surfactant adsorption on surfaces; and the use of micelles and microemulsion droplets as microreactors in which chemical reactions are performed.

In principle, what has just been stated for surfactant micelles also holds for the larger and more complex self-assemblies that surfactants and amphiphilic block copolymers can form: microemulsion droplets, vesicles, and mesophases. The lifetimes of these assemblies are much longer than for micelles, mainly when they involve block copolymers. Nevertheless, exchanges and other processes can also take place. Vesicles and lyotropic mesophases can be considered as permanent objects. However, vesicles can be transformed into micelles, and vice versa. Likewise, a lyotropic mesophase may be transformed into another mesophase or in a micellar solution by an appropriate change brought to the system. The kinetics of these transformations is of basic as well as of practical interest.

The purpose of this book is to present an up-to-date picture of the dynamics aspects of self-assemblies of surfactants and amphiphilic block copolymers, from micelles to solubilized systems, microemulsions, vesicles, and lyotropic mesophases. It is organized as follows. The first chapter introduces amphiphiles, surfactants, and self-assemblies of surfactants and examines the importance of dynamics of self-assemblies in surfactant science. Chapter 2 briefly reviews the main techniques that have been used to study the dynamics of self- assemblies. Chapters 3 and 4 deal with the dynamics of micelles of surfactants and of amphiphilic block copolymers, respectively. The dynamics of microemulsions comes next, in Chapter 5. Chapters 6 and 7 review the dynamics of vesicles and of transitions between mesophases. The last three chapters deal with topics for which the dynamics of self-assemblies is important for the understanding of the observed behaviors. The dynamics of surfactant adsorption on surfaces are considered in Chapter 8. The rheology of viscoelastic surfactant solutions and its relation to micelle dynamics are reviewed in Chapter 9. The last chapter deals with the kinetics of chemical reactions performed in surfactant self-assemblies used as microreactors.

This book summarizes a large number of papers scattered in many scientific journals. It attempts to present a unified view of the dynamics of the systems that are considered. It is hoped that this book will provide readers with the keys to access the dynamic aspects of self-assemblies of surfactants they are investigating or wish to investigate.

Raoul Zana

REFERENCES

1. Lindman, B., Wennerstom, H. *Topics Current Chem.* 1980, 87, 1.
2. Chevalier, Y., Zemb, T. *Rep. Prog. Phys.* 1990, 53, 279.
3. Vinson, P.K., Talmon, Y., Walter, A. *Biophys. J.* 1989, 56, 669.
4. Skoulios, A. *Ann. Phys.* 1978, 3, 421.
5. Blom, A., Duval, F.P., Kovacs, L., Warr, G.G., Almgren, M., Kari, M., Zana, R. *Langmuir,* 2004, 20, 1291.
6. Alaxandridis, P., Lindman, B., Eds. *Amphiphilic Block Copolymers.* Elsevier, Amsterdam, 2000.
7. Aniansson, E.A.G., Wall, S.N., Almgren, M., Hoffmann, H., Kielmann, I., Ulbricht, W., Zana, R., Lang, J., Tondre, C. *J. Phys. Chem.* 1976, 80, 905.
8. Aniansson, E.A.G. *Prog. Colloid Interface Sci.* 1985, 70, 2.

List of Contributors

Rob Atkin
School of Chemistry
University of Bristol
Bristol, U.K.
rob.atkin@bristol.ac.uk

Colin Bain
Physical and Theoretical
  Chemistry Laboratories
University of Oxford
Oxford, U.K.
colin.bain@chem.ox.ac.uk

Julian Eastoe
School of Chemistry
University of Bristol
Bristol, U.K.
julian.eastoe@bristol.ac.uk

Heinz Rehage
University of Dortmund
Dortmund, Germany
heinz.rehage@uni-dortmund.de

Brian H. Robinson
School of Chemical Sciences
  and Pharmacy
University of East Anglia
Norwich, U.K.
thematurbo@aol.com
  and
Dipartiments di Chimica
University di Pisa
Pisa, Italy

Madeleine Rogerson
School of Chemical Sciences and
  Pharmacy
University of East Anglia
Norwich, U.K.
madeleinerogerson@hotmail.
  com

Christian Tondre
Groupe de Chimie Physique
  Organique et Colloïdale
UMR-CNRS 7565
Université Henri Poincaré
Nancy, France
tondre@lesoc.uhp-nancy.fr

Erika Wanless
School of Environmental and
 Life Sciences
University of Newcastle
Callaghan, Australia
ewanless@mail.newcastle.
 edu.au

Raoul Zana
Institut C. Sadron (CNRS)
Strasbourg, France
zana@ics.u-strasbg.fr

1

Introduction to Surfactants and Surfactant Self-Assemblies

RAOUL ZANA

CONTENTS

I. SURFACTANTS AND SELF-ASSEMBLIES OF SURFACTANTS

The word *amphiphile* first introduced by Hartley[1] refers to a large class of compounds whose molecules contains moieties that have affinity for water (or are waterlike) and moieties that have affinity for oil (or are oil-like). Many biological compounds, most notably *phospholipids*, which make up living cell walls, as well as many *drugs* are amphiphilic. So are the so-called *surfactants*, compounds whose molecules are made up of one hydrophilic moiety, referred to as the *head group*, which is covalently bonded to one hydrophobic moiety, generally a single or double alkyl chain also called the *tail*. The head group may be nonionic (a short poly(ethylene oxide) segment or a sugar), ionic (anionic or cationic), or zwitterionic. The alkyl chain may be a hydrocarbon, a perfluorocarbon, or a mixed hydrocarbon/perfluorocarbon group. It may also be partly aromatic.[2,3] Silicon may also partly replace carbon in the hydrophobic moiety (silicon surfactants).[4] Recently new amphiphiles have started to attract much attention. These compounds are block copolymers made up of hydrophilic and hydrophobic blocks.[5–8] The hydrophilic block may be nonionic (poly(ethylene oxide)) or ionic (sodium poly(styrenesulfonate) or poly(acrylate)). The hydrophobic block may be poly(propylene oxide),[5] poly(butylene oxide),[6] or poly(styrene),[7,8] to cite but a few. The blocks being of polymeric nature, these copolymers can be considered to be *macromolecular* surfactants. *Oligomeric* (dimeric (gemini), trimeric, etc.) and *polymeric* surfactants are also attracting much interest. The repeating unit in these compounds is amphiphilic.[9] In this chapter surfactant is used to refer to conventional surfactants as well as to amphiphilic block copolymers.

The dual character of surfactants is responsible for their peculiar behavior in the presence of water. Consider a system made up of water and of a small amount of surfactant solubilized in the water. Alkyl chain/water contacts are energetically unfavorable with respect to water/water contacts.[10–12] In order to avoid such contacts, some surfactants tend to locate at the air/water interface, with the head group in water and the tail in the air side of the interface, thereby forming an adsorbed layer of surfactant. The *adsorption* of surfactants at the air/solution interface reduces the surface tension of water, thus the generic name surfactants. Similarly, surfactants adsorb on the solid walls of the flask containing the solution. The structure of an adsorbed layer of surfactant on a solid surface depends on the nature of the surface and of the surfactant (see Section II).[13,14]

As the surfactant concentration in the solution is increased, the amount of surfactant adsorbed at the air/solution interface (and on the walls of the flask containing the solution) increases up to the point where the interface becomes saturated by absorbed surfactant. From this point, the concentration of the solution may still be further increased and the surfactant dissolved in the aqueous phase may still be in a molecularly dispersed form. However, the free energy of the system increases with the concentration, due to the increasing number of unfavorable alkyl chain/water contacts. At the *critical micellization concentration* (cmc), the surfactant starts to self-associate into *micelles* in order to prevent a further increase of free energy.[1-3,15] This self-association is driven by the *hydrophobic interaction* that arises from the tendency of water molecules to reduce contacts with surfactant alkyl chains. In micelles the alkyl chains are in contact with one another, forming an oily *core* that is coated by the polar head groups (see Figures 1.1A and 1.1B).[1,16] The formation of micelles is a *cooperative* process that is *spontaneous* and *reversible*.[1,10,15] Micelles are *thermodynamically stable* species that are in chemical equilibrium with free surfactants

Compounds that are insoluble or only sparingly soluble in water (oils, aromatic molecules and drugs) can be solubilized into the oily core of micelles, yielding *solubilized systems*.[2,3]

The solubilizing capacity of micellar solutions (expressed as the ratio of the concentrations of solubilized molecules and of micellized surfactant) is usually not large. This capacity can be much enhanced by the addition to the micellar solution of a *cosurfactant* that is in most instances a medium-chain length alcohol (propanol to hexanol). Recall that milky or turbid *emulsions* often occur when adding to a micellar solution more oil than it can solubilize, and then stirring the

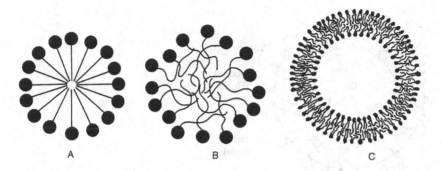

A   B   C

Figure 1.1 Schematic representations of a spherical micelle, (A) and (B) (diameter: 5 nm), and of a spherical vesicle, (C) (diameter 20–1000 nm, thickness, 4–5 nm). (A) is the misnamed Hartley micelle as represented in most chemistry and biochemistry books and papers, (B) is a more realistic representation with disordered alkyl chains. Reproduced from reference 16 with permission of the *Journal of Chemical Education.*

system. Addition of a cosurfactant to such emulsions often results in transparent or translucent systems that are referred to as *microemulsions*.[2,17] Microemulsions are monophasic and thermodynamically stable. Depending on the nature of the components of the system and on its composition, one can generate *oil-in-water* or *water-in-oil* microemulsions. The classical representation of a microemulsion is that of water (or oil) droplets coated by a mixed film of surfactant + cosurfactant suspended in a continuum of oil (or water)[2,18] (see Figure 1.2). Other microemulsion structures have been evidenced (see Section IV).

Some surfactants, as for instance the sodium diethylhexylsulfosuccinate (better known as AOT), are soluble in oil.[2] These organic solutions may contain small surfactant aggregates. They are able to solubilize water, giving rise to *reverse micelles* that have an aqueous core. There is no clear or obvious difference between reverse micelles and water-in-oil microemulsions. As pointed out by Friberg[19,20] there is continuity in phase diagrams between (direct or reverse) micellar solutions, solubilized systems, and microemulsions.

Other surfactants self-assemble into closed bilayers called *vesicles* (or *liposomes* when formed from phospholipids).

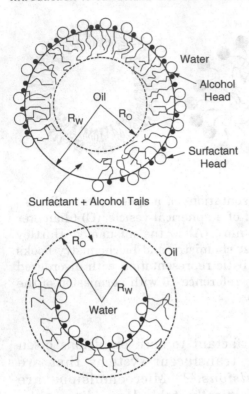

Figure 1.2 Schematic representations of an oil droplet in an oil-in-water microemulsion (*top*) and of a water droplet in a water-in-oil microemulsion (*bottom*). Reproduced from reference 18 with permission of the American Chemical Society.

Vesicles are often spherical (see Figure 1.1C) but can take other shapes and can be unilamellar or multilamellar. In contrast to micelles, vesicles may not be thermodynamically stable.[21] Another important difference between vesicles and micelles is that vesicles have an *inside* (that encloses some of the aqueous phase) and an *outside*. The existence of a critical vesiculation concentration, above which some surfactants would form vesicles, is sometimes mentioned. This is probably incorrect. At very low concentration, sufficiently close to the cmc, surfactants start forming micelles that may turn into vesicles upon increasing concentration. This process takes place at extremely low concentration for surfactants with two long alkyl chains, and micelles may be present in such a narrow range of concentration as to go undetected by the presently available techniques.

As the surfactant concentration is further increased, intermicellar forces come into play and give rise to *liquid crystalline phases*, also called *lyotropic phases* or *mesophases*. The structure of mesophases can be extremely varied (see below).[22,23]

The main qualitative features of the self-assemblies formed by surfactants are considered successively in some detail in the following paragraphs.

II. ADSORBED SURFACTANT LAYERS ON SOLID SURFACES

When a surfactant solution is put in contact with a clean (surfactant-free) solid surface, the surfactant adsorbs onto the surface, thereby forming an adsorbed layer. The structure of the layer depends on the nature of the surfactant and of the surface.

If the surface is hydrophobic, such as polyethylene or polypropylene, for instance, the surfactant contacts the surface through its tail, forming a progressively more compact monolayer on the surface as the amount of adsorbed surfactant increases. In such monolayers the surfactant head groups are oriented toward water. The monolayer is constituted by isolated surfactants at very low surface coverage. It then turns into discrete aggregates, hemi-micelles or admicelles, at higher surface coverage[24] and eventually into a compact monolayer (saturated surface). A different structure has been observed for the adsorbed layer of cationic surfactants on the hydrophobic cleavage plane of graphite, when the surface is saturated. Atomic force microscopy[14] showed that the surfactant is adsorbed under the form of hemi-cylindrical aggregates, with the cylinder axis parallel to a graphite axis of symmetry, irrespective of the nature of the surfactant (see Figure 1.3 bottom left for the dimeric (gemini) surfactant 12-2-12 adsorbed on graphite). The structure of the adsorbed layer is then completely controlled by the nature of the solid surface.

In the case of hydrophilic charged surfaces (mica or silica, for instance), the first adsorption step still involves individual cationic surfactants that bind to oppositely charged sites on

Figure 1.3 Atomic force microscopy images of the dimeric (gemini) surfactants 12-2-12 (*top right*) and 12-4-12 (*top left*) on the cleavage plane of mica, and of 12-2-12 on silica (*bottom right*) and graphite (*bottom left*). Reproduced from references 14 and 25 with permission of the American Chemical Society.

the surface (ion exchange step). The second step is cooperative and corresponds to the formation of surface aggregates. The surface becomes saturated at a bulk surfactant concentration slightly below the cmc. At saturation, the nature of the surface aggregates depends heavily on the nature of the surfactant and of the surface. Thus the adsorbed surfactant layer on the cleavage plane of mica may be constituted by a bilayer or by spherical or cylindrical aggregates, depending on the nature of the surfactant.[13,14,25] This is illustrated for the dimeric (gemini) surfactants 12-2-12 and 12-4-12 in Figure 1.3 (top right and left). One can talk of surface micelles. In general, for cationic surfactants, the adsorbed layer is less organized on silica than on mica; this is probably related to the lower charge density of the former. Thus the dimeric surfactant 12-2-12 adsorbs as a bilayer on mica[14] (Figure 1.3, top right) and as flattened ellipsoidal aggregates on silica[25] (Figure 1.3, bottom right).

The adsorption of surfactants on surfaces is of the utmost importance for a variety of industrial processes involving surfactants. The kinetics of adsorption of cationic surfactants is reviewed in Chapter 8.

III. MICELLES AND SOLUBILIZED SYSTEMS

A. Theoretical Aspects

The driving force for surfactant self-assembly into micelles is the hydrophobic interaction between surfactant alkyl chains.[1,10-12] This interaction has its origin in the strong attractive interaction that exists between water molecules. Even though alkyl chain-water molecule interactions are attractive, they are energetically less favorable than interactions between water molecules.[10-12] As a result, when alkyl chains are immersed into water, the system tends to minimize its free energy by eliminating alkyl chain-water molecule contacts. With surfactants, this results in the formation of micelles where the alkyl chains are in contact with one another, forming the micelle core. The head groups remain at the surface of the core, further reducing alkyl chain-water molecule contacts (see Figure 1.1B).

Several repulsive interactions oppose the formation of micelles.[10,26-28] The main ones are the electrostatic interaction between head groups, the steric interaction arising from the packing of head groups at the micelle surface and of alkyl chains in the micelle core, and an interaction associated with residual alkyl chain-water molecule contacts at the micelle surface. The balance between attractive and repulsive interactions results in micelles of *finite size*.

The molecular thermodynamic theory for micelle formation has been worked out with increasing sophistication following the pioneering work of Israelachvili, Mitchell, and Ninham.[26] The most comprehensive reports on micelle formation are those of Nagarajan and Ruckenstein[27] and of Shiloah and Blankschtein.[28] Many other theoretical approaches have been used in recent years to account for the formation of micelles and their properties: thermodynamics of small systems, the self-consistent field lattice model, the scaled particle theory, and Monte-Carlo and molecular dynamics MD simulations.[29] MC and DC simulations are presently much in favor due to the increased availability of fast computers. A prediction common to all these theories is that micelles represent a thermodynamically stable state and that micellar solutions are *single-phase* systems. Several recent results of MD and (MC) simulations are in agreement with experimental results.[29]

At a concentration very close to the cmc the micelles are spherical or nearly spherical on a time-average basis.[26] The head groups together with a small part of the alkyl chains, a significant fraction of the counterions, and water molecules make up a spherical shell often referred to as *palisade layer*. Figure 1.1B gives a realistic representation of a micelle with disordered alkyl chains (consistent with a quasi-liquid state of the core). The representation of a micelle as in Figure 1.A is misleading because it suggests a very ordered core. This type of representation should be totally banned from publications and textbooks.[16]

As the concentration is increased, the micelles may remain spheroidal or grow and become oblate (disklike) or prolate (or elongated, cylindrical, or rodlike), with the prolate shape much more often encountered than the oblate shape. The micelle shape is determined by the value of the *surfactant packing parameter P* given by[26]

$$P = v/a_0 l \qquad (1.1)$$

where v and l are the volume and length of the hydrophobic moiety (alkyl chain) and a_0 is the optimal surface area occupied by one surfactant at the micelle-water interface.[26] It is important to realize that the value of a_0 is determined by the cross-sectional area of the surfactant head group *and* by the various interactions at play in micelle formation.[26] Surfactants characterized by values of $P < 1/3$ give rise to spherical/spheroidal micelles. Surfactants with $1/3 < P < 1/2$ tend to form elongated micelles that can be considered as the precursor of the hexagonal phase (see below). Surfactants with $1/2 < P < 1$ form disklike micelles that can be viewed as precursors of surfactant bilayers and of lamellar phases (see Section VI). Last, surfactants with $P > 1$ form reverse micelles. Figure 1.4, adapted from the monograph of J. Israelachvili,[30] represents schematically the shapes of the surfactant molecule and of the surfactant self-assembly for different values of the packing parameter. The packing parameter concept is extremely powerful and useful. In particular, it permits one to understand or predict changes of micelle shape induced by changes of experimental conditions. It can also be extended to surfactant mixtures. Besides, the packing parameter of a surfactant is related to the spontaneous curvature of the monolayer that the surfactant can form.[31]

The micelle size increases with the carbon number m of the surfactant alkyl chain. For conventional surfactants, the aggregation number of the *maximum* spherical micelle formed by a surfactant with a hydrophobic moiety assumed to be an alkyl chain C_mH_{2m+1} can be estimated from

$$N_S = 4\pi l^3/3v \tag{1.2}$$

The length l and volume v of the alkyl chain are given by[10]

$$l \text{ (nm)} = 0.15 + 0.1265m \tag{1.3}$$

$$v \text{ (nm}^3) = 0.0274 + 0.0269m \tag{1.4}$$

The micelle growth with the surfactant concentration is well explained by different models, which all assume that the free energy of a surfactant is higher in a spherical micelle than in a rodlike or disklike micelle.[26–28,32] The free energy difference ΔG_{SC} is rather small, $-0.2kT$ to $-0.5kT$ per surfactant, as compared to a free energy of micellization of about $-15kT$ per surfactant having a dodecyl chain. The larger the

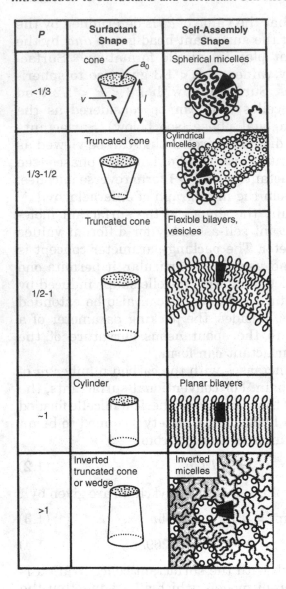

P	Surfactant Shape	Self-Assembly Shape
<1/3	cone	Spherical micelles
1/3-1/2	Truncated cone	Cylindrical micelles
1/2-1	Truncated cone	Flexible bilayers, vesicles
~1	Cylinder	Planar bilayers
>1	Inverted truncated cone or wedge	Inverted micelles

Figure 1.4 Schematic representation of the shapes of the surfactant and of the surfactant self-assemblies for various values of the packing parameter. Adapted from reference 30 with permission of Elsevier Science.

magnitude of ΔG_{SC}, the steeper the increase of micelle size with increasing surfactant concentration.[26-28] Surfactants with $P > 1/3$ show micelle growth. Ionic surfactants with a value of P slightly below 1/3 may still show micelle growth. Indeed as the surfactant concentration increases, so does the ionic strength of the solution. This in turn increases the value of P to a value that may be higher than 1/3. The polydispersity in size of spherical micelles is small, while that of prolate and oblate micelles can be very large.[26-28]

Some models of micelle growth take into account the fact that the surfactants located in the part connecting the cylindrical body of rodlike micelles to the endcaps are at a still higher chemical potential than those in the endcaps.[33,34] This leads one to predict the existence of a second critical concentration, sometimes referred to as second cmc, above which micelles start growing. It also leads one to predict the coexistence of spherical and rodlike micelles, i.e., a bimodal distribution of micelle sizes. This is in contrast to the two-chemical potential approach that predicts a continuous growth of all micelles and a unimodal size distribution curve.

B. Experimental Aspects

Micelles start forming in aqueous solutions of surfactant at concentration $C >$ cmc. The cmc is not a single concentration but rather a narrow range of concentration.[3,15] Its value slightly depends on the method of determination, reflecting differences in the way each measured property weighs micelles and free surfactants (see Figure 1.5). The value of the cmc is determined mainly by the nature of the surfactant (ionic or nonionic) and by the length of its alkyl chain. The cmc decreases exponentially upon increasing number of carbon atoms m in the surfactant alkyl chain according to[3,15]

$$\log \text{cmc} = b - am \qquad (1.5)$$

The value of a corresponds to a decrease of the cmc value by factors 2 and 3 per additional methylene group in the alkyl chain for ionic and nonionic surfactants, respectively.[3,15] In both instances, this corresponds to a free energy of transfer from the micelles to the aqueous phase of $1.1kT$ per methylene

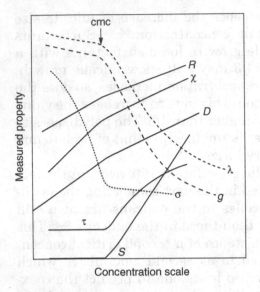

Figure 1.5 Dependence of various properties of a solution of an ionic surfactant with the surfactant concentration C. Variations of the refractive index R, density D, specific conductance χ, turbidity τ and solubility S of a water-insoluble dye (Orange OT) with C (plots in continuous line), of the osmotic coefficient g and of the equivalent conductance λ with $C^{1/2}$ (plots in broken line), and of the surface tension σ with log C (plot in dotted line). Reproduced from reference 15 with permission of Elsevier Publishing Co.

group. For ionic surfactants, the cmc also decreases upon increasing concentration C_e of an added electrolyte having a common ion with the surfactant, according to[3,15]

$$\log \text{cmc} = A - B\log(\text{cmc}_0 + C_e) \qquad (1.6)$$

A and B are two constants and cmc_0 is the cmc in the absence of salt. The value of B is close to the *degree of counterion association* to the micelle.[3,15] Thermodynamic studies showed that the formation of micelles is *entropy-driven*.[1,2,10,15] The large positive value of the entropy of micellization reflects a change affecting water molecules surrounding surfactant alkyl chains when the chains are transferred from the aqueous phase to the micelle core.[10,15]

The micelle oily core is in a *quasi-liquid* state. This is demonstrated by several results:

1. The motions of the methyl and methylene groups within the core are very rapid, and the order parameter of the successive carbon atoms along an alkyl chain decreases progressively from a value around 0.3 to a value close to 0 in going from the micelle surface to its center. This behavior reveals a progressive increase of disorder in alkyl chain packing in the core.[35]

2. The values found for the viscosity of the micelle interior are low.[36]

3. Extensive chain looping occurs in the core, thereby permitting tight packing of the chains.[37] Any methylene group or terminal methyl group in an alkyl chain has a finite probability to be found at a given distance from the micelle center. In fact, a few terminal methyl groups are located close to or at the micelle surface.[37]

Spherical and nonspherical micelles have been directly visualized by means of cryo-TEM (transmission electron microscopy at cryogenic temperature).[38,39] The giant *wormlike* or *threadlike* micelles represent an extreme case of growth into elongated micelles. Figure 1.6 shows cryo-TEM images of *linear* and *branched* wormlike micelles in solutions of the dimeric surfactant 12-2-12.[39] The endcaps of threadlike micelles are slightly larger than hemispheres as they have a larger diameter than the cylindrical body,[39] in agreement with theoretical predictions.[33,34] The same is probably true for the edge of disklike micelles. Figure 1.6 top also shows the coexistence of spherical and elongated micelles that has been theoretically predicted.[33,34] *Ringlike* (or toroidal, or closed loop) micelles have also been visualized (see Figure 1.6).[39,40]

Theoretical[32] and experimental[37] studies have revealed the existence of fluctuations of micelle shape that are rapid (characteristic time of 10^{-10} to 10^{-6} s) and large. Substantial shape deformations can take place at a rather low cost of free energy. For instance, a spherical micelle of diameter 4 nm can

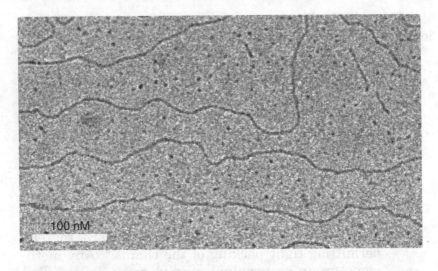

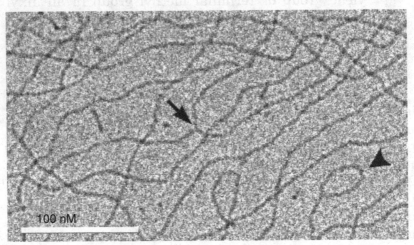

Figure 1.6 Electron micrographs of 0.74 wt% (*top*) and 1 wt% (*bottom*) solutions of the dimeric (gemini) surfactant dimethylene-1,2-bis(dodecyldimethylamonium bromide) showing giant thread-like micelles (*top*), elongated micelles with a branching point (bottom, arrow) and a ringlike micelle (bottom, arrowhead). Reproduced from reference 39 with permission of the American Chemical Society.

be distorted into a prolate micelle 4.8 nm long and 3.6 nm wide at a cost of only about 1 kT.[32]

Besides, micelles are dynamic objects. They constantly exchange surfactant with the bulk phase (*exchange process*), and they form and break down by different pathways (*micelle formation/breakdown*), which are reviewed in Chapter 3.

C.   Solubilized Systems

Most of what has been said earlier about micelles remains qualitatively valid for solubilized systems where the micelles contain solubilized molecules (solubilizates). Nevertheless, the solubilizates may affect the micelle size and/or shape. The most extensive reports concern the effect of alcohols.[41] Thus short-chain alcohols (methanol to propanol) generally decrease the micelle size and progressively transform anisotropic micelles into spherical ones. A sufficiently high concentration of short-chain alcohols can result in the complete disruption of micelles into molecularly dispersed surfactant. Medium-chain length alcohols such as butanol, pentanol, and to a lesser extent hexanol bring about a decrease of micelle size at low alcohol content and an increase of micelle size at higher concentrations. Micellar solubilization of long-chain alcohols (heptanol and longer alcohols) brings about an increase of micelle size and a change of micelle shape.[41]

The location of a solubilizate in micelles and its effect on micelle size depend much on the nature of the solubilizate.[2,3,42] Thus alkanes are solubilized in the micelle core. Aromatic molecules tend to be solubilized first at the micelle surface and then into the core once the surface is saturated. Cosurfactants (alcohols, amines, carboxylic acids, for instance) are solubilized with their head group at the micelle surface and their tail penetrating more or less deeply into the core, depending on its length.

The dynamics of micelles can be considerably affected by the presence of solubilizates (see Chapter 3, Section V).

D.   Water-Soluble Polymer/Surfactant Systems

The micellization of surfactants in aqueous solution is much affected by the presence of a water-soluble polymer, whenever interactions occur between the surfactant and the polymer. In most instances the interaction results in a lowering of the cmc and of the aggregation number of the surfactant, whenever the surfactant concentration is low enough and the micelles formed are bound to the polymer. A widely accepted model of these systems is that of a necklace (the polymer chain) bearing beads (surfactant micelles).[43] Locally, the polymer can be wrapped around the surfactant micelles. The binding of the polymer to the micelles can affect the micelle dynamics (see Chapter 3, Section VI).

IV. MICROEMULSIONS

Microemulsions were first reported by Hoar and Schulman[44] in 1943. These authors were studying emulsions of benzene in micellar solutions of surfactant. In an attempt to increase the solubility of benzene in the system, they added increasing amounts of alcohol and observed that the system clarified at above a certain content of alcohol. The word microemulsion was used to refer to the clarified and transparent or translucent system. Indeed it was believed that the addition of alcohol had brought about a reduction of the size of the benzene droplets in the emulsion to a point that they did not scatter visible light. However, the term "microemulsion" is not well suited to describe these systems. Indeed, it includes the word "emulsion," which corresponds to a thermodynamically unstable system made up of droplets of one liquid dispersed in another liquid. It was later shown that microemulsions are thermodynamically stable systems that can be described as dispersions of one liquid in another liquid but where the droplet structure is not always present. In fact, the most interesting microemulsions have a bicontinuous structure. To circumvent this difficulty, Danielsson and Lindman[45] defined a microemulsions as "a system of oil, water and amphiphile which is a single optically isotropic and thermodynamically stable solution." Friberg[46] suggested replacing "thermodynamically stable" with "formed spontaneously." These two definitions say nothing about structure.

Microemulsions started to attract considerable interest around 1975 because of the oil crisis that occurred during that period. Indeed, one of the main potential uses of microemulsions was for enhanced oil recovery (EOR) from oil fields. Microemulsions of appropriate composition can have an extremely high solubilizing capacity for oil, and the values of the interfacial tension between microemulsion phases and water and/or oil can be extremely low. These two properties are essential in EOR. Microemulsions have not yet been used on a large scale for EOR. Nevertheless, there is still much interest in microemulsions as more and more applications have been found for these systems.

Most microemulsions contain a cosurfactant that has a multiple-faceted role:

1. Increase the amounts of oil and water that can be mixed in the presence of a given amount of surfactant and still give rise to one-phase systems.
2. Shift to higher surfactant content the range of occurrence of lyotropic mesophases.
3. Lower the interfacial tension between the microemulsion phase and excess oil or water to values as low as 10^{-4} mN/m.

Water-in-oil (W/O) microemulsions at low water content and oil-in-water (O/W) microemulsions at low oil content contain droplets of water in an oil continuum or droplets of oil in a water continuum that have been visualized (see Figure 1.7 top).[47] The droplets are coated by a mixed film of surfactant + cosurfactant as represented in Figure 1.2. Such systems are referred to as Winsor IV systems. As the water or oil content is increased, the solubility limit of the water or oil is reached and the system becomes biphasic. It is then made up of a microemulsion phase and, as a second phase, the excess water or oil containing very small amounts of the other components of the system. The phase diagrams of water/surfactant/cosurfactant/oil systems have been much investigated.[48,49] The biphasic systems with an O/W microemulsion in equilibrium with excess oil are the so-called Winsor I systems, whereas Winsor II systems are made up of a W/O microemulsion phase in equilibrium with excess water. In both Winsor I and Winsor II systems, the microemulsion phase has a droplet structure, as in Winsor IV systems. Winsor I and II systems can be

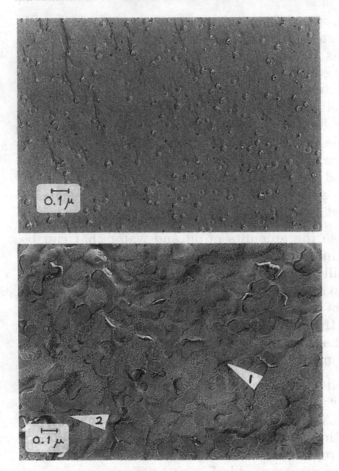

Figure 1.7 Freeze-fracture electron micrographs showing the structure of the microemulsion phase in the system water/n-octane/$C_{12}E_5$. *Top*: microemulsion with a droplet structure that shows the random distribution of the droplets and their small polydispersity. *Bottom*: bicontinuous microemulsion that displays the saddle-shaped structures of the film separating oil domains (grained aspect) from water domains (smooth aspect). Reproduced from reference 47 with permission of Springer Verlag.

modified by the addition of surfactant or cosurfactant and changing the salinity, temperature, or pressure. Such modifications often result in a change of phase behavior. For instance a Winsor I system can give rise successively to a three-phase system (Winsor III system) and then to a Winsor II system, and conversely (Winsor II → Winsor III → Winsor I) depending on the initial system and the type of modification brought to it. The three-phase systems include a middle-phase microemulsion, excess water as lower phase, and excess oil as upper phase, when the density of oil is lower than the density of water.

The three-phase systems are the most interesting ones for applications. Indeed, the interfacial tension between the middle phase of microemulsion and the upper or lower phase is then extremely low.[50] Besides, the middle-phase microemulsion corresponds to maximum amounts of oil and water that can be mixed and still form a one-phase system. The middle-phase microemulsion has a bicontinuous structure[47,51] where the oil and water domains are intimately mixed and extend over microscopic distances (see Figure 1.7 bottom). As in droplet systems, a mixed film of surfactant and cosurfactant separates water and oil domains. This type of structure undergoes large and spontaneous fluctuations. The approach of the compositions associated with the Winsor I → Winsor III and Winsor II → Winsor III phase transitions is critical, with a divergence of many properties of the system such as the intensity of scattered light or the interfacial tension.[50] Also the interactions between droplets in the microemulsion phase of Winsor I or Winsor II systems become increasingly attractive at the approach of the transition.[50]

Reported theories explain the stability of microemulsions. The driving force for the formation of stable microemulsions with a droplet structure is the entropy of dispersion of one liquid (oil, for instance) in the other (water).[52] The presence of the cosurfactant is required for lowering the surface tension to a value low enough (0.01 mN/m) for the stabilization of the system. At such a low value of interfacial tension, the energy of the curvature of the surfactant layer must be taken into account in the calculation of the free energy of the system. In the case of bicontinuous microemulsions, theory shows that a stable middle phase occurs when the average spontaneous curvature of the mixed surfactant + cosurfactant monolayer is close to zero and its bending modulus is close to $1kT$.[53] A unified treatment was recently reported.[18]

Various types of spontaneous processes take place in microemulsions. The surfactant and cosurfactant exchange between the interfacial film separating water and oil domains and the bulk phases. Also collisions between droplets with temporary merging of the collided droplets ("sticky" collisions) have been evidenced. The kinetics of these processes (and of other ones) is reviewed in Chapter 5, Sections VI.F and VIII, and Chapter 10.

V. VESICLES

Aqueous vesicles are commonly observed with phospholipids and with their synthetic analogs, i.e., two-chain surfactants having alkyl chains with 10 or more carbon atoms. Different methods are used for preparing vesicles.[54] For instance, turbid dispersions of appropriate sparingly water-soluble surfactants or phospholipids give rise upon sonication to clear and slightly bluish systems that contain vesicles. Vesicles may also be obtained by solubilizing an appropriate amphiphile in an organic solvent, depositing the solution on a glass plate, evaporating the organic solvent, and exposing the resulting film to water. Vesicle-forming surfactants can also be solubilized in an aqueous solution of a hydrotrope (a water-soluble compound that does not form micelles on its own but that is able to promote the solubilization of water-insoluble compounds in water) or of a micelle-forming surfactant. Dilution of this solution with water gives rise to vesicles. Sonication, extrusion through Millipore filters, and other methods have been used to transform large multilamellar vesicles into small unilamellar vesicles.[54] An image of doubly lamellar vesicles obtained with the dimeric (gemini) surfactant 12-20-20 and visualized by cryo-TEM is shown in Figure 1.8.

Vesicles prepared by the above methods can have long-term stability. Nevertheless, vesicles made up of a single surfactant or phospholipid are unstable and generally revert to liquid crystalline aggregates (lamellar phases, in most instances) after a time that can be quite long. Unilamellar vesicles obtained by mixing aqueous solutions of two surfactants of appropriate structure and of different electrical

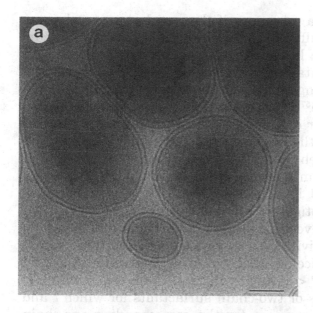

Figure 1.8 Cryo-transmission electron micrograph of a vitrified 1.4 wt% solution of the dimeric (gemini) surfactant 12-20-12 showing doubly lamellar vesicles (bar = 0.1 μm). Reproduced from reference 63 with permission of Academic Press/Elsevier.

charge may be different. Such vesicles appear to form *spontaneously* and reproducibly and can have long-term stability.[55] It has been claimed that vesicles so prepared can be thermodynamically stable. However, a consensus on this claim remains to be reached for the following reason. A surfactant monolayer takes up a curvature equal or close to its spontaneous curvature, whose value depends on the surfactant chemical structure and the interaction between surfactants. Considerations based on the bending energy of a surfactant monolayer lead readily to the conclusion that only flat bilayers (zero curvature) can be thermodynamically stable, and in turn, that vesicles (curved bilayers) formed from a single surfactant or phospholipid cannot be thermodynamically stable.[21] In contrast, Safran et al.[56] showed that the free energy of a system of vesicles, made up from a mixture of two interacting surfactants forming monolayers with large bending constants, can be a minimum if the vesicle inner and outer monolayers have different compositions. The spontaneous curvatures of the two layers are then equal but have opposite signs. Systems containing such vesicles would be thermodynamically stable single-phase systems, like micellar solutions. However, Laughlin[57] pointed out that binary mixtures of surfactants are rather complex because they contain up to five components when the two surfactants are ionic. He concluded that the phase behavior of binary mixtures of water and catanionic surfactants, which are made up of two oppositely charged surfactant ions, in the absence of other small ions, must first be investigated before a conclusion can be reached on the stability of vesicles formed in surfactant mixtures.

Surfactants giving rise to vesicles are also those giving rise to disklike micelles, that is, surfactants with a packing parameter $1/2 < P < 1$. Such values of P are easily obtained with phospholipids or two-chain surfactants for which l and a_0 are nearly the same as for the corresponding one-chain surfactants, whereas v is twice as large. A disklike micelle can be thought of as a fragment of a surfactant bilayer having its edge coated by a half of a cylindrical micelle. Disklike micelles become rapidly energetically unfavorable as they grow in size and may turn into *vesicles* (or *liposomes* when formed from phospholipids) where the edge is eliminated, when mechanical energy is provided to the system. Vesicles can be very large (diameter of several microns) or small (diameter of 20 nm), unilamellar or multilamellar. An extreme case of large multilamellar vesicles (LMV) is encountered in the so-called *onion phase*, where each LMV includes tens to hundreds of concentric vesicles and where the LMVs are in contact as seen in Figure 1.9.[58] The onion phase is used in slow-release formulations of active compounds. Vesicles and liposomes are often spherical, particularly those of small size. Faceted and tubular vesicles have been reported.[59,60]

Packing is tighter in vesicles than in micelles. It is thus likely that water penetration in vesicles is even less than in micelles. The viscosity reported by various probes solubilized in vesicles is not significantly larger than in micelles when the measurements are performed at above the transition tem-

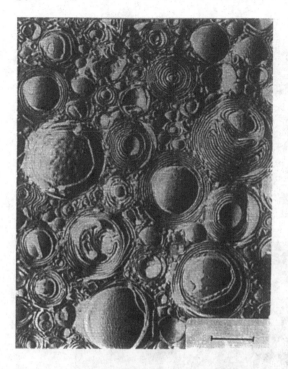

Figure 1.9 Freeze-fracture electron micrograph showing the onion phase of vesicles present in the tetradecyldimethylamineoxide/hexanol/water system (bar = 1 μm). Reproduced from reference 58 with permission of Academic Press/Elsevier.

perature of the system.[61] The interior of bilayers and vesicles is disordered.[62]

Vesicles can be transformed into micelles, and vice versa. In most instances the vesicle-to-micelle transition is induced by the addition of a micelle-forming surfactant[59,63] or a hydrotrope[64] to a vesicular system. The micelle-to-vesicle transition is often induced by mixing two oppositely charged surfactants[55,65,66] or removing a micelle-forming surfactant (or a hydrotrope) from a mixed micellar solution of this surfactant (or hydrotrope) and of a vesicle-forming surfactant. The transition can also be induced by a change of pH, temperature,[67,68] or ionic strength of a micellar solution or of a vesicular system, or by shearing the system.[68]

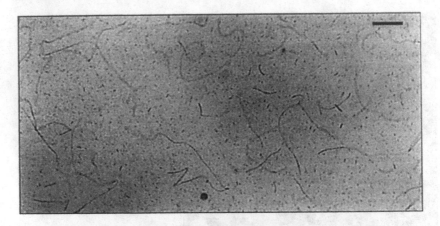

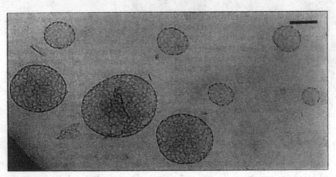

Figure 1.10 Aggregates visualized by cryo-TEM in the course of the vesicle-to-micelle transformation in the glycerol monooleate/cetyltrimethylammonium bromide mixture. *Top*: coexistence of globular micelles, elongated micelles that are probably ribbon-like, and vesicles (lipid mole fraction 0.47) in the presence of 100 mM NaCl. *Bottom*: perforated vesicles (lipid mole fraction 0.64). Reproduced from reference 69 with permission of the American Chemical Society.

Various studies[21,65,66] suggested the existence of structures intermediate between micelles and vesicles. These structures — perforated vesicles, bilayer fragments, giant wormlike micelles, ringlike micelles, and disklike micelles, depending on the investigated system — were visualized by cryo-TEM.[38,59,69] Some of these structures are shown in Figure 1.10. However, for some systems no intermediate structure was evidenced when vesicles were transformed into micelles.[63] The course of the vesicle-to-micelle and micelle-to-vesicle transitions has obviously much to do with the packing parameter of the mixture of the micelle-forming and vesicle-forming surfactants and its variation with composition. In a study where the vesicle-to-micelle transition was induced by an increase of the temperature, the vesicles gave rise to giant wormlike micelles.[68]

The size of vesicles prepared by sonication and other methods is often increased upon the addition of a small amount of micelle-forming surfactant. This increase has also been visualized by cryo-TEM.[59] Reverse vesicles have been evidenced in organic solvents.[70]

Vesicles are also dynamic objects. They exchange surfactant or phospholipid with the surrounding solution. These exchanges are much slower than in micellar solutions essentially because surfactants and lipids making up vesicles are much more hydrophobic than those giving rise to micelles. Besides, since the number of surfactants making up a vesicle is 10 to 1000 times larger than for a micelle, the lifetime of a vesicle must be extremely long and vesicles can probably be considered as "frozen" on the laboratory time scale (weeks to months or years). These aspects are reviewed in Chapter 6, together with the kinetics of the vesicle-to-micelle and micelle-to-vesicle transitions.

VI. LYOTROPIC LIQUID CRYSTALS (MESOPHASES)

The following considerations are largely based on the papers by Israelachvili et al.[26] and Mitchell et al.[23] Only the essential aspects of these papers that are relevant to surfactant phase behavior have been retained and simplified in order to give the reader some markers in the understanding of phase behavior.

The discussion on micelles in Section III considered only the intramicellar interactions that give rise to micelles of finite size and of different shapes. Those forces work parallel to the micelle–water interface.[71] As the surfactant concentration increases further, intermicellar forces come into play.

They are repulsive and work perpendicular to the interface.[71] Repulsive interactions induce micelle ordering at values of the surfactant volume fraction and with symmetry that depend on the micelle shape.[23,71] Spherical micelles close pack into a cubic array, forming the discontinuous cubic phase I_1. Rodlike micelles pack into a hexagonal array, forming the hexagonal phase H_1. Bilayers pack into the lamellar phase L_α.[71] An important point is that this simple approach leads one to predict for a surfactant that forms spherical micelles ($P < 1/3$) the following sequence of phases as the surfactant concentration is increased:

Spherical micelles → cubic phase →
hexagonal phase → lamellar phase

If the surfactant used can form elongated micelles ($1/3 < P < 1/2$) or disklike micelles ($1/2 < P < 1$), then the sequences of phases are, respectively:

Disordered elongated micelles →
hexagonal phase → lamellar phase

Disordered disklike micelles → lamellar phase

However, the intermicellar interactions also include a soft-core repulsion. This additional interaction can be relatively long range and can assist the formation of cubic structures other than of the face-centered type.[23,71] The unfavorable curvature energy can then be compensated by a contribution from intermicellar interactions. Since intermicellar interactions work perpendicularly to the micelle-water interface, any decrease in the distance between micelles (upon increasing concentration) increases the free energy of the system. A way for the system to compensate this effect and to maintain an intermicellar distance as large as possible is to reduce the number of micelles, i.e., to cross over from spherical to cylindrical micelles or from cylinders to lamellae.[71] This renders possible the occurrence of the phase sequences:

Spherical micelles → hexagonal phase → lamellar phase

Disordered elongated micelles → lamellar phase

Figure 1.11 gives a schematic representation of the phase sequence at different curvatures and as a function of the

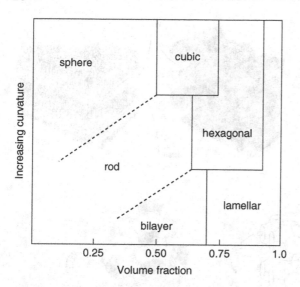

Figure 1.11 Schematic illustration of the evolution of the mesophase structure as a function of the surfactant volume fraction. The micelle shape transitions indicated in dotted lines occur over a range of volume fraction while transitions between mesophases occur at constant volume fraction. Reproduced from reference 23 with permission of the Royal Society of Chemistry.

surfactant volume fraction.[23] This diagram reproduces many of the features of the phase behavior of aqueous surfactants.

The structures of the discontinuous cubic phase, direct hexagonal phase, lamellar phase, bicontinuous cubic phase and of the so-called L_3 or sponge phase are given in Figure 1.12.[72-74] The bicontinuous cubic phase is observed at high concentration, and is usually located between the hexagonal and lamellar phases.[73] The surfactants form a surface that possesses three-dimensional long range order and that separates the water into two unconnected regions. The L_3 phase can be viewed as a disordered bicontinuous cubic phase.[74] Several other lyotropic structures have been observed with binary mixtures of surfactant and water.[22,73]

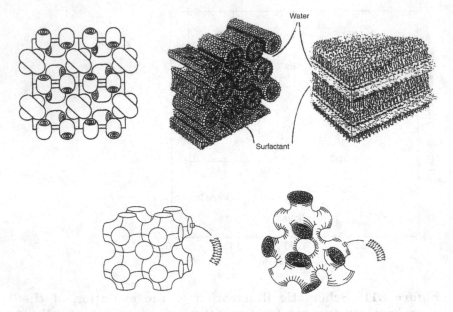

Figure 1.12 Mesophases that are the most often encountered in aqueous surfactant systems. *Top left*: discontinuous cubic phase built up of slightly elongated micelles. *Top center*: hexagonal phase. *Top right*: lamellar phase. *Bottom left*: bicontinuous cubic phase. *Bottom right*: L_3 phase. Reproduced from references 72, 73, and 74 with permission from J. Wiley and Academic Press.

It is possible by using pressure-jump (p-jump), temperature-jump (T-jump), and stopped-flow setups (see Chapter 2) to induce a phase transition from one type of structure to another. This has permitted the study of the kinetics of phase transitions in surfactant (including amphiphilic block copolymer) and lipid systems (see Chapter 7). In these studies, the relaxation of the system is monitored using time-resolved intensity of light (or turbidity), of neutrons, and more particularly of x-rays. Indeed, the powerful x-ray sources from synchrotron radiation now available permit the recording of full diffraction spectra in only a few milliseconds.

VII. CONCLUSION: IMPORTANCE OF DYNAMIC ASPECTS OF SURFACTANT SELF-ASSEMBLIES

This chapter reviewed the main self-assemblies that surfactants can give rise to in the presence of water and pointed out the fact that these self-assemblies are of dynamic character. The dynamics of surfactant self-assemblies are important for the understanding of all phenomena that take place in a time scale comparable to that of either surfactant exchange or assembly formation/breakdown. Here are some examples:

1. Diffusion in micellar systems can be strongly affected by surfactant exchange. This effect may result in an apparent acceleration of the diffusion.[75]
2. Critical behavior of, and electrical conductivity percolation in, microemulsions is much dependent on the dynamics of "sticky" collisions between droplets of oil (water) present in oil-in-water (water-in-oil) microemulsions[76] (see Chapter 5). Sticky collisions refer to collisions that bring assembly cores into contact. During the time of contact matter may be transferred from one assembly to another. Such collisions occur in systems with strong attractive interactions between assemblies.
3. The adsorption behavior of surfactants at interfaces and thus the dynamic surface tension of surfactant solutions may be affected by the lifetime of the micelles that act as reservoirs of surfactants[77,78] (see Chapter 8).
4. The rheological behavior of micellar solutions may be completely controlled by the kinetics of micelle formation/breakdown[79] (see Chapter 9). This occurs when the lifetime of a giant elongated micelle is much shorter than the time required for the reptation of one giant micelle in the network formed by the other micelles.
5. Solubilization, emulsification, wetting, and foaming have all been discussed in terms of micelle dynam-

ics.[80,81] Indeed, all of these effects involve the time required for a surfactant to leave a micelle.

6. The kinetics of reactions performed within surfactant self-assemblies used as microreactors can be completely controlled by the rate of collisions between surfactant assemblies with temporary merging of the collided assemblies. This occurs when the rate of such collisions is much slower than the rate of the investigated chemical reaction (see Chapter 5, Sections VI.F and VIII, and Chapter 10).

REFERENCES

1. Hartley, G.S. *Aqueous Solutions of Paraffinic-Chain Salts. A Study of Micelle Formation*, Herman, Paris, 1936. See also *Micellization, Solubilization and Microemulsions*, Mittal K., Ed., Plenum Press, New York, vol. 1, p. 23.

2. Myers, D. *Surfactant Science and Technology*, VCH Publ. Inc., Weinheim, Germany, 1988.

3. Moroi, Y. *Micelles: Theoretical and Applied Aspects*, Plenum Press, New York, 1992.

4. Ananthapadmanabhan, K.P., Goddard, E.D., Chandar, P. *Colloids Surf.* 1990, 44, 281.

5. Alexandridis, P., Hatton, T.A. *Colloids Surf. A.* 1995, 96, 1.

6. Yu, G., Yang, Z., Ameri, M., Attwood, D., Collett, J.H., Price, C., Booth, C. *J. Phys. Chem. B* 1997, 101, 4394.

7. Jada, A., Hurtrez, G., Siffert, B., Riess, G. *Macromol. Chem Phys.* 1996, 197, 3697.

8. Zhang, L., Eisenberg, A. *Macromolecules* 1996, 29, 8805.

9. Zana, R. *Adv. Colloid Interface Sci.* 2002, 97, 205; *J. Colloid Interface Sci.* 2002, 248, 203.

10. Tanford, C. *The Hydrophobic Effect*, John Wiley and Sons, New York, 1980.

11. Lum, K., Chandler D., Weeks J. *J. Phys. Chem. B* 1999, 103, 4570.

12. Chandler, D. *Nature* 2002, 417, 491.

13. Warr, G. *Curr. Opinion Colloid Interface Sci.* 2000, 5, 88.

14. Manne, S., Schäffer, T.E., Huo, Q., Hansma, P.K., Morse, D.E., Stucky, G.D., Aksay, I.A. *Langmuir* 1997, 13, 6382.

15. Mukerjee, P. *Adv. Colloid Interface Sci.* 1967, 1, 241.

16. Menger, F.M., Zana, R., Lindman, B. *J. Chem. Ed.*, 1998, 75, 115.

17. Kahlweit, M., Strey, R. *Angew. Chem. Int. Ed.* 1985, 24, 654.

18. Nagarajan, R., Ruckenstein, E. *Langmuir* 2000, 16, 6400.

19. Rance, D., Friberg, S.E. *J. Colloid Interface Sci.* 1977, 60, 207.

20. Sjöblom, E., Friberg, S.E. *J. Colloid Interface Sci.* 1978, 67, 16.

21. Lichtenberg, D. In *Biomembranes: Physical Aspects*, Shinitzky, M., Ed., VCH: New York, 1993, chap. 3.

22. Skoulios, A. *Ann. Phys.* 1978, 3, 421.

23. Mitchell, D.J., Tiddy, G.T., Waring, L., Bostock, T., McDonald, P. *J. Chem. Soc., Faraday Trans.* 1983, 79, 975.

24. Rupprecht, H., Gu, T. *Colloid Polym Sci.* 1991, 269, 506.

25. Atkin, R., Craig, V.S., Wanless, E. J., Biggs, S. *J. Phys. Chem. B* 2003, 107, 2978.

26. Israelachvili, J., Mitchell, D. J., Ninham, B.W. *J. Chem. Soc., Faraday Trans.* 1976, 72, 1525.

27. Nagarajan, R., Ruckenstein, E. *Langmuir* 1991, 7, 2934.

28. Shiloah, A., Blankschtein, D. *Langmuir* 1998, 14, 7166 and references therein.

29. Bruce, C.D., Berkowitz, M.L., Perera, L., Forbes M.D. *J. Phys. Chem. B* 2002, 106, 3788 and references therein.

30. Israelachvili, J. *Intermolecular and Surface Forces*, Academic Press, London, 1985.

31. Hyde, S.T. *Prog. Colloid Polym. Sci.* 1990, 82, 236.

32. Halle, B., Landgren, M., Jönsson, B. *J. Phys. France* 1988, 49, 1235.

33. Porte, G., Poggi, Y., Appell, J., Maret, G. *J. Phys. Chem.* 1984, 88, 5713.

34. Eriksson, J.C., Lundgren, S.J. *J. Chem. Soc., Faraday Trans. 2* 1985, 81, 1209.

35. Néry, H., Söderman, O., Canet, D., Walderhaug, H., Lindman, B. *J. Phys. Chem.* 1986, 90, 5802.

36. Turley, W.D., Offen, H.W. *J. Phys. Chem.* 1985, 89, 2933.

37. Cabane, B., Duplessix, R., Zemb, T. *J. Phys. France* 1985, 46, 2161.

38. Talmon, Y. In *Modern Characterization Methods of Surfactant Systems*, Binks, B.P., Ed., M. Dekker Inc., New York, 1999.

39. Bernheim-Groswasser, A., Zana, R., Talmon, Y. *J. Phys. Chem. B.* 2000, *104*, 4005 and references therein.

40. In, M., Aguerre-Chariol, O., Zana, R. *J. Phys. Chem. B* 1999, 103, 7749.

41. Zana, R. *Adv. Colloid Interface Sci.* 1995, 57, 1.

42. Sepulveda, L., Lissi, E., Quina, F. *Adv. Colloid Interface Sci.* 1986, 25, 57.

43. Kwak, J.C.T., Ed. *Polymer-Surfactant Systems*, M. Dekker Inc., New York, 1998.

44. Hoar, T.P., Schulman, J.H. *Nature* 1943, 152, 103.

45. Danielsson, I., Lindman, B. *Colloids Surf.* 1981, 3, 391.

46. Friberg, S.E. *Colloids Surf.* 1982, 4, 201.

47. Strey, R. *Colloid Polym. Sci.* 1994, 272, 1005.

48. Winsor, P.A. *Solvent Properties of Amphiphilic Compounds*, Butterworths, London, 1954.

49. Biais, J., Clin B., Bothorel, P. In *Microemulsions: Structure and Dynamics*, Friberg, S.E., Bothorel, P., Eds., CRC Press, Boca Raton, FL, 1987, p. 1.

50. Bellocq, A.M., Roux, D. In *Microemulsions: Structure and Dynamics*, Friberg, S.E., Bothorel, P., Eds., CRC Press, Boca Raton, FL, 1987, p. 33.

51. Jahn, W., Strey, R. *J. Phys. Chem.* 1988, 92, 2294.

52. Ruckenstein, E., Chi, J.C. *J. Chem. Soc., Faraday Trans. 2* 1975, 71, 1690.

53. De Gennes, P.G., Taupin, C. *J. Phys. Chem.* 1982, 86, 2294.

54. Lasic, D. *Biochem. J.* 1988, 256, 1.

55. Kaler, E.M., Murthy, A.K., Rodriguez B.E., Zasadzinski, J.A. *Science* 1989, 245, 1371.

56. Safran, S.A., Pincus, P., Andelman, D. *Science* 1990, 248, 354.

57. Laughlin, R.G. *Colloids Surf. A.* 1997, 128, 27 and references therein.

58. Hoffmann, H., Munkert, U., Thunig C., Valiente, M.J. *J. Colloid Interface Sci.* 1994, 163, 217.

59. Bernheim-Groswasser, A., Zana, R., Talmon, Y. *J. Phys. Chem. B* 2000, 104, 12192 and references therein.

60. Almgren, M., Edwards, K., Karlsson, G. *Colloids Surf. A* 2000, 174, 3.

61. Turley, W.D., Offen, H.W. *J. Phys. Chem.* 1986, 90, 1967.

62. Czarniecki, M., Breslow, R. *J. Am. Chem. Soc.* 1979, 101, 3675.

63. Danino, D., Talmon, Y., Zana, R. *J. Colloid Interface Sci.* 1997, 185, 84.

64. Campbell, S.E., Yang, H., Patel, R., Friberg S.E., Aikens P.A. *Colloid Polym. Sci.* 1997, 275, 303.

65. Robinson, B.H., Bucak, S., Fontana, A. *Langmuir* 2000, 16, 8231 and references therein.

66. Shioi, A., Hatton, T.A. *Langmuir* 2002, 18, 7341.

67. Majhi, P.R., Blume, A. *J. Phys. Chem. B* 2002, 106, 10753.

68. Mendes, E., Narayanan, J., Oda, R., Kern, F., Candau, S.J., Manohar, C. *J. Phys. Chem. B* 1997, 101, 2256.

69. Gustafsson, J., Orädd, G., Nyden, M., Hansson, P., Almgren, M. *Langmuir* 1998, 14, 4987 and references therein.

70. Kunieda, H., Makino, S., Ushio, N. *J. Colloid Interface Sci.* 1991, 147, 286.

71. Charvolin, J., Sadoc, J. F. In *Micelles, Membranes, Microemulsions and Monolayers*, Gelbart, W.M., Ben Shaul, A., Roux, D., Eds., Springer-Verlag, New York, 1994.

72. Jönsson, B., Lindman, B., Holmberg, K., Kronberg, B. *Surfactants and Polymers in Aqueous Solution*, J. Wiley and Sons: New York, 1998, pp. 71–73.

73. Laughlin, R.G. *The Phase Aqueous Behavior of Surfactants*, Academic Press, London, 1994.

74. Porte, G.J. *Phys. Condens. Matter* 1992, 4, 8649.

75. Kato, T., Terao, T., Tsukuda, M., Seimiya, T.J. *Phys. Chem.* 1993, 97, 3910.

76. Jada, A., Lang, J., Zana, R., Makhloufi, R., Hirsch, E., Candau, S.J. *J. Phys. Chem.* 1990, 94, 387.

77. Eastoe, J., Rankin, A., Wat, R., Bain, C.D. *Int. Rev. Phys. Chem.* 2001, 20, 357.

78. Eastoe, J., Dalton, J.S. *Adv. Colloid Interface Sci.* 2000, 85, 103.

79. Oelschlager, Cl., Waton, G., Candau, S.J. *Langmuir* 2003, 19, 10495 and references cited.

80. Patist, A., Kanicky, J., Shukla, P., Shah, D.O. *J. Colloid Interface Sci.* 2002, 245, 1.

81. Patist, A., Oh, S.G., Leung, R., Shah, D.O. *Colloids Surf. A.* 2000, 176, 3.

2

Methods for the Study of the Dynamics of Surfactant Self-Assemblies

RAOUL ZANA

CONTENTS

I.   INTRODUCTION

Some processes that occur in systems containing surfactant self-assemblies — most particularly the exchange process involving surfactant, cosurfactant, or solubilizates (see the preface and Chapter 3) — can be extremely rapid. They cannot be investigated by methods used in conventional kinetic studies that involve mixing reactants and monitoring the formation of products. Indeed, fast mixing apparatuses usually have a dead time of 1–10 ms and, therefore, do not permit studies of kinetics of reactions with half-time of reaction shorter than this limit. This fact led to the use of chemical relaxation methods in the early studies of micellar kinetics. These methods were developed in the 1950s.[1] In chemical relaxation methods one starts from a mixture of reactants and products in a state of thermodynamic equilibrium and perturbs this equilibrium by *very rapidly* changing one of the external parameters: pressure p or temperature T, for instance. The system responds to the perturbation by shifting toward a new state of equilibrium imposed by the new value of the modified external parameter. The evolution of the system with time is referred to as *chemical relaxation*. It can be described in terms of one (or several) time constant(s), the *chemical relaxation time*(s), which characterizes the ability of the system to follow the perturbation. The relaxation time(s) depends on the rate constants of the investigated reaction(s).

As studies of surfactant self-assemblies developed during 1970 to 1980, other methods, such as time-resolved luminescence quenching (and the related methods of flash photolysis and pulse radiolysis), nuclear magnetic resonance (NMR), electron spin resonance (EPR), and rheology, were used or introduced for studying the kinetics of specific processes involving surfactant self-assemblies. These methods cover a wide time scale. In the 1980s the use of chemical relaxation methods was extended to studies of the dynamics of transitions between surfactant or lipid lyotropic phases.

The purpose of this chapter is to review the methods that have been used for studies of dynamics of surfactant self-assemblies in solution. The emphasis is mostly on the principles underlying each method and on its possibilities. No mathematical developments are presented. The methods used for the study of the dynamics of surfactant adsorption on surfaces are reviewed in Chapter 8.

II. CHEMICAL RELAXATION METHODS

A. Principle

In relaxation studies, the perturbation applied to the system is always of very small amplitude, in such a way that the system remains close to equilibrium during the entire course of its evolution. Nevertheless, the perturbation must be sufficient for generating a relaxation signal large enough with respect to noise. Chemical relaxation methods permit the study of chemical processes, with half-time of reaction ranging from minutes to nanoseconds. Several books and review papers on chemical relaxation methods have been published.[1–10] Only the main features of chemical relaxation methods are presented in this section.

The change of external parameter that induces the relaxation of the system can be of different types: stepwise, rectangular, or harmonic.[1,2,5,7–9] Consider the simplest case, that of a *stepwise* perturbation as in temperature-jump (T-jump), pressure-jump (p-jump), or electrical field-jump (E-jump) studies. Figure 2.1A shows the time-dependence of the perturbation, that is, the imposed variation, here the temperature T(t) and the response of the system, P(t). The response is in fact the variation with time of a property P of the system measured as the system shifts toward its new state of equilibrium. The property P is selected as to be proportional to

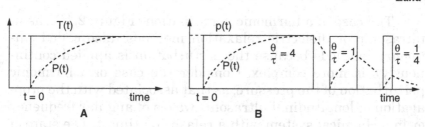

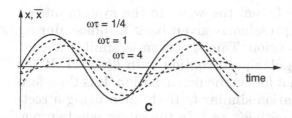

Figure 2.1 Shape of the perturbation (in full line) used in chemical relaxation methods and response of the system (dashed line); (A) step perturbation as in T-jump; (B) rectangular perturbation as in shock-tube (θ = duration of the perturbation); (C) harmonic perturbation as in ultrasonic absorption relaxation.

the concentration of a reactant or product. It can be the electrical conductivity, the absorbance of light by the solution, the intensity of fluorescence or the intensity of scattered radiation (light, x-ray, neutron), for instance. The relaxation signal P(t) may be single- or multiexponential. Its analysis yields the relaxation time(s) and the relaxation amplitude(s) of the system.

Figure 2.1B shows the case of a rectangular perturbation, as used in shock-tube and E-jump methods (see below). In these methods, the perturbation is applied only during a certain time, θ, which must be longer than the relaxation time(s) τ of the system, say θ/τ > 4. Indeed, Figure 2.1B shows that when this condition is not fulfilled, the system does not reach its new state of equilibrium during the duration of the perturbation. This may result in a larger error on the fitted value(s) of the relaxation time(s). The use of step or rectangular perturbations (*transient* methods) gives direct access to the value(s) of the relaxation time(s).

The case of a harmonic perturbation (Figure 2.1C), as in ultrasonic or dielectric relaxation methods (also called *stationary* methods because the perturbation is applied continuously), is more complex. Consider the case of a harmonic perturbation of the pressure, as that associated with the propagation of longitudinal ultrasonic waves of angular frequency ω, in a chemical system with a relaxation time τ. The state of the system shifts back and forth at the frequency of the wave. If $1/\omega \gg \tau$, the system can follow the perturbation and all of the energy transferred from the wave to the system during the half-wave of compression is given back in phase during the half-wave of expansion. This situation is similar to that when using a rectangular perturbation with $\theta/\tau \gg 1$. If $1/\omega \ll \tau$, the system cannot follow the perturbation. It is therefore not perturbed, a situation similar to that when using a rectangular perturbation with $\theta/\tau \ll 1$. In the intermediate range of values of ω, part of the energy transferred from the wave to the system during the half-wave of compression is dissipated as heat in the system because the system lags behind the wave owing to the finite value of its relaxation time. The propagation of the wave results in its progressive *absorption* associated with the chemical process. A determination of the absorption of the wave as a function of frequency provides a measure of the relaxation time of the system as the absorption coefficient is given by an expression with a denominator of the form $(1 + \omega^2\tau^2)$. The same reasoning applies to dielectric absorption studies, where the perturbation is induced by a harmonic electrical field. The perturbed chemical equilibrium must involve a change of dipole moment. Obviously, the determination of relaxation times by stationary methods is more time consuming than by transient methods because it implies measurements of the absorption coefficient at as many discrete frequencies as possible.

B. Relaxation Times

1. Single-Step Processes

For single-step reactions (*elementary* reactions), the expression of the relaxation time is readily obtained from the rate equation, taking into account the stoichiometry of the reaction and the appropriate mass-balance equation. The rate equations are differential equations, with the concentration changes as variables. The equations are linearized on the assumption that the concentration changes induced by the perturbation are extremely small.[1,2,7,9] This explains why in chemical relaxation measurements one always tries to keep the perturbation as small as possible but still consistent with a reasonable signal/noise ratio. Three examples are considered.

a.   Unimolecular equilibrium

$$A \underset{k_{-1}}{\overset{k_1}{\rightleftarrows}} B \tag{2.1}$$

$$1/\tau = k_1 + k_{-1} \tag{2.2}$$

Unimolecular reactions are mostly conformational changes of small molecules (isomerism), proteins, and other macromolecules.

b.   Bimolecular equilibrium

$$A + B \underset{k_{-1}}{\overset{k_1}{\rightleftarrows}} C \tag{2.3}$$

$$1/\tau = k_{-1} + k_1\{[A] + [B]\} \tag{2.4}$$

Bimolecular reactions include ion-pairing, proton transfer, complexation, and dimerization, to cite a few.

c.   Catalyzed equilibrium

$$A + D \underset{k_{-1}}{\overset{k_1}{\rightleftarrows}} B + D \tag{2.5}$$

$$1/\tau = (k_1 + k_{-1})[D] \tag{2.6}$$

In Equation 2.1 to Equation 2.6, k_1 and k_{-1} are the forward and backward rate constants. [A], [B], and [C] are the equilibrium concentrations of reactants or products A, B, and C. [D] is the concentration of the catalyst D.

2. Multistep Processes

Most chemical processes including micellar kinetics involve several steps and are characterized by several relaxation times (relaxation spectrum). The maximum number of observable relaxation times is equal to the number of independent rate equations that can be written for the system investigated. This number is equal to that of chemical species minus the number of mass-balance equations.[2,7,9,11]

The expressions of the relaxation times for a system of n independent coupled reactions can be obtained using a matrix method to solve the system of n linearized differential equations. One of the easiest methods is that proposed by Castellan.[11]

Thus far it has been implicitly assumed that the simple or complex equilibrium giving rise to the observed relaxation processes has been identified prior to the relaxation experiments. In fact, this can be also achieved through the use of chemical relaxation methods. Indeed, it is clear from Equation 2.2, Equation 2.4, and Equation 2.6 that the concentration dependence of the relaxation time depends on the type of process under investigation. A unimolecular process, such as reaction (1), is characterized by a concentration-independent relaxation time, while the relaxation time of a bimolecular process, such as reaction (2), depends on the concentrations of the reactants. Thus the concentration dependence of the measured relaxation time permits one to know what type of process is responsible for the observed relaxation. In general, the determination of the variation of each of the relaxation times with appropriate parameters (concentration, ionic strength, pH, T, p, E,...) permits the assignment of the observed relaxation processes to specific reactions involving the chemical species present in the investigated system. A reaction scheme is then assumed for which expressions of the relaxation times are derived and fitted to the experimental results. A satisfactory fit is usually considered as supporting the assumed reaction scheme. If the quality of the fit is not good, the reaction scheme must be modified and the procedure repeated.

The rate constants and equilibrium constants for the various steps are obtained from such fits. Their variation with temperature or pressure permits the determination of the activation energy or volume for each step.

C.   Relaxation Amplitudes

The amplitude of the relaxation signal depends on the thermodynamic quantity associated with the type of perturbation used. Take, for instance, the case of a p-jump. A change of pressure perturbs a chemical system only if the system involves a chemical reaction with an associated volume change. Equations relate the relaxation amplitudes to the associated parameters characterizing the chemical reactions. The amplitudes of the relaxation signals in p-jump and T-jump experiments permit the determination of the values of the volume change and enthalpy change, respectively, associated with the investigated reaction.

The derivation of expressions for the relaxation amplitudes is fairly straightforward in the case of single-step reactions.[2,5,7,9,11] The calculations are more complex for multistep reactions but well within the reach of present computers, as demonstrated in a recent study of micellar equilibria.[12]

The use of the relaxation amplitudes permits in some cases a choice between two possible reaction schemes or between two different sets of assumptions in the analysis of chemical relaxation data. For instance, a controversy arose about the validity of an assumption made in deriving the expression of the relaxation time associated with the micelle formation/breakdown process for surfactant solutions. This controversy was solved by fitting the expressions of the amplitudes derived with both sets of assumptions to the amplitude data and retaining the set giving the best fit.[13-16]

D. Chemical Relaxation Techniques

This section is an actualized version of a review of chemical relaxation techniques that was published in 1987.[10] Detailed descriptions of chemical relaxation setups can be found in older reports.[2–5,7,9]

Figure 2.2 shows the block diagram of a chemical relaxation setup. At the outset it must be pointed out that in transient chemical relaxation methods the time required to achieve the perturbation must be much shorter (say by a factor of 10) than the shortest relaxation time being measured. When this condition is not met, the response of the system is convoluted by the perturbation function of the apparatus and must be corrected.[2] Extraction of the relaxation time from the relaxation signal requires the knowledge of the perturbation function. The latter can be obtained from an experiment involving a chemical system with a relaxation time much shorter than the time required for perturbing the system.[17] Acid-base equilibria of indicators or metal ion complexation reactions are generally used for this purpose. Computer analysis of the relaxation signal then yields the relaxation time. The T-jump, p-jump, shock-tube, E-jump, ultrasonic absorption, and stopped-flow techniques are considered successively.

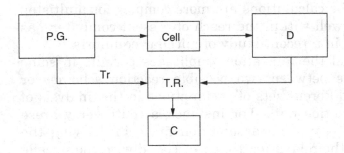

Figure 2.2 Block diagram for chemical relaxation setups. P.G., perturbation generator; D, detector monitoring the change of property of the system contained in the cell; Tr, trigger line; T.R., transient recorder (now replaced by a computer card); C, computer.

1.   T-jump

This method can be applied to chemical reactions that have an enthalpy of reaction $\Delta H°$ that is not too small. This is the case for most reactions and explains the wide use of the T-jump method in chemical relaxation studies.

The jump of temperature (ΔT) results in a change of the equilibrium constant K of the reaction given by Equation 2.7 and thus in a shift of the equilibrium state of the system. This shift is usually monitored by the change of an optical property of the system (see below).

$$\delta \ln K = (\Delta H°/RT^2)\Delta T \qquad (2.7)$$

Two methods are still in use for achieving a fast heating. The first one, referred to as Joule heating, involves the discharge of a capacitor (capacitance C = 0.01 to 0.05 μF; charging voltage, 10 to 50 kV) into the solution.[2,7,18] The solution is contained in a cell between the two electrodes used for the discharge. The cell resistance R must be low. Indeed, the rise time of the T-jump (heating time) is RC/2. The second method uses a pulse of laser light of proper wavelength to ensure its partial absorption by the investigated solution. Raman-shifted neodynium lasers with an output wavelength of 1.41 μm[19,20] or 1.89 μm[21] and iodine lasers (λ = 1.31 μm)[22–24] are now available for direct heating of aqueous solutions, thus avoiding the shortcomings of the earlier laser T-jump setups.[25]

T-jump studies of surfactant solutions have made much use of Joule heating because it was in the only commercially available apparatus when such studies were started. Laser heating has two distinct advantages over Joule heating. First, the heating can be achieved in 10 ns or less against 1 μs with Joule heating. Second, it can be used with both conducting and nonconducting systems, whereas Joule heating is restricted to sufficiently conducting systems. Thus Joule heating cannot be used with surfactant solutions in organic solvents, aqueous solutions of nonionic surfactants, or even dilute aqueous solutions of ionic surfactants, without the addition of salt, which may affect the kinetics of the system being investigated. The main disadvantages of laser heating lie in its high cost and technical complexity. Also, too-short laser pulses may give rise to transient effects in the solution, which perturb measurements on the short time scale. The experiments require about 1 ml (Joule heating) or even less (laser heating) of solution. T-jumps of a few degrees can be obtained by both methods. The magnitude of the T-jump can be varied by changing the values of the charging voltage and capacitance in Joule heating, and the laser output in laser heating.

The upper limit of relaxation times that can be measured by the T-jump method is about 1 s. This value corresponds to the time after which the cooling-down effect of the thermostatic system of the cell becomes detectable.

Several properties have been used to monitor relaxation processes in surfactant solutions: optical density, fluorescence intensity, circular dichroism, and intensity of scattered radiation.[2,6,7,18,26–28] Optical density has been by far the most widely used, owing to the very high sensitivity of the detecting devices.[2,18] The spectrophotometric detection implies the presence in the surfactant molecule of a chromophoric group whose spectrum depends on whether the surfactant is free or micellized. The T-jump method can then be used to study both the exchange process and the micelle formation/breakdown. In most instances, however, the surfactant contains no chromophoric group. A dye whose binding to, or solubilization into, micelles results in a large spectral change, which can then be added to the system to study the kinetics of micelle formation/breakdown.[29] Eosine and acridine orange can be used for the study of cationic and nonionic surfactants and anionic surfactants, respectively.[29] *N*-alkylpyridinium and *N*-alkyl-4-cyanopyridinium halides with a sufficiently long alkyl chain can be used for concentrated solutions of surfactant irrespective of its nature.[30] It is essential to use a [surfactant]/[dye] molar ratio sufficiently high, say above 200, in order that the perturbation introduced by the dye be negligible.[29,30] At any rate, it is suggested that the relaxation time τ_2 associated to the micelle formation/breakdown process be measured as a function of the dye concentration at constant surfactant concentration. Valid measurements are those performed in conditions where τ_2 is independent of the dye concentration.

Light scattering is increasingly used for the monitoring of T-jump relaxation signals.[12,26,31] Indeed, recent studies involve solutions of surfactant with giant micelles and of amphiphilic block copolymers. In both systems, the micelle size can be very large and temperature-dependent. More recently with the availability of strong sources of x-rays and neutrons, the intensity of the scattered x-rays[27] and neutrons[28] has been used to monitor relaxation processes arising from vesicle-to-micelle transformations (see Chapter 6) and from transitions between lyotropic phases (see Chapter 7).

A Joule heating T-jump apparatus with a heating time below 8 µs and an adjustable jump of temperature up to 7°C is commercially available from U.K. Hi-Tech Scientific Ltd. (U.K.) Figure 2.3 shows a block diagram of the setup.

2. p-jump

The p-jump method applies to chemical reactions with a non-zero volume of reaction $\Delta V°$. The jump of pressure, Δp, brings about a change of the equilibrium constant K given by Equation 2.8 and thus a shift of the equilibrium.

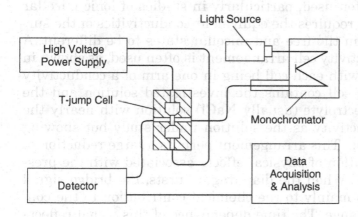

Figure 2.3 Block diagram of the T-jump apparatus commercially available from Hi-Tech Scientific Ltd. (U.K.).

$$\delta \ln K = -(\Delta V_i^0 / RT) \Delta p \qquad (2.8)$$

with

$$\Delta V_i^0 = \Delta V_T^0 - \theta \Delta H^\circ / dC_p \qquad (2.9)$$

ΔV_i^0 and ΔV_T^0 are the isentropic and isothermal volume changes associated with the reaction; θ, d, and C_p are the coefficient of thermal expansion, the density and the specific heat at constant pressure of the solution.[2]

In the commercially available version of this apparatus[32] (Dialog, West Germany) the perturbation is a step decrease of pressure, Δp, from the pressure initially applied to the investigated solution to the atmospheric pressure. The cell containing the solution is connected to an autoclave through an impermeable elastic membrane (Figure 2.4). One wall of the autoclave is a thin metal diaphragm, which bursts when the pressure in the autoclave reaches 50 to 100 kg/cm². The rupture of the diaphragm and the ensuing pressure drop occur in about 100 μs.

Pressure-jump setups with conductivity[32] and optical[33] detection have been described. The conductivity detection is the most often used, particularly in studies of ionic micellar solutions. It requires the equivalent conductivities of the surfactant ion in the free and micellar states to be different. A dual-conductivity cell arrangement is often used, as shown in Figure 2.4, with each cell being in one arm of a conductivity bridge. One cell contains the investigated solution and the other an electrolyte (usually NaCl) solution with nearly the same conductivity as the solution under study but showing no relaxation. This arrangement permits a large reduction of the contribution of physical effects associated with the pressure change. When the diaphragm bursts, the bridge signal corresponds mainly to the chemical contribution to the conductivity change. The time dependence of this signal reflects the relaxation of the system. The conductivity detection has a very high sensitivity. However, its use is restricted to ionic solutions of low overall conductivity.

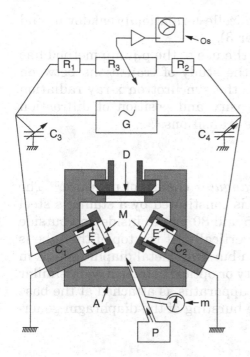

Figure 2.4 Block diagram of the p-jump apparatus with conductivity detection and the twin cell arrangement from Dialog (Germany). A, autoclave; C_1 and C_2, conductivity cells; E, electrodes; M, elastic membrane; D, metal diaphragm; P, pressure pump; m, manometer; G, 40 kHz generator driving the conductivity bridge; C_3 and C_4, tunable capacitors; R_1 and R_2, helipot resistances; R_3, potentiometer; Os, oscilloscope (now replaced by a computer).

The p-jump method can be used to monitor processes with relaxation times in the range between 100 μs and minutes. The method applies to aqueous solutions, where the perturbation occurs mainly through the p-jump (ΔV_T^0 term in Equation 2.9, as the prefactor of the $\Delta H°$ term is close to zero). It also applies to solutions in organic solvents, where the perturbation occurs mainly through the temperature change associated with the essentially adiabatic pressure change ($\Delta H°$ term in Equation 2.9). In the case of micellar solutions, the p-jump method has been used for the study of the exchange process, the micelle formation/breakdown, and other processes (see Chapter 3).

Just as for the T-jump, the use of the p-jump method has recently been extended to the study of transitions between lyotropic mesophases, using the synchrotron x-ray radiation to monitor changes of intensity and position of diffraction peaks associated with these transitions.[27]

3.   Shock Tube

This method involves a *rectangular* change of pressure.[34] The main part of the apparatus is constituted by a stainless steel tube, 2 to 3 m-long and of 35 and 80 mm in inside and outside diameters, respectively, set vertically. At the top of the tube is attached an autoclave with a bursting metal diaphragm, as in p-jump. A cell for conductivity or optical detection, very similar to those used in the p-jump apparatus, is attached at the base of the tube (Figure 2.5). The bursting of the diaphragm gener-

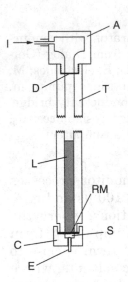

Figure 2.5   Shock-tube apparatus. A, autoclave; I, inlet for pressurizing nitrogen; T, stainless steel tube; L, filling liquid; RM, rubber membrane; C, conductivity cell; D, metal diaphragm E, electrode; S, investigated solution.

ates a p-jump of 50 to 100 kg/cm^2, which propagates from top to bottom through the tube, which is half-filled with water or ethanol. The trailing edge of the pressure front propagates faster than the leading edge, as the velocity of the wave is larger at higher pressure. Thus the pressure front sharpens as it propagates. The length of the liquid column in the tube is sufficient for the rise time of the pressure at the bottom of the tube to be well below 1 μs.[34] The reflections of the pressure-jump at the bottom of the tube (liquid-metal interface) and at the top of the liquid column (liquid-air interface) result in a rectangular pressure change whose duration is twice the time taken by the wave to travel over the length of the liquid column (i.e., about 3 ms). The values 1 μs and 3 ms set the limits of use of the shock tube for relaxation time measurements. As such, it complements, toward the short-time scale, studies performed by p-jump. It has been much used for studies of the kinetics of the exchange process in micellar solutions. It has the disadvantage (with respect to p-jump) of measuring rate constants at a pressure of about 100 kg/cm^2.

4.   E-jump

When an electrical field is applied to a solution, different processes take place. For instance, induced and permanent dipoles tend to orient parallel to the applied field (orientation polarization), droplets of water in oil-in-water microemulsions tend to become elongated in the direction of the field (structural polarization), and chemical reactions that are accompanied by a change of dipole moment can be perturbed (chemical polarization), if the field is maintained for a sufficiently long time. This requires the use of solutions of low to very low electrical conductivity. As a result of the polarization, the solution becomes optically anisotropic and the relaxation effects arising when turning on and turning off the field can be detected by following the time-dependence of the induced electrical bire-fringence (known as the *Kerr effect*).

The E-jump technique has been recently used for the study of the dynamics of processes occurring in water-in-oil microemulsions.[35,36]

Reference 37 describes an E-jump apparatus with electrical birefringence detection capable of measuring relaxation times ranging from about 0.2 to 900 µs.

5. Ultrasonic Absorption

The relaxation times are obtained from the measurement of the ultrasonic absorption α/ω^2 (α = absorption coefficient; ω = angular frequency) at as many ultrasonic frequencies as possible and in a large frequency range, covering two to three decades. Equation 2.10 is then fitted to the data.

$$\frac{\alpha}{\omega^2} = \sum \frac{A_i}{1 + \omega^2 \tau_i^2} + B \tag{2.10}$$

In Equation 2.10 B is a constant and the A_i's represent the relaxation amplitudes that are proportional to the square of the isentropic volume change ΔV_i^0 given by Equation 2.9 and that contains both ΔV_T^0 and $\Delta H°$. Indeed, the propagation of ultrasonic waves in fluids gives rise to harmonic changes of p and T. The investigated equilibria are shifted periodically by these two perturbations, thus the ultrasonic relaxation amplitude dependence on ΔV_i^0.

The use of ultrasonic absorption techniques has decreased in recent years because the experiments are time consuming. The advent of the laser T-jump, which covers a large part of the range of relaxation times explored by ultrasonic relaxation techniques, has also contributed to this situation. The apparatuses still in use are now fully automated and computer-driven, and the time required for data acquisition has been greatly reduced.[38–40]

Many apparatuses for measuring ultrasonic absorption have been reported[40–46] and the state of the art in ultrasonic absorption measurements can be found in Reference 44. Only the swept-frequency resonator and the pulse technique are still in use. The resonator[40–45] consists of two parallel X-cut quartz crystal plates set at a fixed distance at the top and bottom of a cell filled with the investigated solution (Figure 2.6 top). The input transducer, driven by a swept-frequency

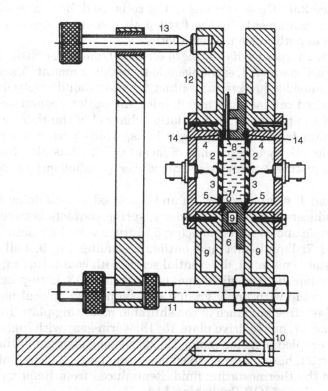

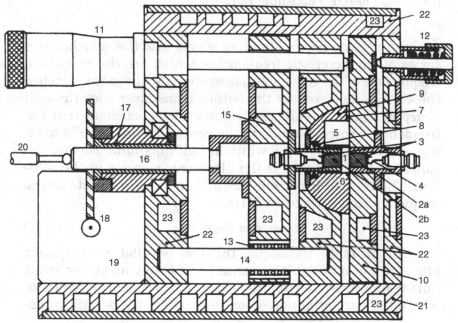

Figure 2.6   Cross sections of the cells used for ultrasonic absorption measurements by the fixed path resonance method (*top*) and variable path pulse method (bottom).
Top: 1, sample cavity; 2, piezo-electric transducer disk; 3, flexible electrical contact; 4, exchangeable transducer mount; 5, sealing ring with embedded electrical contact for the ground electrode; 6, thermostatted cell jacket; 7 and 8, inlet and outlet, respectively, for the liquid sample cavity; 9, circulation channel of the thermostat fluid; 10, main frame and 11, screw bolts, which permit a variation of path length by an exchange of jacket 6; 12, adjustable frame plate; 13, adjustment screws for transducer parallelism; 14, sealing O-ring.
Bottom: 1, sample cavity; 2a and 2b, fused quartz delay lines with transducers 3 on the backside; 4, spring contacts between the hot electrode and coaxial connector; 5, storage vessel for liquid overflow; 6 and 7, liquid inlet and outlet; 8, sealing lip; 9, ball joint; 10, adjustable plate; 11, differential screw with counteracting spring 12 for the parallelism adjustment of delay lines; 13, high-precision ball-bush guide with finely ground and lapped cylindrical pin 14 for a backlash-free guidance of the shiftable mounting plate 15; 16, spindle and 17, nut to drive plate 15; 18, worm gear with high reduction ratio coupled to a step motor 19; 20, sensing element of the distance meter; 21, base plate; 22, fixed mounting plate; 23, circulation channel of the thermostating fluid. Reproduced from Reference 44 with permission of IOP Publishing Ltd.

generator, gives rise to standing waves into the solution at a series of characteristic frequencies, which are the resonance frequencies of the cell + solution system. At these frequencies, the voltage delivered by the output transducer goes through sharp maxima. The ultrasonic absorption coefficient α at the frequency ω corresponding to the central frequency of a maximum is related to its full width at half-maximum $\delta\omega$ through Equation 2.11, where c is the ultrasonic velocity in the solution. The resonator permits absorption measurements in the frequency range 0.1 to 15 MHz.

$$\alpha = \delta\omega/c \qquad (2.11)$$

In the pulse technique, the two parallel X-cut quartz plates are at a variable distance.[39,40,46] One transducer sends pulses of ultrasound of frequency ω through the solution, where they are attenuated before reaching the second transducer (Figure 2.6 bottom). The attenuation is measured as a function of the transducer separation distance using a comparison method. The absorption coefficient at frequency ω is obtained from the attenuation versus distance plot. The pulse technique permits measurements in the range 5 to 500 MHz.

Overall, these two techniques permit the determination of relaxation times in the range 0.5 µs to 0.3 ns. They have been used extensively at the early stage of kinetic studies of micellar solutions for the study of the exchange process of surfactants with a relatively short alkyl chain and, more recently, for the study of novel surfactants, including gemini (dimeric) surfactants. They have also been used for the study of phase transition in vesicle systems.

Other methods that have been described for absorption measurements below 100 kHz[47] are practically no longer in use. Indeed, such measurements are tedious and require a considerable amount of solution. Besides, this time window is covered by the p-jump and shock tube techniques.

Many of the reported ultrasonic absorption apparatuses are home-built. The companies Dispersion Technology Inc. (in the U.S.) and Heath Scientific Co. Ltd. (in the U.K.) manufacture apparatuses.

The principle of the dielectric absorption relaxation technique is very similar to that of the ultrasonic relaxation. The coupling between electrical waves and the chemical equilibrium occurs through the change of dipole moment associated with the chemical reaction. The drawback of this method lies in its extreme sensitivity to all rotational motions occurring in the system and which all give rise to relaxation signals. The chemical contributions are then more difficult to extract and analyze. Nevertheless, the dielectric relaxation has been recently used for the study of the kinetics of processes occurring in systems of threadlike micelles.[48,49]

6. Stopped-Flow

Stopped-flow apparatuses permit the fast mixing of two solutions, one of reactant A and the other of reactant B, and the monitoring of the course of reaction A + B $\rightarrow$ C. Thus the stopped-flow method deals with irreversible reactions and not equilibria and is not a real chemical relaxation method. Nevertheless, it can be made to work in conditions close to those prevailing in chemical relaxation methods. For example, two micellar solutions, one slightly below cmc, the other slightly above cmc, can be mixed and the micelle breakdown monitored. Similarly, one can induce a jump of ionic strength or pH. One can also mix two populations of identical micelles, vesicles, or microemulsion droplets containing reactive solubilizates and follow the course of the reaction between these solubilizates, thereby obtaining information on the rate of collisions between aggregates and/or on the rate of exit of the solubilizate from the aggregates. The main interest in the stopped-flow method is that it permits perturbations through composition jumps.

The solutions to be mixed are placed in two syringes, the pistons of which are actuated by a pneumatic system. The solutions flow into a mixing chamber, where intimate mixing is achieved in 1 to 5 ms, and then into an observation chamber, where the reaction is monitored by a conductivity or optical detection (changes of circular dichroism, optical density, optical rotation and fluorescence intensity, for instance have been used for this purpose).[6,8] A mixing system where the mixing time has been reduced to 10 μs has been reported.[50]

The stopped-flow method has been used to study the kinetics of micelle formation/breakdown in surfactant solutions (see Chapter 3), of the exchange process in micellar solutions of amphiphilic block copolymers (see Chapter 4, Sections IV and V), and also of collisions between droplets in microemulsions (see Chapter 5, Section VI.F). It has been also used to study the kinetics of the vesicle-to-micelle transformation (see Chapter 6) and of various types of chemical reactions performed in micelles or microemulsion droplets (see Chapter 10). The stopped-flow method has also been used to study the rate of dissolution of oil or water in microemulsions (see Chapter 5, Section VII.C). In such studies the syringe that contains the oil or water to be solubilized is of a much smaller diameter than that containing the microemulsion.

This permits mixing with a volume ratio of the mixed solutions as large as 100.

Recently stopped-flow devices have been inserted in synchrotron small angle x-ray scattering[27,51] and in small angle neutron scattering[52] machines to monitor kinetics of micelle breakdown,[51] of transitions between mesophases,[27] and formation and growth of vesicles.[52]

Stopped-flow apparatuses are commercially available from Applied Photophysics Ltd. (in the U.K.) or Hi-Tech Scientific (also in the U.K., this company also commercializes a setup for work under high pressure), and Bio-logic (France), to name a few. The mixing time can vary between 0.5 and 10 ms. Modern stopped-flow apparatuses are equipped with a diode array detector, which permits the acquisition of the full absorption spectrum of the system in a very short time. Thus the kinetics can be followed simultaneously at several wavelengths where different reactants or products absorb light.

In some apparatuses, very similar to stopped-flow the flow of reactants is continuous (continuous flow methods). The observation of the reaction is integrated. An apparatus based on this principle has been described that allows the study of systems with half-time of reaction as short as 5 μs.[53]

III. TIME-RESOLVED LUMINESCENCE QUENCHING

This method has been used extensively to measure the rate constant of exit of solubilizates that are phosphorescent or fluorescent from micelles. For this purpose the solubilizate is excited by an appropriate means, generally a pulse of laser light of appropriate wavelength, and brought to its triplet state. This state is usually long-lived and has a lifetime longer than the solubilizate residence time in the micelle. If the intermicellar solution contains a water-soluble quencher of the triplet state, then an increase of quencher concentration will result in a decrease of the measured triplet state lifetime until the point where all excited solubilizates exiting the micelles are quenched. Almgren et al.[54] showed that the measured lifetime is then equal to the reciprocal of the solubilizate exit rate constant k⁻. Several conditions must be met for the method to apply:

1. The solubilizate must be mainly located in the micelles.
2. The quenching of the triplet state of the solubilizate must take place in the aqueous phase.
3. The solubilizate concentration must be sufficiently low in order to avoid self-quenching.
4. The dynamics of exchange of the triplet state must be the same as that of the ground state. This condition has been checked.[55]

This method is illustrated in Figure 2.7, which shows the variation of the phosphorescence decay rate of 1-bromonaphthalene solubilized in sodium dodecylsulfate (SDS) micelles in the presence of an increasing concentration of sodium nitrite in the aqueous phase. The 1-bromonaphthalene triplet

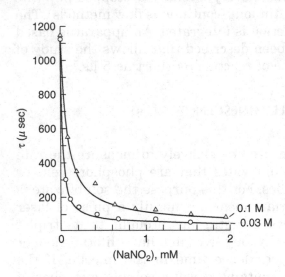

Figure 2.7 Variation of the phosphorescence lifetime of 1-bromonaphthalene with the sodium nitrite concentration and at SDS concentration as indicated on the plots. Reproduced from reference 54 with permission of the American Chemical Society.

is quenched by the nitrite ion.[54] The analysis of the results yielded the value of the exit rate constant of bromonaphthalene from SDS micelles.

Extensions of this method are possible. For instance the quencher may be micelle-solubilized (biphenyl) and the molecule giving rise to the triplet may be water-soluble (acetone).[56] Conditions similar to those described earlier must apply for the method to be valid in such a situation. Also, both the probe (1-bromonaphthalene) and the quencher (pyrene) can reside mainly in the micelles (SDS).[57]

Another extension of the method makes use of fluorescent rather than phosphorescent solubilizates (the excited state is a singlet state instead of a triplet state). Given that fluorescence lifetimes are much shorter than phosphorescence lifetimes, fluorescent probes or quenchers can be used only for studies of extremely fast exchanges. The method has also been extensively used to study collisions with temporary merging of micelles[58] and of microemulsions droplets.[59]

The principle of the method is briefly described. Consider a micellar solution (or a microemulsion) with a micelle-solubilized fluorescent probe. The fluorescence decay plot is linear and its slope yields the fluorescence lifetime τ_0 of the probe in its micellar environment (plot 1 in Figure 2.8). Add a

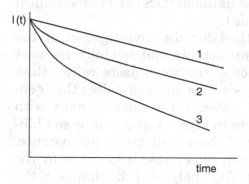

Figure 2.8   Variation of the intensity of fluorescence with time for a micelle-solubilized probe in the absence of quencher (plot 1); in the presence of a quencher that is insoluble in the intermicellar solution (plot 2); and in the presence of a quencher that is partitioned between micelles and intermicellar solution (plot 3).

quencher Q. If the quencher resides in the micelles a long time compared to τ_0, the probe lifetime is not affected by the presence of the quencher, but the time dependence of the fluorescence intensity is modified and becomes biphasic (plot 2 in Figure 2.8). The fast quenching process occurs in micelles containing one probe and one (or more) quencher(s). If the quencher is partitioned between micelles and intermicellar solution, a micelle containing an excited probe and no quencher at the time of excitation will have a finite probability to incorporate a quencher during the probe lifetime. This results in a fast quenching and in a variation of the slope of the long time portion of the fluorescence decay curve (plot 3 in Figure 2.8). From this variation one can extract the exit rate constant of the quencher from a micelle or microemulsion droplet, or the rate of collision with temporary merging of microemulsion droplets, depending on the system investigated.[60-64]

The above methods involve luminescent probes in the excited state, singlet or triplet. They make use of laser light for exciting triplet states or of more conventional light sources for exciting singlet states. Laser flash photolysis setups are commercially available, such as those from Applied Photophysics (U.K.). Fluorescence quenching studies require photon-counting apparatuses that are available from EGG Ortec and Photon Technology International (U.S.) or Photochemical Research Associates (Canada).

A closely related method for the investigation of the exchange process of both surfactants and solubilizates uses radicals of the surfactant or of the solubilizate rather than excited states. Pulse radiolysis uses electron pulses that generate hydroxyl radicals in water. These radicals react with the surfactant or solubilizate to give a surfactant or solubilizate radical.[65] A scavenger of these radicals — for instance, $Fe(CN)_6^{3-}$ — present in the aqueous phase is bleached by the radical exiting the micelles. The analysis of the change of the apparent absorption of the solution with the time at different quencher concentrations permits one to obtain the exit rate constant of the surfactant or solubilizate radical. In another pulse radiolysis study, the presence of a scavenger was not required as the surfactant adsorbed in the near UV.[66] The method is unfortunately restricted to the availability of a source of high-energy electrons (several MeV), not often available on campuses or research laboratories.

IV. NUCLEAR MAGNETIC RESONANCE (NMR)

Several properties probed in NMR spectroscopy can be used to study the rate of chemical processes. This section is restricted to the most easily accessible properties: the NMR chemical shifts and the line width of resonance signals.[67]

If a nucleus has a finite lifetime in a given state, the NMR resonance signal corresponding to this nucleus is broadened. The broadening, characterized by the width $\delta\omega$ of the band, is related to the lifetime τ of the nucleus in this state by: $\delta\omega = h/2\pi\tau$. If the nucleus exists in two states (for instance, micellar M and nonmicellar F states) and exchanges between the two states (as surfactants in micellar solutions), the NMR spectra depend on the rate at which the nucleus is exchanged and the lifetimes τ_M and τ_F of the nucleus in the two states, as compared to the NMR time scale.[67] The NMR time scale depends on the nucleus used and on the characteristics of the NMR spectrometer, mainly its frequency of operation.[68]

Suppose that the exchange process between the two states is extremely slow on the NMR time scale — that is, both τ_M and τ_F are much larger than $(\omega_F - \omega_M)^{-1}$, ω_F and ω_M being the resonance frequencies of the nucleus in states F and M. The NMR spectrum of the system then consists of two bands centered at frequencies ω_F and ω_M and broadened by the nucleus exchange between the two states. If the two signals are well separated, their broadening can be analyzed with existing theories to extract the lifetime of each state.

If the exchange is very fast — that is, if τ_M and τ_F are much smaller than $(\omega_F - \omega_M)^{-1}$ — the two resonance signals collapse into a single one centered at an intermediate frequency given by

$$\omega_0 = p_F\omega_F + p_M\omega_M \qquad (2.12)$$

with

$$p_F = 1 - p_M = \tau_F/(\tau_M + \tau_F) \qquad (2.13)$$

p_F and p_M represent the fractions of nuclei in states F and M (or more generally in the two exchanging states).

If the exchange is slow enough to contribute to the width of the signal but is still well beyond the rate corresponding to the slow exchange (separated signals), the signal is not completely collapsed. Its analysis also yields information on the lifetimes in the two states.

In the special case where $p_F = p_M$ — that is, equal populations in the two states — the equations become particularly simple as $\tau_M = \tau_F = \tau$. For micellar solutions the situation of equal populations corresponds to a surfactant concentration equal to twice the critical micellization concentration.[69] At this concentration $1/\tau = \Delta\Delta\delta\omega/2$, where $\Delta\Delta\delta\omega$ is the difference between the linewidths of the monomer signal at twice the cmc (broadened by the exchange) and below the cmc (no exchange).[69] The coalescence of the two signals into a single one then occurs when $\tau = 2^{0.5}/(\omega_F - \omega_M)$. The NMR method provides an easy access to the rate constant for surfactant and solubilizate exchange. It has been recently used to demonstrate that gemini (dimeric) surfactants exchange much more slowly between micelles and bulk phase than conventional surfactants of the same chain length.[69] Figure 2.9 illustrates the NMR spectra obtained with such surfactants. The presence of two broad signals in the spectra of micellar solutions of the gemini surfactants 14-2-14 and 18-2-8 indicates slow surfactant exchange. The single broad signal observed with the shorter chain gemini surfactant 12-2-12 is indicative of fast exchange on the NMR time scale.

The advent of spectrometer operating at increasingly higher frequencies and the variety of nuclei that can be used to probe exchange processes in micellar solutions and microemulsions will undoubtedly lead to an increased use of NMR for the study of these processes.

The self-diffusion coefficients of the different chemical species present in a solution can be measured using the so-called pulsed field gradient Fourrier-transform NMR (PFG-FT NMR) spectroscopy.[70] The diffusion coefficients do not give direct access to the dynamics of the processes of exchange or micelle formation-breakdown. They can nevertheless be use-

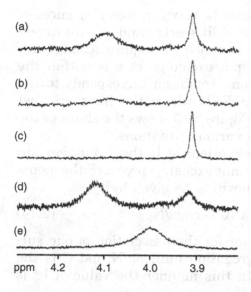

Figure 2.9 Part of the 400 MHz ^{1}H NMR spectra of gemini surfactants in D_2O at 25°C. The signals observed are those of the $N^+CH_2CH_2N^+$ protons (3.92 ppm in the bulk phase and 4.03–4.13 ppm in micelles). (a), (b) and (c): surfactant 14-2-14 at 0.25 mM, 0.167 mM and 0.125 mM; (d) surfactant 18-2-8 at 0.5 mM; (e) surfactant 12-2-12 at 1.25 mM. Reproduced from Reference 69 with permission of the Royal Society of Chemistry.

ful for revealing certain dynamic aspects of the investigated micellar solutions.

V. ELECTRON SPIN RESONANCE (ESR)

Just like NMR, ESR is sensitive to exchange phenomena. This also results in a broadening and shift of the ESR lines. However, since electrons rather than nuclei are now involved, the time scale of processes that can be probed by ESR is shifted to much shorter values. Typically exchange processes with characteristic times between 10 µs and 1 ns can be probed by ESR.[71] The main problem in using ESR is that it requires spin-labeled surfactants. The process investigated is then the exchange of the spin-labeled surfactant and not that of the surfactant of interest.

The exchange of surfactants between free and micellar states can be observed in the ESR spectra and quantitatively analyzed provided that interference with spin exchange can be excluded and that the spin exchange rate is within the range 10^5–10^7 Hz. The second condition corresponds to the slow exchange condition in ESR where the lines are just broadened but not shifted. Figure 2.10 shows the shape of the spectra of an ESR probe in various situations.[72]

Using the same approximations as in the Aniansson and Wall treatment of the surfactant exchange process[73] the monomer lifetime τ_M has been shown to be given by[74]

$$1/\tau_M = k^+(C - cmc)/N \qquad (2.14)$$

where k^+ is the rate constant for the association of one surfactant to a micelle of aggregation number N and C is the surfactant concentration. In this manner the value of k^+ is

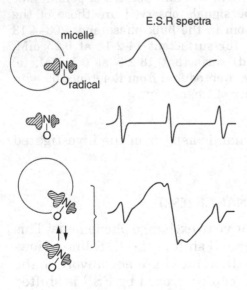

Figure 2.10 ESR spectra of a nitroxide radical solubilized in micelles (*top*), dissolved in water (*bottom*), and exchanging between micelles and water (*middle*). Reproduced from Reference 72 with permission of Springer Verlag.

easily obtained from measurements of line broadening as a function of the surfactant concentration. One can then obtain all the dynamic characteristics of the investigated surfactant. As noted earlier, the ESR method only yields information on spin-labeled surfactants. This information can be easily compared to that for unlabeled surfactants, and one can then check that both unlabeled and labeled surfactants behave similarly.

VI. RHEOLOGY

The use of rheological methods as tools for the study of the dynamics of surfactants self-assemblies is illustrated by the following two examples.

The first example concerns surfactants that give rise in aqueous solutions to very elongated micelles, the so-called "giant" micelles, which are rather flexible and can be micron-long (see Chapter 1, Section III.B). The solutions of giant micelles possess specific rheological properties. For instance, they can display shear-induced viscoelasticity, i.e., they become viscoelastic at above a critical shear rate (see Chapter 9, Section VI). They are often viscoelastic, and an applied stress may relax with a single relaxation time, a very surprising result in view of the large polydispersity of giant micelles (see Chapter 9, Sections III and IV). This time has been related to the time after which a giant micelle breaks reversibly into two daughter micelles.[75,76] In the case of systems showing shear-induced structuring, rheological methods may also be able to provide information on the formation of ringlike micelles.[77] The second example is that of "associative polymers," which are hydrophilic polymers with some pendent hydrophobic grafts. Associative polymers can give rise to extremely viscous or viscoelastic aqueous solutions and even to gels because hydrophobic grafts from several polymers can self-associate, thereby forming hydrophobic microdomains and giving rise to a network. Rheological studies of viscoelastic solutions of associative polymers have shown that they also sometimes relax with a single time constant.[78] This time has been related to the time required for the exit of one hydrophobic graft from the microdomain to which it belongs.[78] The rate of exit of one hydrophobic graft from a micelle then becomes an important characteristic of the rheological behavior of the system. The rheology of these systems is an extremely important property when it comes to practical aspects such as their handling and uses. It is therefore not surprising that their rheological properties have been extensively investigated with the aim of identifying the processes responsible for their special behavior and extracting information on their dynamics.

Rheological studies show some similarities with chemical relaxation studies. For instance, a rectangular shear rate is applied and the relaxation of the stress is monitored. This directly yields the stress relaxation time(s). One can also apply a sinusoidal deformation or strain of angular frequency ω. The response of the system is a two-component sinusoidal shear stress. The first component is in phase with the strain and corresponds to the elastic (storage) properties of the system. The second component is out of phase with the strain with a phase angle δ, and corresponds to the viscous loss in the system. These quantities give access to the storage (elastic) modulus $G'(\omega)$ and to the loss (viscous) modulus $G''(\omega)$, with $G''(\omega)/G'(\omega) = tg\delta$. As in the case of chemical relaxation methods with harmonic perturbation, the variations of $G'(\omega)$ and $G''(\omega)$ with ω yield the relaxation time(s) of the system.

A large number of rheological instruments (viscometers or rheometers) are commercially available from many companies: Paar (Austria), Rheometrics and Carrimed (U.S.), Contraves (Switzerland), and Haake (Germany), to name a few. These instruments differ by their geometry (capillary arrangement coaxial cylinder, cone-plate, and rolling-ball), the type of perturbation, and the range of frequency available in the case of sinusoidal perturbation. These instruments measure quantities from which one can calculate the main rheological parameters such as shear stress or strain, viscosity, and modulus. The commercial apparatuses permit one to cover the time range between, say, 1 ms and hundreds of seconds. More details on rheological methods can be found in Chapter 9, Section IV.B.

VII. MISCELLANEOUS METHODS

This section briefly introduces two techniques that seem to have potential for the study of micellar kinetics.

A. Capillary Wave Propagation

The propagation of capillary waves at the surface of surfactant solutions is influenced by the adsorption of the surfactant. The damping of the waves yields information on the mechanism of surfactant adsorption. If the surfactant adsorption occurs from a micellar solution and if the adsorption is somehow coupled to the micelle formation/breakdown, then the study of the damping of capillary waves may provide information on the kinetics of this process. Recent results suggest that such is the case.[79] An apparatus for the generation of capillary waves and measurement of their damping has been described.[79] The fitting of the data to the available theories yielded values of the relaxation time for the micelle formation/breakdown in fair agreement with the values measured by p-jump or T-jump. The main drawback of the method is its enormous sensitivity to the presence of hydrophobic impurities that all adsorb at the surface and greatly perturb the damping of capillary waves. Its advantage over measurements of dynamic surface tension measurements for the study of micelle kinetics is that the perturbation brought to the system for the generation of capillary waves is extremely small and that the response of the system can be dealt with using the linear approximation, as in chemical relaxation. Dynamic surface tension measurements usually involve a strong perturbation of the system under study and lead to the appearence of nonlinear hydrodynamic phenomena that may complicate the analysis of the data in view of extracting information on micelle kinetics.

B. Fluctuation Spectroscopy

The fluctuations of the electrical potential between two electrodes placed in two cells filled with a micellar solution and connected by a capillary tube are expected to be affected by the micelle lifetime. Indeed, the fluctuations should differ

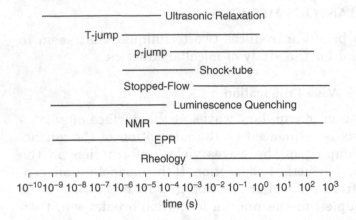

Figure 2.11 Time scales of the methods reviewed in this chapter.

depending on whether the micelle lifetime is shorter or longer than the time required for diffusion through the tube.[80] Thus the technique can in principle be used to study micelle kinetics.

VIII. CONCLUSIONS

Figure 2.11 concludes this chapter by schematically representing the time ranges covered by the methods reviewed in this chapter.

REFERENCES

1. Eigen, M. *Disc. Faraday Soc.* 1954, 17, 194.

2. Eigen M., De Maeyer, L. In *Techniques of Organic Chemistry*, Friess, S.L., Lewis, E.S., Weissberger, A., Eds., Wiley-Interscience, New York, 1963, p. 895.

3. Kustin, R., Ed. *Methods in Enzymology*; vol. XVI, Academic Press, New York, 1969.

4. Hague, D. *Fast Reactions*, Wiley-Interscience, London, 1971.

5. Hammes, G.C., Ed. *Techniques of Organic Chemistry*, vol. VI, part II, *Investigations of Elementary Steps in Solution and Very Fast Reactions*, Wiley-Interscience, New York, 1974.

6. Wyn-Jones, E., Ed. *Chemical and Biological Applications of Relaxation Spectrometry*, D. Reidel, Dordrecht, Holland, 1975.

7. Bernasconi, C.F. *Relaxation Kinetics*, Academic Press, New York, 1976.

8. Gettins, W.J., Wyn-Jones, E., Eds. *Techniques and Applications of Fast Reactions in Solution*, D. Reidel, Dordrecht, Holland, 1979.

9. Czerlinski, G. *Chemical Relaxation*, M. Dekker, New York, 1965.

10. Lang, J., Zana, R. In *Surfactant Solutions. New Methods of Investigation*, R. Zana, Ed., M. Dekker, New York, 1987, p. 405.

11. Castellan, G. *Ber. Bunsenges. Phys. Chem.* 1963, 67, 898.

12. Goldmints, I., Holzwarth, J.S., Smith, K.A., Hatton, T.A. *Langmuir* 1997, 13, 6130.

13. Aniansson, E.A.G. *Ber. Bunsenges. Phys. Chem.* 1978, 82, 981.

14. Chan, S., Herrmann, U., Ostner, W., Kahlweit, M. *Ber. Bunsenges. Phys. Chem.* 1977, 81, 396.

15. Aniansson E.A.G., Wall, S.N. *Ber. Bunsenges. Phys. Chem.* 1977, 81, 1293.

16. Chan, S., Kahlweit, M. *Ber. Bunsenges. Phys. Chem.* 1977, 81, 1294.

17. Friess, S.L., Lewis, E.S., Weissberger, A., Eds. *Techniques of Organic Chemistry*, vol. VIII, Wiley-Interscience, New York, 1963, p. 980.

18. French, T., Hammes, G.C. In *Methods in Enzymology*, K. Kustin, Ed., vol. XVI, Academic Press, New York, 1969, p. 3.

19. Turner, D., Ryan, R., Flynn, G., Sutin, N. *Biophys. Chem.* 1974, 2, 385.

20. Turner, D., Flynn, G., Sutin, N., Beitz, J. *J. Am. Chem. Soc.* 1972, 94, 1554.

21. Ameen, S. *Rev. Sci. Instrum.* 1975, 46, 1209.

22. Holtzwarth, J. In *Techniques and Applications of Fast Reactions in Solution*, W.J. Gettins and E. Wyn-Jones, Eds., D. Reidel, Dordrecht, Holland, 1979, p. 47.

23. Holtzwarth, J., Eck, V., Genz, A. In *Spectroscopy and the Dynamics of Molecular Biology Systems*, Bayley, P.M., Dale, R.E., Eds., Academic Press, London, Dordrecht, 1985, p. 315.

24. Holtzwarth, J., Schmidt, A., Wolff, H., Volk, R. *J. Phys. Chem.* 1977, 81, 2300.

25. Hoffmann, H., Yeager, E. *Rev. Sci. Instrum.* 1968, 39, 649.

26. Waton, G., Porte, G. *J. Phys. II France* 1990, 3, 515.

27. Bras, W., Ryan, A.J. *Adv. Colloid Interface Sci.* 1998, 75, 1.

28. Simmons, B., Agarwal, V., Singh, M., McPherson, G., John, V., Bose, A. *Langmuir* 2003, 19, 6329.

29. Tondre, C., Lang, J., Zana, R. *J. Colloid Interface Sci.* 1975, 52, 372.

30. Lang, J., Zana, R. Unpublished results.

31. Candau, S.J., Merikhi, F., Waton, G., Lemaréchal, P. *J. Phys. France* 1990, 51, 977.

32. Knoche, W., Wiese, G. *Chem. Instrum.* 1973–1974, 5, 91.

33. Knoche, W., Wiese, G. *Rev. Sci. Instrum.* 1976, 47, 220.

34. Hoffmann, H., Yeager, E. *Rev. Sci. Instrum.* 1968, 39, 1151.

35. Schelly, Z. *Current Opin. Colloid Interface Sci.* 1997, 2, 37 and references therein.

36. Runge, F., Schlicht, L., Spilgies, J.H., Ilgenfritz, G. *Ber. Bunsenges. Phys. Chem.* 1994, 98, 506.

37. Tekle, E., Schelly, Z. *J. Phys. Chem.* 1994, 98, 7657.

38. Menzel, K., Rupprecht, A., Kaatze, U. *J. Phys. Chem.* 1985, 89, 2896.

39. Nishikawa, S., Kotegawa, K. *J. Phys. Chem.* 1997, 101, 1255.

40. Kuramoto, N., Ueda, M., Nishikawa, S. *Bull. Chem. Soc. Jap.* 1994, 67, 1560.

41. Eggers, F., Funck, T. *Rev. Sci. Instrum.* 1973, 44, 969.

42. Eggers, F., Funck, T., Richmann, K. *Rev. Sci. Instrum.* 1976, 47, 361.

43. Kaatze, U., Wehrmann, B., Potel, C.H. *J. Phys. E.: Sci. Instrum.* 1987, 20, 1025.

44. Eggers, F., Kaatze, U. *Meas. Sci. Technol.* 1996, 7, 1.

45. Labhardt, A., Schwarz, G. *Ber. Bunsenges. Phys. Chem.* 1976, 80, 83.

46. Kaatze, U., Kuhnel, V., Menzel, K., Schwerdtfeger, S. *Meas. Sci. Technol.* 1993, 4, 1257.

47. Hayakawa, R., Shintani, H., Wad, Y. *Jap. J. Appl. Phys.* 1974, 13, 787.

48. Imai, S., Shikata, T. *Langmuir* 1999, 15, 8388.

49. Shikata, T., Imai, S. *Langmuir* 2000, 16, 4840.

50. Regenfuss, P., Clegg, R., Fulwyler, M., Barrantes, F., Jovin, T. *Rev. Sci. Instrum.* 1985, 56, 283.

51. Eastoe, J., Dalton, J.S., Downer, A., Jones, G., Clarke, D. *Langmuir* 1998, 14, 1937.

52. Grillo, G., Kats, E.I., Muratov, A.R. *Langmuir* 2003, 19, 4573.

53. Holzwarth, J. In *Techniques and Applications of Fast Reactions in Solution*, Gettins, W.J., Wyn-Jones, E., Eds., D. Reidel, Dordrecht, Holland, 1979, p. 13.

54. Almgren, M., Grieser, F., Thomas, J.K. *J. Am. Chem. Soc.* 1979, 101, 279.

55. Kowalczyk, A.A., Vecer, J., Hodgson, B.W., Keene, J.P., Dale, R.E. *Langmuir* 1996, 12, 4358.

56. Leigh, W.J., Scaiano, J.C. *J. Am. Chem. Soc.* 1983, 105, 5652.

57. Gläsle, K., Klein, U.K., Hauser, M. *J. Mol. Struct.* 1982, 84, 353.

58. Malliaris, A., Lang, J., Sturm, J., Zana, R. *J. Phys. Chem.* 1987, 91, 1475.

59. Jada, A., Lang, J., Zana, R., Makhloufi, R., Hirsch, E., Candau, S.J. *J. Phys. Chem.* 1990, 94, 387.

60. Infelta, P.P. *Chem. Phys. Lett.* 1979, 61, 88.

61. Zana, R. In *Surfactant Solutions. New Methods of Investigation*, Zana, R., Ed., Marcel Dekker, New York, 1987, p. 241.

62. Almgren, M. *Adv. Colloid Interface Sci.* 1992, 41, 9.

63. Gehlen, M.H., De Schryver, F.C. *Chem. Rev.* 1993, 93, 199.

64. Tachiya, M. *Chem. Phys. Lett.* 1975, 33, 289.

65. Almgren, M., Grieser, F., Thomas, J.K. *J. Chem. Soc., Faraday Trans.* I 1979, 75, 1674.

66. Henglein, A., Proske, Th. *J. Am. Chem. Soc.* 1978, 100, 3706.

67. Kaplan, J.I., Frenkel, G. *NMR of Chemically Exchanging Systems*, Academic Press, New York, 1980.

68. Bryant, R.G. *J. Chem. Ed.* 1983, 60, 933.

69. Huc, I., Oda, R. *Chem. Comm.* 1999, 2025.

70. Lindman, B., Söderman, O. In *Surfactant Solutions. New Methods of Investigation.* Zana, R., Ed., M. Dekker, New York, 1987, p. 295.

71. Atherton, N.M. *Electron Spin Resonance*, Ellis Horwwod, Chichester, 1973.

72. Nakagawa, T., Jizomoto, H. *Colloid Polym. Sci.* 1979, 257, 502.

73. Aniansson, E.A.G., Wall, S.N. *J. Phys. Chem.* 1974, 78, 1024 and 1975, 79, 857.

74. Schmidt, D., Gahwiller, Ch., Von Planta, C. *J. Colloid Interface Sci.* 1981, 83, 191.

75. Kern, F., Lequeux, F., Zana, R., Candau, S.J. *Langmuir* 1994, 10, 1714.

76. Rehage, H., Hoffmann, H. *Molecular Phys.* 1991, 74, 933.

77. Cates, M.E., Candau, S.J. *Europhys. Lett.* 2001, 55, 887.

78. Annable, T., Buscall, R., Ettelaie, R., Whittlestone, D. *J. Rheol.* 1993, 37, 695.

79. Noskov, B.A., Grigoriev, D.O. *Prog. Colloid Polym. Sci.* 1994, 97, 1 and *Langmuir* 1996, 17, 3399.

80. Green, M.E., Krishnamurthi, M. *J. Colloid Interface Sci.* 1989, 127, 295.

3

Dynamics in Micellar Solutions of Surfactants

CONTENTS

I. INTRODUCTION

In aqueous solutions, surfactants self-associate into micelles
at a concentration above the critical micellization concentration (cmc). The formation of micelles is a cooperative process
that is well described by the mass action law. Micelles can be
considered as chemical species that are in equilibrium with
free (nonmicellized) surfactant. These various aspects are
briefly reviewed in Chapter 1.

Much was known about the equilibrium properties of
micellar solutions of surfactants in 1964, but very little was
known about the dynamics of these systems. In this chapter
dynamics is mainly used to refer to the rate at which micelles
form or break down, or the time a surfactant remains in a
micelle. At that time only qualitative statements could be
made about these processes, namely that the rate at which
micelles form or break down was rather fast, or in other words,
the lifetime of a micelle was short. Indeed, it had been realized
that when a micellar solution was diluted below the cmc, the
micelles disappeared instantaneously at the time scale of this
experiment, i.e., seconds. Besides, ESR experiments using
spin-labeled surfactants had shown that the residence time
of labeled surfactants in micelles was short. In 1964 the chemical relaxation techniques (see Chapter 2, Section II), which
had been in full use for the previous ten years for the study
of the kinetics of fast elementary steps in chemical reaction
mechanisms,[1] started to be applied to micellar solutions. The
stopped-flow (SF) technique with conductance readout was
the first one used to probe the kinetics of micelles.[2] The experiments showed that the rate of micelle breakdown was fast,
with a half-life of reaction shorter than 10 ms (dead time of

the SF apparatus used). The same year, NMR experiments showed that a single resonance peak characterized micellar solutions, indicating that the exchange of the surfactant between free and micellized states was fast, in the submilli-second time scale.[3] In 1965 studies of micellar solutions by pressure-jump (p-jump)[4] and SF[5] provided further evidence for the fast breakdown of micelles but yielded no reliable values of the relaxation times. It was only in 1966, with the first investigation performed by means of the temperature-jump (T-jump) technique, that a relaxation process associated with micellar equilibria was clearly observed and values of relaxation times reported.[6] This process was assigned to the reversible exit (or dissociation) of one surfactant from the most probable micelle A_N (A refers to a surfactant and N = micelle aggregation number) according to

$$A_N \rightleftarrows A_{N-1} + A \qquad (3.1)$$

This reaction can also be considered as that by which one surfactant is exchanged between the two micelles A_N and A_{N-1}. It is often referred to as *surfactant exchange*.

In the years that followed until around 1974, many chemical relaxation studies of micellar systems were reported. A number of studies of ionic and nonionic surfactants were performed by T-jump, p-jump, and SF.[7-13] The observed relaxation process (relaxation time > 50 µs) was assigned to reaction (3.1) or to a micelle reorganization. A second group of studies of ionic surfactants used ultrasonic absorption relaxation techniques (Chapter 2, Section II.D.5).[14-19] The observed relaxation processes were characterized by relaxation times much shorter than 1 µs and assigned to either the surfactant exchange[14-16] (i.e., same assignment as in References 6–12) or reactions of counterion association/dissociation to/from micelles.[18,19] The confusion that existed until 1974 concerning the dynamics of micelles and the origins of the observed relaxations is summarized in a short report presented at a NATO Advanced Institute held in 1974.[20]

The situation dramatically changed between the end of 1973 and early 1975. First, a p-jump and shock-tube (ST) study[21] showed unambiguously that micellar solutions are characterized by *two* well separated relaxation processes whereas, thus far, all studies had reported the observation of only one process. It was also shown that the slow process is much affected by the presence of impurities. Second, several authors[21-24] independently assigned the fast process to reaction (3.1), as in References 14–16, and the slow process to the micelle formation/breakdown that can be schematically represented by the single-step reaction (3.2). Nakagawa[23] was the first to associate the relaxation processes with shifts of the distribution curve of the micelle aggregation number N_s versus s (s = number of surfactants per micelle) along the s-axis, induced by the applied perturbation. Last but not least, Aniansson and Wall[24] derived the first analytical expressions of the relaxation times characterizing the two processes based on this assignment. Micelles were assumed to form and break down through a series of steps involving one surfactant at a time, as in reaction (3.1).

$$NA \rightleftarrows A_N \qquad (3.2)$$

Extensive chemical relaxation data for solutions of ionic surfactants were first interpreted on the basis of this theory in 1976.[25] The conclusions reported in this study still constitute the basis for the present understanding of the dynamics of micelles. The treatment of Aniansson and Wall was later refined and extended to ionic surfactant micelles, taking into account the presence of the counterions and of added electrolyte.[26,27] It was also extended to include fragmentation/coagulation (or fission/fusion) reactions (3.3) by which a micelle A_s can break into two daughter micelles A_i and A_j (with s = i + j), and conversely.[28]

$$A_s \rightleftarrows A_i + A_j \qquad (3.3)$$

The contribution of reactions (3.3) to the micelle kinetics becomes important for systems where collisions between aggregates are possible. These very useful extensions permitted a more thorough analysis of the chemical relaxation results but did not affect the main conclusions reached in 1976. The theory was also extended in the direction of mixed micelles.[27,29-31] It predicted the existence of three relaxation processes: two fast ones associated with the entry/exit of the two surfactants into/from mixed micelles and a slow process due to the formation/breakdown of the mixed micelles.

The very large number of chemical relaxation studies of micellar solutions of conventional surfactants reported since 1976, including some studies of surfactants in nonaqueous solvents, all confirmed the conclusions reached in Reference 25. The most recent studies of micellar dynamics concern aqueous solutions and two new types of surfactants: the dimeric (gemini) surfactants (see Sections III.B.7 and III.C) and the amphiphilic block copolymers (see Chapter 4). These surfactants deviate from the general behavior reported for conventional surfactants.

Thus far only processes involving motion of the surfactant as a whole have been mentioned. Other processes may occur in micellar solutions: internal motion of the surfactant alkyl chains within the micelles; exchange of counterions between free and micelle-bound states; and fast changes of micelle shape, among others. Also in the case of solubilized systems, i.e., micellar solutions that have solubilized compounds that are sparingly soluble in water, the solubilizate may exchange between micelles and the intermicellar solution. The dynamics of the exchange of counterions and of solubilizates are reviewed later. The dynamics of internal motions of the surfactant alkyl chains are not dealt with in this chapter, but some information and references can be found in Chapter 5, Section V. Some information on the fluctuations of micelle shapes can be found in Chapter 1, Section III.B.

The chapter is organized as follows. Section II briefly recalls the theoretical aspects of micellar dynamics and the expressions of the relaxation times characterizing the main relaxation processes (surfactant exchange, micelle formation/breakdown). Section III reviews studies of micellar kinetics of various types of surfactants: conventional surfactants with a hydrocarbon chain, surfactants with a fluorinated chain, and gemini (dimeric) surfactants. Section IV deals with mixed micellar solutions. Section V considers the dynamics of solubilized systems. Section VI reviews the dynamics of surfactant aggregates bound to polymers. Section VII deals with the dynamics of miscellaneous processes. Section VIII considers the relevance of micellar kinetics in practical aspects of surfactant solutions. Section IX concludes this chapter.

The dynamics of micelles of amphiphilic block copolymers (macromolecular surfactants) of even relatively small size is reviewed in Chapter 4.

II. SURFACTANT EXCHANGE AND MICELLE FORMATION/BREAKDOWN: THEORETICAL ASPECTS

This section recalls the main aspects of the derivation of the expressions of the relaxation times for the surfactant exchange process and for the micelle formation/breakdown as done by Aniansson and Wall[24] in 1974 and 1975, and the main extensions of this theory by Kahlweit et al.[26,28] and Hall[27] in the years that followed.

A. Aniansson and Wall Theory of Micellar Kinetics

1. Basic Assumptions

The authors start from the experimentally well-established fact that relatively dilute micellar solutions are characterized by two well-separated relaxation processes. They attribute the fast process to the exchange of a surfactant A between aggregates (micelles) A_s and A_{s-1} as in reaction (3.4), with the rate constants of association (entry), k_s^+, and dissociation (exit), k_s^-:

$$A_{s-1} + A \underset{k_s^-}{\overset{k_s^+}{\rightleftarrows}} A_s \qquad (3.4)$$

They also assign the slow process to the micelle formation/breakdown (global reaction (3.2)) and assume that this reaction takes place via a series of stepwise reactions (3.4).

Reactions (3.3) of fragmentation/coagulation are excluded. The contribution of the counterions is not included.

The authors proceed by using a model of micellar solution at surfactant concentration C > cmc, which is based on the result of thermodynamic calculations of the aggregate size distribution curve for fairly dilute micellar systems.[32-34] This curve is represented in Figure 3.1 as a plot of $[A_s]$, concentration of aggregates A_s, against s. It shows a minimum and a maximum, the latter occurring at a value of s close to the micelle aggregation number N that can be measured by means of classical methods. The range around s = N corresponds to the micelles proper.[24] The deep minimum in the concentration of the aggregates between the oligomers (s = 1, 2, ...) and the

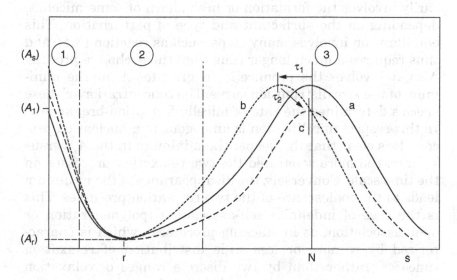

Figure 3.1 Distribution curve of the micelle aggregation number and its modifications after a very rapid step perturbation. (a) distribution curve before perturbation with a maximum at s = N and a minimum at s = r; (b) state of pseudo-equilibrium of the system reached upon equilibration of the micelles proper at constant number of micellar species; (c) final distribution curve reached after equilibration between micelles and oligomers. Regions 1 and 3 correspond to oligomers and micelles proper. Region 2 is that of the species around the minimum of the distribution curve.

micelles proper is responsible for the two-step relaxation behavior of micellar systems. Indeed, upon perturbation of such a system, the micelles proper (and oligomers) undergo a fast equilibration according to exchange reaction (3.4), where each micellar species gains or loses a small (fractional) number of monomers. This process qualitatively results in a shift of the range of micelles proper of the distribution from the initial curve **a** to curve **b** in Figure 3.1 and leaves the number of micelles constant. However, the distribution curve **b** corresponds to a state of pseudo-equilibrium, as the equilibrium between oligomers and micelles proper did not have time to be reached during the fast exchange process. This equilibration leads to the equilibrium distribution curve **c,** which corresponds to the final state of the system and necessarily involves the formation or breakdown of some micelles, depending on the surfactant and type of perturbation. This equilibration involves many steps, such as reaction (3.4), and thus requires a much longer time than the exchange process. Also, it involves the premicellar aggregates A_r at the minimum of the size distribution curve. The concentration of these species determines the rate of micelle formation-breakdown. In this respect micellization is analogous to a nucleation process. It is clear that the deeper the minimum in the A_s versus s curve, the more separated the two relaxation processes on the time scale. Conversely, the disappearance of the minimum leads to the coalescence of the two relaxation processes. This is the case of indefinite self-association (polymerization or open association, as in stacking processes), which is characterized by a more or less wide distribution of relaxation times[35–37] rather than by two discrete ranges of relaxation times.

Aniansson and Wall[24] pointed out the analogy existing between the response of a micellar system to a sudden perturbation and heat conduction (through diffusion) in a system constituted by two metal blocks (oligomers and micelles proper) connected by a thin wire (aggregates around the distribution curve minimum). When heat is provided to the system, a fast thermal equilibration occurs within each block, owing to their large heat conductivity. Then the thermal equilibrium between the two blocks is slowly established via heat flow through the connecting wire.

In fact, Aniansson and Wall showed that the analogy between micellar kinetics and diffusion phenomena goes quite far by introducing into the rate equations corresponding to reaction (4) the relative concentration change ξ_s upon perturbation, and the quantity J_s, which is analogous to a flux, given by

$$\xi_s = \{[A_s] - [A_s]^{eq}\}/[A_s]^{eq} \tag{3.5}$$

$$J_s = -k_s^- [A_s]^{eq}[\xi_s - \xi_{s-1}(1 + \xi_1) - \xi_1] \tag{3.6}$$

where $[A_s]^{eq}$ is the equilibrium concentration of A_s. On the assumption of small perturbations (relaxation condition) and of a size distribution curve large enough for s to be treated as a continuous variable, the rate equations corresponding to reactions (4) become

$$J_s = -k_s^- [A_s]^{eq}\left(\frac{\delta\xi_s}{\delta s} - \xi_1\right) \text{ and } -\frac{\delta J_s}{\delta s} \cong [A_s]^{eq}\frac{\delta\xi_s}{\delta t} \tag{3.7}$$

These equations are formally identical to Fick's laws for diffusion in a tube. The space coordinate, the diffusion coefficient, the concentration, and the section of the tube identify to s, k_s^-, ξ_s and $[A_s]^{eq}$.

2. Relaxation Time for the Exchange Process

The following assumptions were made by Aniansson and Wall:

(1) The amplitude of the perturbation is small.
(2) There is no exchange of surfactants between oligomers and micelles proper during the exchange process.
(3) The shape of the distribution curve around the maximum is Gaussian, that is

$$[A_s] = [A_s]^0 \exp[-(N - s)^2/2\sigma^2] \tag{3.8}$$

where σ^2 is the variance of the distribution. This assumption is consistent with theoretical calculations of the size distribution function for dilute solutions:[34]

(4) The rate constants k_s^+ and k_s^- are independent of s and equal to k^+ and k^- in the range of micelles proper. This assumption reduces the spectrum of relaxation times associated with the fast process to a single relaxation time, given by

$$1/\tau_1 = (k^-/\sigma^2) + (k^-/N)a \qquad (3.9)$$

In Equation 3.9 N is the average micelle aggregation number, that is, the average number of surfactants making up a micelle, a quantity that is experimentally accessible by a number of techniques. a is the reduced surfactant concentration given by

$$a = (C - [A_1]^{eq})/[A_1]^{eq} \qquad (3.10)$$

To a first approximation, the free surfactant concentration $[A_1]^{eq}$ can be taken as the cmc. Thus, $1/\tau_1$ should increase linearly with C provided that all other quantities in Equation 3.9 are independent of C. From such plots, one can obtain k^- and σ^2, provided that N is known. Notice that $T_R = N/k^-$ is the *average residence time of a given surfactant in the micelle* whereas $1/k^-$ is the residence time of any surfactant in the micelles. The variance σ^2 characterizes the micelle polydispersity. Moreover, since $[A_1]^{eq} \cong$ cmc, k^+ can be obtained from

$$k^+ \cong k^-/cmc \qquad (3.11)$$

Most of the assumptions, particularly (3) and (4) under which Equation 3.9 was derived, were later relaxed.[38] Thus Equation 3.9 was shown to remain valid if k_s^+ and k_s^- vary linearly with s.[38] Numerical calculations showed that very large deviations from a linear dependence of the two rate constants on s are required in order that the fast process be characterized by more than one experimentally accessible relaxation time.[38–40] An expression of τ_1 in the case of not very small perturbations was also derived.[24]

Criticism[41] concerning the validity of assumption (2) was shown to be unfounded.[28,39,42] Nevertheless, it is noteworthy that Equation 3.9 strictly applies only to nonionic surfactants and dilute micellar solutions.[27,43–45]

Aniansson[46] further derived the expression of the surfactant residence time T_R in a micelle of aggregation number N:

$$T_R = N/k^- = (l_b^2/D)\exp(\varepsilon/k_B T) \qquad (3.12)$$

where D is the diffusion coefficient of the free surfactant, l_b is a molecular distance (0.1–0.2 nm) and ε is the free energy of dissociation of one surfactant from the micelle. The model used suggested that ε should vary with the surfactant alkyl chain length (characterized by its carbon number m) exactly as the free energy of transfer ΔG_{tr}^0 of the surfactant alkyl chain from the micelle core to water, that is, linearly with m. The free energy increment per methylene group should have the same value for ε and ΔG_{tr}^0.

3. Relaxation Time for the Micelle Formation/Breakdown

The main assumption underlying Aniansson and Wall treatment is that micelles form or break down only via a series of stepwise reactions (4).[24] Assumptions (2)–(3) are also made in the calculations.

The aggregation space (s-axis) is divided in three regions (Figure 3.1). Region 1 ($1 \leq s \leq s_1$) corresponds to the oligomers; region 2 ($s_1 < s < s_2$) is that of the premicellar aggregates that are present at very low concentration; and region 3 ($s \geq s_2$) corresponds to the micelles proper. The authors assume that the species in region 2 contribute negligibly to the mass-balance equation and that the flux J_s in region 2 is independent of s (see Equation 3.7). Under such assumptions, at concentrations above, say, 2 cmc, the following expression was obtained:

$$1/\tau_2 = N^2\{R[A_1]^{eq}[1 + (\sigma^2/N)a]\}^{-1} \qquad (3.13)$$

with

$$R = \sum_{s_1+1}^{s_2}(k_s^-[A_s]^{eq})^{-1} \qquad (3.14)$$

R is similar to a resistance opposed by the system to the transfer of monomers between regions 1 and 3. The value of R determines that of τ_2. The dependence of R on the concentration of the species in region 2 suggests that τ_2 can be used to obtain information on these species, which are present at a very low level.

Changes in $1/\tau_2$ with C in situations corresponding to various positions of the minimum of the distribution curve and lengths of region 2 have been simulated.[25] The simulations have shown that, in the case of a long narrow passage and for σ^2/N around 1 (as experimentally found, see Section III.B.6), $1/\tau_2$ goes through a maximum with increasing concentration, a very unusual variation for a reciprocal relaxation time (see Figure 3.2). The maximum disappears and is replaced by a decrease of $1/\tau_2$ with increasing C in the case of a short narrow passage at s = 20 and for $\sigma^2/N = 1$. Simulations confirmed that very little material is transferred between regions 1 and 3 in the course of time during the fast relaxation process.[39,42]

The last contribution of Aniansson[47] to micellar dynamics before his untimely death in 1984 was the derivation of the relationship between the micelle lifetime T_M and τ_2 in the case of dilute micellar solutions:

$$T_M = N\tau_2 a/[1 + (\sigma^2/N)a] \qquad (3.15)$$

As stated by Aniansson, "except close to the cmc $a/[1 + (\sigma^2/N)a]$ is of the order of one so that generally the order of magnitude of the micelle lifetime is determined by the product $N\tau_2$."[47] Since N is often close to 100, the micelle lifetime is therefore much longer than τ_2.

B. Extension of the Theory to Ionic Surfactants with or without Added Salt

1. Exchange Process

If only reactions (4) contribute to the exchange process, the expression of τ_1 for ionic surfactants in the presence of added salt is given by Equation 3.16:

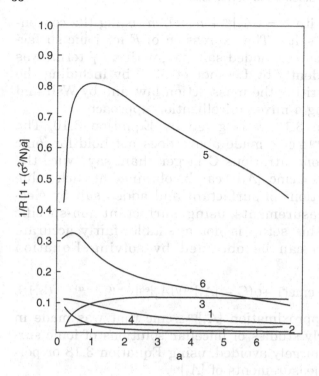

Figure 3.2 Variation of $1/R[1 + (\sigma^2/N)a] \propto 1/\tau_2$ with a for a short region 2 around $s = 20$ (low value) and $\sigma^2/N = 1$ (curve 1) and 5 (curve 2); a short region 2 around $s = 80$ (high value) and $\sigma^2/N = 1$ (curve 3) and 5 (curve 4); and a long region 2 between $s = 20$ and $s = 80$ and $\sigma^2/N = 1$ (curve 5) and 5 (curve 6). The value $\sigma^2/N = 1$ characterizes most micellar solutions at concentrations close to the cmc, where the micelles are nearly monodisperse; the value $\sigma^2/N = 5$ corresponds to systems of higher polydispersity (high surfactant concentration and/or presence of added salt). Adapted and reproduced from Reference 25 with permission of the American Chemical Society.

$$1/\tau_1 = (k^-/\sigma^2) + (k^-/N)aF \qquad (3.16)$$

with

$$F = 1 + (1 - \alpha)^2/(1 + \alpha a) + 2\gamma[A_1]^{eq} \qquad (3.17)$$

This expression of $1/\tau_1$ is very similar to Equation 3.9 except for the correcting term F. In Equation 3.17 α is the experimentally accessible micelle ionization degree and $\gamma =$

$0.587/b(1 + b)^2$ with $b = ([A_1]^{eq} + m_s)^{0.5}$, m_s being the concentration of added salt.[27] The expression of F for ionic surfactants in the absence of added salt (i.e., with no γ term) was derived independently by Lessner et al.,[28] by including the counterions in writing the mass action law, and by Wall and Elvingson,[30] using a mixed micellization approach.

In Equation 3.16, a is given by Equation 3.10. The assumption $[A_1]^{eq} \cong cmc$ made above does not hold for ionic surfactants at concentrations C larger than, say, twice the cmc, where $[A_1]^{eq} < cmc$. $[A_1]^{eq}$ can be obtained at each value of the concentrations of surfactant and added salt by electrochemical measurements using surfactant ion-specific electrodes.[48] If this setup is not available, fairly accurate values of $[A_1]^{eq}$ can be obtained by solving Equation 3.18:[26,27,49]

$$[A_1]^{eq} = cmc\{1 + \alpha(C - [A_1]^{eq})/[A_1]^{eq}\}^{(1 - \alpha)/(2 - \alpha)} \qquad (3.18)$$

Thus the approximation $[A_1]^{eq} \cong cmc$ that was made in many of the early studies of micellar solutions of ionic surfactants can be largely avoided, using Equation 3.18 or performing direct measurements of $[A_1]^{eq}$.

2. Micelle Formation/Breakdown

a. Micelle Formation/Breakdown Through Stepwise Reactions (3.4) Only

Expressions of τ_2 for ionic surfactant solutions in the presence of added electrolyte have been reported.[26–28,43,45] Reference 26 reports the following expression

$$1/\tau_2 = N^2(1 + \mu)/\{R[A_1]^{eq}[1 + (1 + \mu)(\sigma^2/N)a]\} \qquad (3.19)$$

where the correcting term $\mu = (1 + \alpha)^2[A_1]^{eq}/[X]$, [X] is the concentration of free counterions.

b. Micelle Formation/Breakdown via Fragmentation/Coagulation Reactions (3.3) and Stepwise Reactions (3.4)

The inclusion of fragmentation/coagulation reactions (3.3) to the theory of the dynamics of the micelle formation/breakdown process has been carried out by Kahlweit et al.,[28,44,50] after an earlier attempt by Coleen.[51] The contribution of such reactions becomes significant in all systems where the intermicellar interactions are attractive or very weakly repulsive such as nonionic surfactants, ionic surfactants at high surfactant concentration and/or ionic strength, or in the presence of divalent counterions and water-in-oil microemulsions.

The treatment of Kahlweit et al.[28,44,50] makes use of the fact that since the mass-action law is unaltered, the contributions of reactions (3) and (4) to the relaxation of the system add in the same manner as parallel resistances:

$$1/\tau_2 = 1/\tau_{21} + 1/\tau_{22} \qquad (3.20)$$

In Equation 3.20 τ_{21} corresponds to reactions (4) and is given by Equation 3.13, and τ_{22} corresponds to reactions (3), also referred to as fusion-fission processes, and is given by

$$1/\tau_{22} = \beta N a/[1 + (\sigma^2/N)a] \qquad (3.21)$$

In Equation 3.21 β is a measure of the mean dissociation rate constant for reactions (3). Its expression is obtained using DLVO theory. The authors showed that

$$1/\tau_2 = Q_1\left([X]/[A_1]^{eq}\right)^{-q_1} + Q_2\left([X]/[X]_0\right)^{q_2} \qquad (3.22)$$

Q_1 and Q_2 are two constants; $[X]_0$ is the concentration of free counterions at the onset of coagulation; and q_1 and q_2 are two positive numbers. As a result, the change in $1/\tau_2$ with the concentration of surfactant and/or added salt is expected to be V-shaped and to level off as $[X]$ tends toward $[X]_0$. This behavior that has been repeatedly reported for concentrated surfactant solutions and at high salt concentration (see Section III.C). A similar expression of $1/\tau_2$ was derived by Hall[31] using the formalism of the thermodynamics of irreversible processes.

c. Additional Treatments of Coagulation/Fragmentation Processes

The kinetics of fragmentation/coagulation reactions is extremely important for the understanding of the rheological properties of solutions of the so-called wormlike or threadlike micelles (see Section III.D and Chapter 9). In their calculations, Kahlweit et al. assumed that the probability of micelle breakdown was independent of the surfactant concentration, that is, of the micelle length. This assumption has been criticized.[52,53] Besides Kahlweit assumed that reactions (3) involve only the premicellar aggregates in region 2 in Figure 3.1.

Turner and Cates[54] assumed an exponential distribution of micelle lengths and a rate constant of breakdown of a micelle into two daughter micelles that is proportional to the mean micelle length <L>. The resulting relaxation time is given by

$$1/\tau_2 = 2k_b<L> \qquad (3.23)$$

In Equation 3.23 k_b is the rate constant of micelle fragmentation (or fission) expressed in $(s \times unit\ length)^{-1}$.

Waton[53] derived the general expression (3.24) for this relaxation time. This expression applies to different types of distribution of micelle lengths.[53]

$$1/\tau_2 = k_f/[d\log N/d\log(C-cmc)] \qquad (3.24)$$

In this equation k_f is the mean rate constant of fission. Equation 3.24 yields Equation 3.23 in the case of an exponential distribution of micelle lengths.

C. Relaxation Amplitudes

Expressions for the amplitudes of the fast and slow relaxation processes characterizing micellar solutions have been derived for the ultrasonic relaxation techniques for the fast exchange process[55,56] and for jump techniques (T-jump, p-jump, SF) for the two relaxation processes.[26,28,57–65] In an effort to save space, these expressions are not given here. Nevertheless, it is recalled that the analysis of the amplitudes of the relaxation signals can be useful in several respects (see Chapter 2, Section II.C). For instance, if the relaxation process responsible for the observed relaxation signal is identified, the analysis of the relaxation amplitudes can provide the values of thermodynamic quantities characterizing the process. For micellar solutions, the amplitude of the ultrasonic relaxation yields the value of the isentropic volume change associated to the transfer of one surfactant from water to micelle[55,56] (see Chapter 2, Section II.D.5). Likewise, the relaxation amplitude of the slow process as obtained in p-jump and T-jump can be used to determine the values of the volume change and enthalpy change, respectively, associated with this process (see Chapter 2, Sections II.D.1 and II.D.2). Further, the analysis of the relaxation amplitudes measured for the slow process by T-jump, p-jump, and SF were shown to provide information on the changes of the micelle aggregation number with temperature,[63] pressure,[26] and concentration,[64] respectively.

Just as importantly, the variation of the relaxation amplitude with relevant parameters can permit a choice between two possible sets of assumptions used in deriving expressions of the relaxation times and amplitudes. This has been the case in the early stage of studies of micellar solutions by chemical relaxation methods. A controversy arose about an important approximation made in analyzing the relaxation due to the micelle formation/breakdown.[59,60] Chan et al.[58,60] assumed that during the slow process the micelle aggregation number retained the value reached at the end of the fast relaxation, contrary to Aniansson and Wall.[24,59] It was later shown for several surfactants that the relaxation amplitude of this process, as measured in p-jump experiments with a conductivity readout, goes through the value zero twice as a = (C − cmc)/C is increased.[61,65] Teubner et al.[61] recognized that this behavior is correctly predicted with the set of assumptions used in Aniansson and Wall calculations, but not with that in Chan et al.[57,58]

D. Extension to Mixed Micelles

Aniansson[29] showed that mixed micellar solutions of binary surfactant mixtures are characterized by three relaxation processes: two fast processes associated with the exchange of the two surfactants between mixed micelles and bulk phase (τ_{11} and τ_{12}) and a slow process associated with the formation/breakdown of the mixed micelles (τ_2). The basic assumptions and the methods used to derive the expressions of the three relaxation times were very similar to those used for single-component micelles.[24] In particular, the size distribution curve of the mixed micelles was assumed to be doubly Gaussian. Simulations yielded variations of the relaxation times qualitatively similar to those for one component micelles: $1/\tau_{11}$ and $1/\tau_{12}$ increase nearly linearly with concentration, whereas the variations of $1/\tau_2$ are complex and depend much on the characteristics of the narrow passage between oligomers and micelles proper.

The theoretical aspects of the kinetics of two-component micelles based on the stepwise association model were later extended to high surfactant concentrations.[30,66] This model applies to binary mixtures of nonionic surfactants as well as to one ionic surfactant, the surfactant ion and its counterion then being the two components of the system.

The thermodynamics of irreversible processes was also used to derive very general expressions of the relaxation times for mixed micellar solutions (two surfactants).[27,31] Unfortunately, these expressions are not easy to handle.

E.   Other Theoretical Treatments of Micellar Kinetics

The first theoretical treatment of the exchange process was of phenomenological character and assumed a simultaneous exchange of several surfactants between micelles and bulk phase.[16] The reported expression of τ_1 bears much resemblance to Equation 3.9. Sams et al.[67] derived an expression of τ_1 on the assumption that the rate constants for the association/dissociation of one surfactant to/from an aggregate A_i are proportional to i. This assumption was later shown to be not valid. Inoue et al.[68] have tried to incorporate the features of Aniansson and Wall treatment and those of Sams et al. in a unified theory of the fast exchange. This treatment suffers from the same shortcomings as that of Sams et al. and yields unrealistic values of the rate constants k^+ and k^-.

Inoue et al.[69] also derived an expression of τ_2 by using the stepwise association/dissociation reactions (4), including the counterions and assuming all reaction steps to be fast except an intermediate step that is slow and rate-limiting.

This treatment was used to explain the effect of salt on τ_2 at a time where the treatments in Section II.B.2.b had not been reported.

The relaxation of micellar solutions has been recently simulated.[70] The simulated behavior was found to be in good agreement with that predicted by Aniansson and Wall.[24]

III. SURFACTANT EXCHANGE AND MICELLE FORMATION/BREAKDOWN: EXPERIMENTAL ASPECTS

This section reviews the main experimental results for the exchange process and micelle formation/breakdown for solutions of pure surfactants, in the absence of additives other than salts. The surfactants considered are the conventional surfactants (one head group/one hydrocarbon chain and one head group/two hydrocarbon chains), the perfluorinated surfactants (one head group/one perfluorocarbon chain), and the gemini (dimeric) surfactants (two head groups/two alkyl chains).

A. General Observations

Two relaxation processes were evidenced for a large number of surfactants by using two relaxation techniques or, in favorable instances, a single one. Examples of conventional surfactants: sodium dodecylsulfate (shock tube (ST)[25] or ultrasonic relaxation (USR)[18] and p-jump[11,13]); sodium tetradecyl- and hexadecylsulfates (ST and p-jump[25]); alkylpyridinium sulfates (ST and p-jump[71]); alkylpyridinium halides (ST and p-jump[72]); alkylammonium halides (ST and p-jump[73]); surfactants with divalent counterions (T-jump, p-jump, SF[74]); two-chain surfactants (USR,[75] SF, and p-jump[76]); nonionic surfactants (T-jump[77,78]). Examples of perfluoronated surfactants: perfluorononanoate with various counterions (p-jump, T-jump, and SF[79]). Examples of dimeric surfactants: the alkanediyl-α,ω-bis(dodecyldimethylammonium bromide) (p-jump and ST[80]).

In all instances where it could be measured, the relaxation time for the surfactant exchange process was found to obey Equation 3.9, that is, $1/\tau_1$ increased linearly with the surfactant concentration C close to the cmc. Deviations showed up at higher concentration. The variations of $1/\tau_1$ with C are illustrated in Figure 3.3 for sodium tetradecylsulfate (STS). The linear variation at low C is obvious, as well as the departure from linearity at higher C. The origin of this deviation probably lies in the fact that the results are plotted as $1/\tau_1$ against C whereas, according to Equation 3.9, $1/\tau_1$ should be plotted against a, that is $C/[A_1]^{eq}$ (see Equation 3.10). For ionic surfactants $[A_1]^{eq}$ decreases upon increasing C.[48] Thus the curvature toward the ordinate axis seen in Figure 3.3 is expected to be decreased or eliminated if $1/\tau_1$ is plotted against $C/[A_1]^{eq}$. Unfortunately, this was not done at the time the results were reported. Other effects may contribute to a nonlinear variation of $1/\tau_1$ at high C: change of micelle size and shape, interactions, etc. In the case of anionic surfactants such as the alkylcarboxylates with alkali metal counterions[15,67] and alkylsulfates with tetraalkylammonium counterions[81] $1/\tau_1$ depends only little on the nature of the counterion. However $1/\tau_1$ clearly depends on the nature of the counterion when the counterion is lyophobic or can be considered as a short surfactant ion.[82,83] This is, of course, due to the fact that one is then dealing with systems that are somewhat similar to mixed micellar systems. $1/\tau_1$ decreases rapidly upon increasing length of the surfactant alkyl chain, as would be expected for the surfactant exchange process.[15,25] It has been noticed that for a series of homologous surfactants — for instance, the sodium alkylsulfates or the potassium alkylcarboxylates — the $1/\tau_1$ vs C plots for the different alkyl chains nearly fall on a single line when plotted in the same graph.[25,84] $1/\tau_1$ increases moderately with temperature (see Figure 3.3 and Reference 17) and in the presence of added salt with the same counterion as the surfactant.[26,85,86] All these variations were shown to be consistent with an assignment of the fast relaxation process to the surfactant exchange between micelles and bulk phase. As a general rule, $1/\tau_1$ varies in the same manner as $[A_1]^{eq} \cong$ cmc with most of the system parameters, such as the surfactant chain length and concentration, the nature of the counterion, and the salt concentration. $1/\tau_1$ is little affected by the presence of a small amount of impurities.

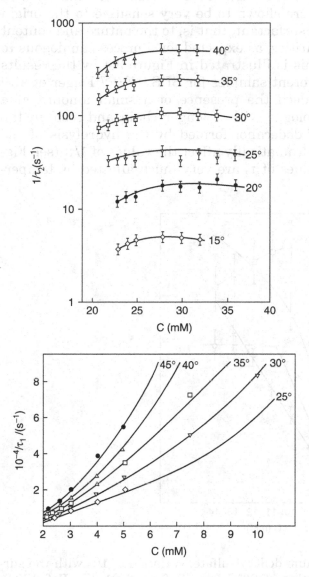

Figure 3.3 Sodium tetradecylsulfate (STS): variations of $1/\tau_1$ and $1/\tau_2$ with the surfactant concentration at different temperatures. Reproduced from Reference 25 with permission of the American Chemical Society.

The reported variations of the relaxation time τ_2 with concentration were shown to be very sensitive to the origin of the sample of surfactant, that is, to the nature and content of possible impurities, as expected for a process analogous to a nucleation. This is illustrated in Figure 3.4 by the results reported for different samples for SDS.[11,13,21,87] Folger et al.[21] showed that indeed the presence of a small amount of a surfactant homologue with a longer chain and also, in the case of SDS, of dodecanol formed by the hydrolysis of the surfactant, can dramatically affect the values of $1/\tau_2$ (see Figure 3.5). The values of τ_2 are very much affected by temper-

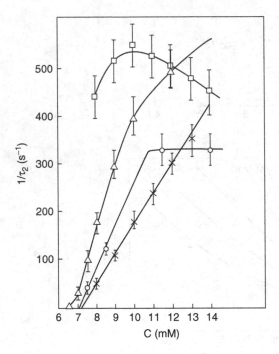

Figure 3.4 Sodium dodecylsulfate: variation of $1/\tau_2$ with the surfactant concentration at 20°C from Reference 21 (□); Ref. 13(O); Reference 11 (×), and Reference 87 (△). Reproduced from Reference 21 with permission of Verlag Chemie GmbH.

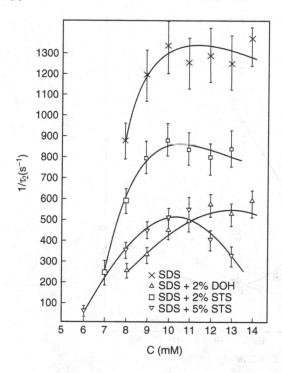

Figure 3.5 Sodium dodecylsulfate: variation of $1/\tau_2$ with the surfactant concentration at 25°C in the presence of additives simulating impurities (STS: sodium tetradecylsulfate; DOH: dodecanol). Reproduced from Reference 21 with permission of Verlag Chemie GmbH.

ature, as can be seen in Figure 3.3. It is noteworthy that in spite of the great sensitivity of $1/\tau_2$ to the parameters just discussed, several studies showed that different relaxation techniques measure the same value of $1/\tau_2$ when the perturbation is small enough.[62,77,88–90] Also, when monitoring the slow relaxation using a micelle-solubilized dye and a spectrophotometric detection (see Chapter 2, Section II.D.1), the same value of τ_2 is measured than when using a different technique, provided that the [dye]/[surfactant] ratio is small enough.[78,89,90] The variations of $1/\tau_2$ are also dramatically affected both qualitatively and quantitatively by additions of salt, as is illustrated in Figure 3.6 for additions of KBr to dodecylpyridinium bromide.[88] The rapid variations of $1/\tau_2$ in

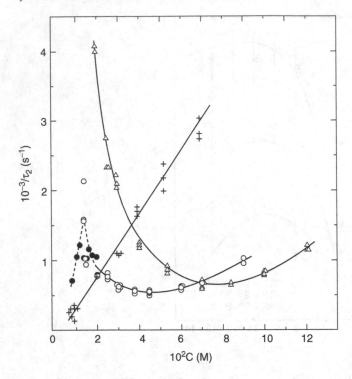

Figure 3.6 Dodecylpyridinium chloride: Variation of $1/\tau_2$ with the surfactant concentration at 25°C in the absence of KBr ($\triangle$, T-jump); in the presence of 0.10 mM KBr (o, T-jump; ● p-jump); and in the presence of 50 mM KBr (+, T-jump). Reproduced from Reference 88 with permission of the American Chemical Society.

Figure 3.6 can be partly explained by changes of length of region 2 in Figure 3.1 and of the position of the minimum of the distribution curve (see the simulations represented in Figure 3.2) and/or, more likely, by the increased contribution of fragmentation/coagulation reactions (3) to the micelle formation/breakdown at high concentrations of surfactant and/or added salt.

Three types of studies confirm the existence of this contribution. First, for the water/SDS/NaClO$_4$ system, the different $1/\tau_2$ vs C plots obtained at various NaClO$_4$ contents collapsed into a single plot when $1/\tau_2$ was plotted against the total counterion (Na$^+$) concentration in the system (see Figure 3.7), as

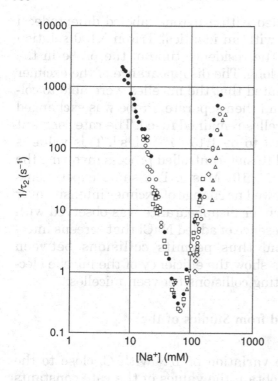

Figure 3.7 Sodium dodecylsulfate: variation of $1/\tau_2$ with the counterion (Na^+) concentration at 25°C in pure water (•) and in the presence of sodium perchlorate at concentration (O) 10 mM; (□) 20 mM; (∇) 50 mM; (◊) 100 mM and (△) 200 mM. Reproduced from Reference 44 with permission of Elsevier/Academic Press.

expected on the basis of Equation 3.22.[44] Second, correlations were shown to exist between the variations of the micelle diffusion coefficient D (inversely proportional to micelle size, obtained from dynamic light scattering experiments), of the variance v of the micelle distribution (indicative of the micelle polydispersity, also obtained from dynamic light scattering experiments), of the rate constant k_{exch} characterizing the transfer of material between micelles through collisions (obtained from time-resolved fluorescence quenching; see Chapter 2, Section III) and of $1/\tau_2$ in the case of aqueous surfactant solutions[91] and of ternary surfactant/alcohol/water systems.[92] The third result comes from a stopped-flow study of excimer disappearance when a micellar solution of the nonionic surfactant Triton

X100 with micelles loaded with a pyrene-labeled diheptadecyl triglyceride was mixed with an identical Triton X100 solution containing no probe.[93] The residence time of the probe in the micelles was extremely long. The disappearance of the excimer band after mixing indicated that the micelles were able to collide, fuse temporarily, and then separate. Probe was exchanged during the time the micelles remained fused. The rate constant for fusion was estimated to be 1.15×10^9 $M^{-1}s^{-1}$. This value is much lower than for a diffusion-controlled process involving the micelles (estimated at 6×10^9 $M^{-1}s^{-1}$). The same experiments performed with SDS showed no change of excimer intensity over several months, but excimer disappearance was observed with SDS solutions in the presence of added NaCl that screens intermicellar repulsion and thus permits collisions between micelles.[94] These results show the efficiency of the micelle electrical charge in preventing collisions between micelles.

B. Information Gained from Studies of the Exchange Process

The linear part of the variation of $1/\tau_1$ with C, close to the cmc, has been used to obtain the values of the rate constants k^+ and k^- for the association and dissociation of one surfactant to/from a micelle and of the distribution width σ (see Equation 3.9), using known or estimated values of the mean micelle aggregation N. Many values of k^+ and k^- have been reported.[25,26,62,71–76,78–83,85,86,95–103] Table 3.1 lists values for representative anionic, cationic (including dimeric), nonionic and zwitterionic surfactants. Note that the values listed in Table 3.1 have been obtained from an analysis of the $1/\tau_1$ vs. C data using Equation 3.9 or Equation 3.16, which is more correct for the ionic surfactants considered here. The two types of analysis yield data than can differ by much more than the experimental error, as noted in several studies.[71,101,102] Also the errors involved in the different methods used for measuring τ_1 can be very different. The main conclusions are as follows:

1. For all the listed surfactants, except the entries 38–40 that refer to long-chain dimeric surfactants discussed below, the values of k^+ fall, say, between 2

TABLE 3.1 Values of the Micelle Size Distribution Width σ, of the Rate Constants k^+ and k^-, and of the cmc of Surfactants at 25°C

Surfactant	σ (N)[a]	k^- (s^{-1})	k^+ (M^{-1}s^{-1})	cmc (mM)	Reference
1. Sodium hexylsulfate[b]	6 (17)	1.3×10^9	3.2×10^9	420	95
2. Sodium heptylsulfate[b]	10 (22)	7.3×10^8	3.2×10^9	220	95
3. Sodium octylsulfate[b]	(27)	1×10^8	0.77×10^9	130	67
4. Sodium nonylsulfate[b]	(41)	1.4×10^8	2.3×10^9	60	95
5. Sodium decylsulfate	14 (50)	6.8×10^7	2.1×10^9	34	65
6. Sodium dodecylsulfate (SDS)	13 (64)	1×10^7	1.2×10^9	8.2	25
7. Cobalt dodecylsulfate	11 (108)	2.6×10^6	1.1×10^9	4.9	74
8. Nickel dodecylsulfate	13 (97)	3.9×10^6	1.6×10^9	5.0	74
9. Sodium tetradecylsulfate (STS)	16 (80)	9.6×10^5	0.47×10^9	2.05	25
10. Tetramethylammonium tetradecylsulfate[c]	13 (80)	7.1×10^5	0.5×10^9	1.35	81
11. Tetraethylammonium tetradecylsulfate[c]	11 (80)	6.1×10^5	0.6×10^9	0.94	81
12. Tetrapropylammonium tetradecylsulfate[c]	11 (80)	4.5×10^5	0.75×10^9	0.59	81
13. Sodium hexadecylsulfate (30°C)	11 (100)	6×10^4	0.13×10^9	0.45	25
14. Dodecylpyridinium iodide	18 (64)	1.6×10^7	3×10^9	5.25	72
15. Tetradecylpyridinium chloride	14 (88)	1.9×10^7	4.5×10^9	4.2	73
16. Tetradecylpyridinium sulfate	14 (60)	2.2×10^6	1.1×10^9	2.0	71
17. Tetradecylpyridinium bromide	12 (80)	6.3×10^6	2.5×10^9	2.55	72
18. Tetradecylpyridinium iodide (35°C)	30 (72)	3.1×10^6	2×10^9	1.5	72
19. Hexadecylpyridinium chloride	11 (100)	6×10^5	0.63×10^9	0.9	72
20. Hexadecylpyridinium bromide	11 (100)	3.3×10^5	0.55×10^9	0.68	72
21. Hexadecylpyridinium bromide	15 (100)	5.2×10^5	0.8×10^9	0.68	62
22. Octyltrimethylammonium bromide	6 (25)	1×10^9	3.6×10^9	280	86
23. Decyltrimethylammonium bromide	6 (38)	1.7×10^8	2.6×10^9	66.3	86
24. Dodecyltrimethylammonium bromide	9 (49)	3.2×10^7	2.2×10^9	14.6	85

25. Tetradecyltrimethylammonium bromide	3.9 (66)	3.2×10^6	0.86×10^9	14.6	96
26. Octadecyltrimethylammonium bromide	9 (125)	6.4×10^5	0.96×10^9	0.245	62
27. Tetradecylpyridinium methylsulfonate	10 (50)	6.6×10^6	1.4×10^9	4.7	83
28. Tetradecylpyridinium ethylsulfonate	9 (48)	3.9×10^6	0.96×10^9	4.1	83
29. Tetradecylpyridinium propylsulfonate	9 (53)	2.7×10^6	0.87×10^9	3.1	83
30. Tetradecylpyridinium butylsulfonate	9 (56)	1.7×10^6	0.72×10^9	2.4	83
31. Tetradecylpyridinium pentylsulfonate	7 (49)	7.3×10^5	0.47×10^9	1.55	83
32. Lithium p-dodecylbenzenesulfonate	16 (75)	2.3×10^5	0.27×10^9	0.82	83
33. sodium p-6-dodecylbenzenesulfonate	3.6 (21)	4.2×10^5	0.17×10^9	2.5	83
34. Dihexylmethylammonium chloride	3.7 (6)	6.2×10^8	1.9×10^9	335	75
35. Diheptylmethylammonium chloride	3 (9)	1×10^8	1×10^9	99	75
36. 8-3-8	5 (13.5)	8.8×10^7	1.6×10^9	55	97
37. 8-6-8	4 (13.5)	1.6×10^8	2.3×10^9	72	97
38. 12-2-12	7 (31)	3.6×10^4	4×10^7	0.84	80
39. 12-3-12	8.5 (31)	7.2×10^4	8×10^7	0.96	80
40. 12-4-12 (15°C)	6 (30)	9.5×10^4	8.7×10^7	1.1	80
41. Triton X100	10 (48)	1.1×10^6	3.7×10^9	0.3	78
42. $C_8H_{17}(OCH_2CH_2)_8OH$	9 (72)	8.7×10^7	8.3×10^9	10.4	98
43. Octylglucoside	15 (92)	1.4×10^8	5.6×10^9	25.1	99
44. Octyl(dimethylammonio)propanesulfonate	4 (23)	4.3×10^8	1.6×10^9	265	100
45. Decyl(dimethylammonio)propanesulfonate	8 (44)	1.7×10^8	4.5×10^9	38	100
46. Sodium perfluorooctanoate	5 (12)	3.2×10^7	1×10^9	32	79
47. Sodium perfluorooctanoate	6 (27)	3.2×10^7	1×10^9	31.4	103
48. Tetraethylammonium perfluorooctanesulfonate	32 (140)	9.2×10^5	1×10^9	0.95	79

[a] In parentheses: value of the micelle aggregation number

[b] Values calculated from the results reported in References 67 and 95

[c] Values calculated from the reported $1/\tau_1$ vs C data using Equation 3.9

[d] Dimeric surfactants $C_mH_{2m+1}(CH_3)_2N^+(CH_2)_sN^+(CH_3)_2C_mH_{2m+1}$, 2Br referred to as *m-s-m*

$\times 10^8$ and 4×10^9 M^{-1}s^{-1}, irrespective of the nature of the surfactant head group (ionic, nonionic, zwitterionic), chain length (from C_8H_{17} to $C_{18}H_{37}$, in the case of the alkyltrimethylammonium bromides, entries 22–26, for instance), nature of the chain (hydrocarbon or perfluorocarbon, as in entries 46–48), and number of chains (entries 33–35 refer to two-chain surfactants). Such values are close to those calculated for a diffusion-controlled process. This means that the rate of association of a surfactant to a micelle proper is nearly equal to the rate of collisions between free surfactants and micelles. The energy barrier to association is low, some 1–3 kT depending on the surfactant.

2. The values of k$^-$ depend mainly on the surfactant chain length characterized by its carbon number m, just like the cmc of the surfactant (see entries 1–5, 9, and 13; entries 22–26; entries 36 and 39; entries 44 and 45). As a general rule k$^-$ decreases by a factor close to 3 per additional methylene group in the alkyl chain for single-chain surfactants, as can be seen from the results in Table 3.1 for series of homologous surfactants having hydrocarbon chains. Recall that k$^-$ is directly related to the *residence time of a surfactant in a micelle* $T_R = N/k^-$, that is, to the reciprocal of the slope of the plot of $1/\tau_1$ vs. a. Since the energy barrier to surfactant association is low, Equation 3.12 predicts a linear variation of logT_R with the alkyl chain carbon number m that is well verified by the results plotted in Figure 3.8. The lines for several surfactant series run parallel. Their slopes yield a free energy increment per CH_2 that is close to the free energy of transfer of a CH_2 from micelles to water, obtained from a log cmc versus m plot.

Another important parameter is the nature of the alkyl chain: for the same m value perfluorocarbon chains are characterized by much lower cmc and k$^-$ values than hydrocarbon chains (compare entries 2 and 46 or 3 and 48). Recall that in terms of values of the cmc and of free energy of transfer from water to

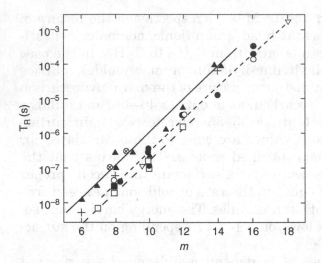

Figure 3.8 Variation of the residence time of surfactants in micelles with the carbon number m of the alkyl chain at 25°C for sodium alkylsulfates (▲); sodium alkylsulfonates (⊗); sodium alkylcarboxylates (■); potassium alkylcarboxylates (□); alkylammonium chlorides (+); alkyltrimethylammonium chlorides (▽); alkylpyridinium chlorides (O); alkylpyridinium bromides (•); Adapted from Reference 84.

micelle, one CF_2 group is equivalent to 1.5 CH_2 group. However, a perfluorocarbon surfactant and a hydrocarbon surfactant having nearly the same cmc value also have the same k^- values within a factor 2 (see entries 5, 23, 45, and 47 and results in Reference 103). The effect of the nature of the counterion on the value of k^- is small (see, for instance, entries 6–8, 9–12 and References 95 and 103). However, some differences can be seen when the counterion is a small surfactant ion as in entries 27–31: k^- and the cmc decrease by factors close to 10 and 5, respectively, in going from the methyl to the pentyl sulfonate counterion.

3. The dynamics of the surfactant exchange is little affected upon increasing the ionic strength of the solution. Thus k^+ is slightly increased and k^- is slightly decreased in the presence of a salt having the same counterion as the surfactant.[25,26,74,85,86,101]

4. The substitution of H_2O by D_2O has little effect on the surfactant exchange.[49,104] For instance, for STS, N decreased from 44 to 31, the cmc decreased from 2.21 to 1.94 mM, and k^- decreased from 7×10^5 to 5.76 $\times 10^5$ s^{-1} in going from H_2O to D_2O.[49] The rate constant k^+ is also decreased because the viscosity of D_2O is larger than for H_2O, as expected for a diffusion-controlled process.[104]

5. The enthalpy of association of one surfactant with a micelle proper, obtained from the dependence of k^- on temperature, is negative, indicating an exothermal process.[25]

6. The values of σ, width of the aggregation number distribution curve permit the calculation of the micelle polydispersity, defined as the ratio of the weight-average (N_w) and number-average (N_n) micelle aggregation numbers, from: $N_w/N_n = 1 + (\sigma/N)^2$. For most of the entries in Table 3.1 the micelle polydispersity is low. This is because most of the measurements refer to surfactant concentrations close to the cmc, at which the micelles are generally small and spheroidal and, thus, of rather low polydispersity.[34] For these systems it has been observed that σ^2/N is close to unity.[25]

7. The behavior of dimeric surfactants is quite interesting. The short-chain dimeric surfactants 8-3-8 and 8-6-8 (entries 36 and 37) show k^+ values in the 10^9 $M^{-1}s^{-1}$ range like most conventional surfactants. However, the long-chain dimeric surfactants (entries 38–40) have k^+ values much lower than for a diffusion-controlled process. The association is no longer controlled by diffusion, and the barrier to association is of about $5kT$. Another interesting feature is the significant increase of both k^+ and k^- as the length of the polymethylene spacer $(CH_2)_s$ is increased. This variation has been attributed to the increased conformational freedom of the spacer.[80]

8. The knowledge accumulated on the exchange process in aqueous surfactant solutions has been used to investigate the conformation of the alkyl chain of

bolaform surfactants of the alkanediyl-α,ω-bis (trimethylammonium bromide) in the micelles.[105] It was tentatively concluded that the alkanediyl chains making up the micelle core are folded.

9. Data for many other surfactants have been reported: sodium alkylsulfates;[106,107] sodium hexylsulfate and hexylammonium chloride in brine solution;[108] octylammonium chloride, octyltrimethylammonium chloride and bromide, and sodium and potassium caprylates;[109] sodium p-(1-propylnonyl)benzenesulfonate;[110] dodecylpyridinium chloride and bromide;[111] heptylammonium chloride;[112] nonionic surfactants C_mH_{2m+1} $(OCH_2CH_2)_nOH$ with various m and n values;[113,114] perfluorinated anionic surfactants;[115] and cationic surfactants with a perfluoroalkylcarboxylate counterion.[116,117] Some of these data must be reanalyzed using Equation 3.9 or, if the required data are available, Equation 3.16, in order to extract meaningful rate constants.

10. Other methods than chemical relaxation methods have been used to determine the rate constants k^+ and k^-. ESR (see Chapter 2, Section V) has been used to investigate the exchange of spin-labeled surfactants[118] and lipids.[119] The spin-labeled surfactant was found to have values of the cmc and k^- that did not differ much from those for the corresponding unlabeled surfactant.[118] The values of k^+ for spin-labeled lecithins with decyl and dodecyl chains were found to be 4.8×10^8 and 1.2×10^8 $M^{-1}s^{-1}$, respectively.[119] These values are somewhat smaller than for the two-chain surfactants, entries 34 and 35 in Table 3.1. The difference may be due to the very large size of the lecithin head groups. The value of the exit rate constant k^- of surfactant radicals (dodecylsulfate radical[120] and 1-(6-dodecylbenzenesulfonate) radical[121]) from micelles have been determined by pulse radiolysis (see Chapter 2, Section III) and found to be in good agreement with those from chemical relaxation methods. Time-resolved phosphorescence quenching (see Chapter 2, Section III) yielded values

of k$^+$ and k$^-$ for naphthalene-labeled decyltrimethy-lammonium bromide that are comparable to those obtained by chemical relaxation methods for the unlabeled surfactant considering the bulkiness of the phosphorescent naphthoyl group label.[122] ^{19}F NMR has been used to determine the exit rate constant of the perfluorooctanesulfonate anion from its micelles.[123] The result is in good agreement with that for comparable surfactants obtained from chemical relaxation studies.[79,103,115] A similar study has been performed for perfluoroheptanoate salts, perfluoro-heptanoic acid and its ethoxylated amines,[124] and also for hybrid surfactants with separate hydrocarbon and perfluorocarbon chains.[125] In this last study the authors were able to determine the residence time of the terminal perfluoromethyl group and of the α-CF$_2$ group in the micelles. In another ^{19}F NMR study of the hybrid surfactant $C_6F_{13}C_6H_4CH(C_5H_{11})OSO_3^-Na^+$ the residence time of the surfactant in the micelle, 2 ms, was claimed to be unusually long[126] However the cmc of this surfactant, 0.34 mM, is smaller than for sodium hexadecylsulfate and the value 2 ms is not that unexpected.

11. The literature shows two studies that report values of k$^-$ and k$^+$ that are several orders of magnitude smaller than for a diffusion-controlled process. The first study concerns gangliosides, complex surfac-tants with two very long alkyl chains and a very large sugar head group. The reported values of the cmc and k^{-127} yield k$^+$ values between 10^4 and 10^6 M^{-1}s^{-1}. The fact that the chains of the investigated gangliosides melt at about 20°C may explain the reported results.[128] The second study concerns the perfluori-nated surfactants $C_mF_{2m+1}C(O)NH(CH_2CH_2O)_nH$.[129] ^{19}F NMR experiments indicated a very slow exchange between free and micellized surfactants and yielded a value of about 1.1 s for the mean lifetime of the surfactant with $m = 6$ and n = 2. These surfactants were later shown to form very large aggregates, per-

haps a cubic phase, characterized by a very slow surfactant exchange process.[124]

12. The chemical relaxation studies of micellar kinetics have been extended to surfactant solutions in water + glycerol[130] and in pure organic solvents such *n*-alkanes[131,132] and formamide.[133] In all instances the characteristics of the exchange process were found to be similar to those in water with $1/\tau_1$ increasing linearly with C. The analysis of the results on the basis of Equation 3.9 or (16) yielded values of k^+, indicating a nearly diffusion-controlled association and values of k^- that correlated well with those of the cmc in the investigated solvents. In organic solvents the changes of k^- with chain length are much smaller than in water. This is readily explained on the basis of Equation 3.12. Indeed the free energy of transfer per CH_2 group from the micelles to the organic solvent is smaller than for the transfer to water.

NMR has been used to determine the exchange rates and lifetimes of phospholipids in reverse micelles formed in organic solvents in the presence of water. The residence time of the unsaturated dilinoleylphosphatidylcholine molecule in a micelle was found to be close to that of a monomer in bulk solution in benzene, around 30 ms.[134] For the saturated dipalmitoylphosphatidylcholine, the residence time in a micelle was about twice as long.

13. There have been some studies of the dynamics of self-association of drugs by the USR method as the processes are very rapid.[135,136] The results were analyzed in terms of the Aniansson and Wall theory and indicated a diffusion-controlled association.

14. The surfactant exit rate constant from the aggregates present in the discontinuous (micellar) cubic phase formed by some surfactants at high concentration has been determined by NMR self-diffusion.[137,138] The reported values of k^- compare well with those obtained in the case of micellar solutions.

C.   Information Gained from Studies of the
Micelle Formation/Breakdown

As for τ_1, the relaxation time τ_2 for micelle formation/break-down has been measured for a large number of surfactants, under a variety of experimental conditions, using mainly the T-jump, p-jump, and SF techniques.[25,26,28,49,58,62,69,71–74, 76–80,87–89,106,113,115–117,139] The results show that τ_2 can vary widely, from tenths of milliseconds to minutes, depending on the composition of the system (nature and concentration of the surfactant and additives), the temperature, the pressure, and the concentration of added salt (see Figure 3.9). Since the micelle lifetime T_M is proportional to $N\tau_2$ (see Equation 3.15), it means that T_M also varies in a wide range, from tens of milliseconds to tens of minutes, depending on the experimental conditions. Very long lifetimes have been inferred for dimeric (gemini) surfactant micelles.[80]

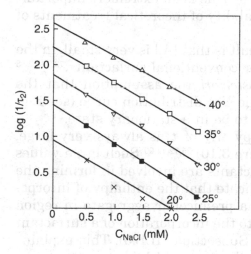

Figure 3.9   Sodium tetradecylsulfate (STS): variations of $1/\tau_2$ with the concentration of added NaCl at different temperatures. Reproduced from Reference 25 with permission of the American Chemical Society.

As pointed out in Subsection II.A.3, the values of τ_2 can be used to obtain information on the premicellar aggregates A_r at the minimum s = r of the micelle size distribution curve (see Figure 3.1), provided the measurements are performed on dilute solutions, where micelle formation/breakdown proceeds via stepwise reactions (4). The data were generally analyzed under the simplifying assumption that the concentration of the aggregates A_r is much smaller than that of the other species in the narrow passage and $k_r^- \approx k^-$ [25] Equation 3.14 then reduces to

$$R \approx 1/k^-[A_r] \propto \{[A_1]^{eq}\}^{-r} \qquad (3.25)$$

On these assumptions, the aggregation number r at the minimum in Figure 3.1 was obtained from the values of τ_2 by various independent methods, particularly from the variation of τ_2 with the concentration of added salt (see Figure 3.9).[25,73,74] For all of the surfactants investigated, r was found to be in the range 7 to 10.[25,73,74,82,83,139] This is an extremely important finding that confirms the validity of theoretical treatments of micellization.[32–34,140]

Another important result is that $[A_r]$ is very small, in the range of 10^{-10} to 10^{-14} M, for conventional surfactants.[73,74,82,83] This result supports *a posteriori* the assumption that the aggregates in region 2 of the size distribution curve (see Figure 3.1) can be considered to be in a stationary state.[25]

The activation enthalpy ΔH_2^+ of τ_2 is always very large, up to 200 kJ/mol (see Figure 3.10).[25,72,74,90] Such large values reflect the fact that r surfactants are involved in forming the aggregate A_r They also indicate that the enthalpy of incorporation of a surfactant into a premicellar aggregate in region 2 is endothermic, contrary to the incorporation of a surfactant into a micelle proper (see Subsection III.B.5). This explains why the overall heat of micellization, which is an average over all association steps, is often small.[25,82,83]

The $1/\tau_2$ versus C plots usually show a plateau at very low C, close to the cmc (see Figure 3.3), then a decrease upon increasing C[21,25,65,88,106] in solutions of conventional ionic surfactants as well as in solutions of ionic gemini (dimeric) surfactants.[80] However, at a higher value of C, which depends on

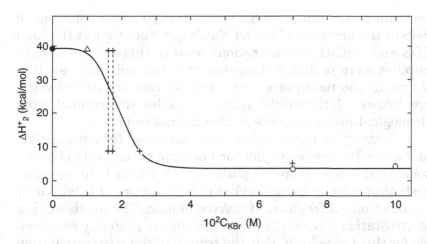

Figure 3.10 Tetradecylpyridinium bromide: effect of additions of KBr on the value of ΔH_2^+, activation enthalpy of $1/\tau_2$, at surfactant concentration 8 mM (+); 2 mM (O) and 1.2 mM ($\triangle$). (•) Value in the absence of KBr. The broken vertical lines indicate the limiting values of ΔH_2^+ obtained from nonlinear variations of $1/\tau_2$ with $1/T$. Reproduced from Reference 90 with permission of Elsevier/Academic Press.

the surfactant and on the experimental conditions, $1/\tau_2$ goes through a minimum, then increases with C.[26,28,43,44] The minimum is shifted to lower values of C upon increasing ionic strength[26,28,43,44] or with the addition of divalent or trivalent counterions.[74] The representation of the results in a doubly logarithmic plot yields a V-shaped curve for the intermediate-to-high range of concentration (see Figure 3.7). The increase of $1/\tau_2$ observed at high C was attributed to the fact that micelle formation/breakdown is now occurring via fragmentation/coagulation reactions (3) rather than via stepwise reactions (4). The overall variation of $1/\tau_2$ with C is well predicted by Equation 3.22. The large decrease of ΔH_2^+, activation enthalpy of $1/\tau_2$, upon addition of salt (see Figure 3.10) supports the above explanation.[74,90] The fitting of the plots in Figure 3.7 to Equations 3.20–3.22 is generally very good and yields an estimate of the Hamaker constant of the expected order of magnitude.[28] Besides, the fitting of the increasing branch of the plot yields a value of the mean breakdown rate constant β (see Equation 3.21) of about 10 s^{-1} for solutions of SDS in the presence of added NaClO$_4$.[28] Note that at the high SDS and NaClO$_4$ concentrations used in this study, the SDS micelles were probably elongated and thus polydisperse. It is shown in the next paragraph that the rate constant for the breakdown of threadlike (giant) micelles determined from rheological measurements is also in this range.

Contrary to the complex variation noted for ionic surfactants, the 1/τ$_2$ versus C plot for nonionic surfactants shows a somewhat linear increase with C.[9,10,77,78] This behavior suggests that micelle formation/breakdown occurs at least partly via reactions of fragmentation/coagulation. The much weaker electrostatic repulsion between micelles is probably responsible for this behavior.[44] Also, the trough in the size distribution curve may not be as deep as for ionic surfactants.

The effect of temperature on the values of τ$_2$ can be quite strong, as noted earlier. In contrast, the effect of pressure appears to be rather weak.[141,142] For instance, for sodium tetradecylsulfate at 25°C, 1/τ$_2$ increased from 53 to 89 s^{-1} as the pressure was increased from 10 to about 7000 Pa.[142]

The amplitudes of the relaxation associated with the micelle formation/breakdown have been used to study the dependence of the aggregation number of several surfactants on temperature,[63] pressure,[26] and concentration of surfactant or added salt.[64] The values of δlogN/δp and of δlogN/δT obtained in this manner agreed with reported values obtained from completely different measurements.

D. Dynamics of Giant Micelles

Many surfactants can form giant micelles also called wormlike or threadlike micelles (see Chapter 1, Section III.B). In some conditions the giant micelles can be branched or in the form of closed rings. The solutions of giant micelles are often viscoelastic and sometimes display shear-induced viscoelasticity. These two aspects have been much investigated using rheological methods, and the results are reviewed in Chapter 9. In this section we simply recall some important results and compare them to the ones discussed in the preceding section.

We also include some T-jump studies specifically performed on systems of giant micelles.

Several rheological studies have reported values of the lifetime of wormlike micelles, that is, the time during which a wormlike micelle retains its length before breaking into two daughter micelles. These values generally range between, say, 0.1 and 50 s.[143,144] their lower range these values are of the same order as those evaluated for the average rate constant of micelle fission.[44]

Coagulation/fragmentation reactions (3) in solutions of giant micelles of SDS in the presence of NaCl[145] and of CTAB in the presence of KBr[146] has been investigated by T-jump. Relaxation times in the range of 10 ms to 1 s were reported. The rate of the reaction of coagulation of two micelles was found to be in the 10^6 $M^{-1}s^{-1}$ range, which is much slower than for a diffusion-controlled process.[145] The fitting of the relaxation data using different models of micelle size distribution led to support a model of elongated micelles where the diameter of the cylindrical body of the elongated micelles is smaller than that of the endcaps.[145] The results also showed that when the micelle size becomes sufficiently large the rate of reversible micelle breakdown controls the rheological properties of the systems.[146]

The shear-thickening effect observed with some solutions of giant micelles is still not fully explained, but it has much to do with the dynamics of the system. The effect is generally observed after a latency time and is a maximum in terms of increase of viscosity at a concentration C* close to that where entanglements start to occur. The latency time depends on the history of the sample.[147–49] Flow birefringence studies showed that the shear-thickening transition is associated with the transformation of the initially anisotropic solution in the quiescent state into a strongly oriented phase.[150] The return to equilibrium (the disappearance of the anisotropy after cessation of the perturbation) can be very slow, hundreds of seconds and more. The effect has been attributed to the formation of bundles of elongated micelles, mediated by counterions.[151] Indeed, cryo-TEM showed the formation of such aggregates upon shearing (see Figure 3.11).[152] The critical shear rate would correspond to the time for debundling. A more fascinating explanation rests on the existence of interlinked ring micelles

Figure 3.11 Cryo-TEM image of a solution of the dimeric surfactant 12-2-12 after shear showing giant wormlike micelles and bundles of such micelles indicated by arrows. These bundles were not seen before shearing. Bar = 1 μm. Reproduced from Reference 152 with permission of the American Chemical Society.

in the quiescent state at concentrations below C*.[153] Upon shearing the ring micelles would separate, open, and align in the flow, rendering the system anisotropic. They may also grow. The return to equilibrium is, of course, slow as it implies the reverse of all these processes. The critical shear rate would now be equal to the time for delinking.

Much work remains to be done on the dynamics of systems of giant micelles, both from the experimental and theoretical viewpoints.

IV. DYNAMICS OF AQUEOUS MIXED MICELLAR SOLUTIONS

A. Preliminary Remarks

In ideal mixed micellar solutions of two different amphiphiles (a surfactant or a cosurfactant 1 and a surfactant 2 of longer alkyl chain), the theory predicts the existence of two fast relaxation processes, with the relaxation times τ_{11} and τ_{12}, associated with the exchanges of the surfactants 1 and 2 from the mixed micelles, and of a slower process associated with the mixed micelle formation/breakdown (see Section II.C).[27,29–31,40,66]

In the case where the dissociation rate constants of surfactants 1 and 2 fulfill the condition $k_1^- \gg k_2^-$, τ_{11} is associated with the exchange of surfactant 1. $1/\tau_{11}$ is given by an equation similar to Equation 3.9 where σ and N are now the variance and aggregation number of surfactant 1 in the mixed micelles and $a = (C_1 - C_{1,free})/C_{1,free}$, C_1 being the total concentration of surfactant 1 and $C_{1,free}$ the concentration of free surfactant 1. The expression of the relaxation time τ_{12} associated with the exchange of surfactant 2 is more complex. It contains the variance, the aggregation number, the total concentration and concentration of free surfactants 1 and 2, and also a term that measures the correlation between the concentrations of surfactants 1 and 2 in the mixed micelles.[29,30,40,66] Under appropriate experimental conditions and provided the condition $k_1^- \gg k_2^-$ is fulfilled, the available expressions of τ_{11} and τ_{12} can be fitted to the experimental data from chemical relaxation studies to yield meaningful information on the dynamics of mixed micelles. In the case where k_1^- and k_2^- are comparable, the expressions of the relaxation times are extremely complicated.[29,30,40] Moreover, this corresponds to an experimental situation where the values of the relaxation times can be extracted from the raw data only with rather large errors. Unfortunately, several of the reported studies correspond to this situation, owing to the lack of an adequate theory to use as a guide at the time the measurements were performed.

B. Mixed Systems with Two Surfactants

Studies have been performed using the ultrasonic absorption relaxation for the study of the surfactant exchange. Mixtures of sodium nonylsulfate with sodium hexyl- and pentylsulfates showed more than one relaxation process.[154] However, the experimental conditions did not permit a determination of the relaxation times associated with the exchange of the two surfactants. Only one relaxation process was observed in mixtures of sodium octyl- and nonylsulfates[154] because in this case k_1^- is rather close to k_2^- and the exchanges of the two surfactants are strongly coupled (see Table 3.1).

Mixtures of dodecyl- and hexadecyltrimethylammonium bromides are characterized by one relaxation time,[155] associated with the exchange of the dodecyl surfactant from the mixed micelles. Indeed, in this case $k_1^- \approx 10^2 k_2^-$, and the technique used probed only the faster of the two exchange processes. Unfortunately, these data were not analyzed quantitatively. Similarly, Folger et al.[21] observed a single fast relaxation process in SDS solutions containing small amounts (mole fraction of 1 to 4%) of STS. The addition of STS increased the value of τ_1. However the addition of the cationic surfactant dodecylpyridinium chloride to SDS resulted in the occurrence of a third relaxation process with a very long relaxation time while the values of τ_1 and τ_2 changed only little. No explanation was given for this result. It is known that mixtures of oppositely charged surfactants give rise to very large aggregates, often vesicles. The exchange from vesicles may be extremely slow because of the attractive interaction between one surfactant and the oppositely charged ones (see below). Besides vesicles are much more tightly packed than micelles, and this may further slow down the exchange.

The ability of alkylpyridinium surfactants (used with the chloride counterion) to quench the pyrene fluorescence has been used to obtain the residence time T_R of these surfactants in various micelles. The residence time of a dodecylpyridinium ion (DPy$^+$) in micelles of hexadecyltrimethylammonium chloride was found to be about three times longer than in dodecylpyridinium chloride micelles.[156] Although this difference is nearly within the experimental error, it gives supports to a theory that predicts a decrease of k^- and thus an increase of T_R upon increasing micelle size.[157] The residence time of the decylpyridinium ion in micelles of $Na^+, C_8 H_{17}(OCH_2 CH_2)_5 OCH_2 CO_2^-$ is about 20 times larger than in micelles of decylpyridinium chloride.[158] The difference reflects the effect of electrostatic interactions between the decylpyridinium cation and the micelles, attractive with anionic micelles and repulsive with cationic micelles. The residence time of DPy$^+$ was found to be much larger in lithium dodecylsulfate micelles than in lithium perfluorooctanesulfonate micelles.[159] The difference reflects the lesser affinity of the quencher hydrocarbon chain for a perfluorocarbon chain than for a hydrocarbon chain.

Studies concerning the micelle formation/breakdown in mixed micellar solutions are few. Folger et al.[21] showed that small amounts of STS (mole fraction 2–5%) significantly affected the value of τ_2 for SDS, particularly at C close to the cmc (see Figure 3.5). This is expected since micelle formation/breakdown is similar to a nucleation process. Patist et al.[160] reported that the slow relaxation process in solutions of SDS became considerably slower upon the addition of alkyltrimethylammonium bromides. The largest effect was obtained with dodecyltrimethylammonium bromide, and the authors interpreted the results in terms of chain compatibility. Measurements of the slow relaxation time have been used to show that solutions of mixtures of some hydrocarbon and perfluorocarbon surfactants contain two types of mixed micelles, one rich in hydrocarbon surfactant, the other rich in perfluorocarbon surfactant.[161]

The rate of formation of mixed micelles of two gangliosides was found to be very slow, with a time constant of about 3 hours.[162] In these experiments, two micellar solutions of two different gangliosides were mixed, and the change of scattered intensity with time reflected the hybridization of the micelles. Such very large values of τ_2 are in line with the very small values of k$^-$ reported for gangliosides[127,128] (see Section III.B.11). When one of the two micellar solutions of ganglioside was replaced by a solution of Triton X100, a conventional nonionic surfactant, the rate of hybridization was very fast.

C. Mixed Systems with One Surfactant and One Cosurfactant

In most instances the cosurfactant was a medium-chain length alcohol. The topic has been recently reviewed,[163] and only the main features are presented here. Note that studies of the dynamics of ternary water/surfactant/cosurfactant systems were a prerequisite for the understanding of the dynamics of the quaternary water/surfactant/cosurfactant/oil microemulsions, where the cosurfactant is most often a medium-chain length alcohol. Most studies were performed using chemical relaxation methods (ultrasonic relaxation, shock-tube, p-jump, and T-jump). One study used pulse radiolysis.[120]

Studies that addressed specifically the exchange of the alcohol used surfactants with an alkyl chain containing 14 or 16 carbons atoms and alcohols not longer than hexanol.[155,164-171] Studies that addressed the effect of the alcohol on the surfactant exchange used fairly short-chain surfactants (with m = 8-12) and alcohols not longer than hexanol.[172,173,174,175]

Other studies considered specifically the effect of the alcohol on the kinetics of micelle formation/breakdown.[92,166,176-182] Spectacular changes of the relaxation time τ_2 were evidenced. Note, however, that the three relaxation processes expected for a mixed surfactant + alcohol system were evidenced and investigated for only the tetradecyltrimethylammonium bromide (TTAB)/1-pentanol system.[166]

1. Studies of Surfactant and Cosurfactant Exchanges

The relaxation time τ_{11} associated with the exchange of the alcohol was found to be nearly independent of the nature of the surfactant when using fixed concentrations of alcohol and surfactant.[165] The value of $1/\tau_{11}$ was found to increase nearly linearly with the cosurfactant concentration as theory predicts (see Figure 3.12).[164-171] Unfortunately, many of the data reported in the early studies were analyzed using inadequate theories or the analysis did not yield values of the rate constants.[155,164,167-170] Even in studies where the equilibrium data required for a full analysis of the kinetic data were available, the interpretation was complicated by the fact that the activity of the alcohol in the micelles varies rapidly with the number of alcohol per mixed micelle. The studies where the data were fully analyzed with existing theories showed behaviors very similar to those for conventional surfactants. Thus, the values of the association rate constant k_1^+ are in the range of

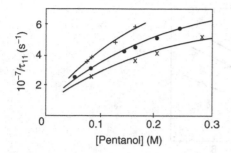

Figure 3.12 TTAB/pentanol systems: variation of the relaxation time associated with the pentanol exchange with the pentanol concentration at constant [pentanol]/[TTAB] molar concentration ratio (+) 0.5; (•) 1.0; and (×) 2. Reproduced from Reference 166 with permission of Elsevier/Academic Press.

TABLE 3.2 Values of the Rate Constants for the Alcohol Exchange

Surfactant[a]	Alcohol	$10^{-7} k_1^-$ (s^{-1})	$10^{-9} k_1^+$ (M^{-1}s^{-1})	Reference
SDS	pentanol	1.9	14	120
SDeS[b]	pentanol	1.2	4.2	172
DTAB[b]	pentanol	1.7	7.0	172
CTAB[b]	pentanol	0.84	7.3	172
CTAB[b]	hexanol	4	15	171
DeTAB[c]	propanol	49	22	173
DeTAB[c]	butanol	35	54	173
DeTAB[c]	pentanol	17	82	173
TTAB[c]	pentanol	100	6	166

[a] SDeS and SDS: sodium decyl and dodecyl sulfate; DeTAB, DTAB, TTAB and CTAB: decyl, dodecyl, tetradecyl and hexadecyl trimethylammonium bromides
[b] Interpretation of the kinetic data according to Hall's theory[27]
[c] Interpretation of the kinetic data according to Aniansson theory[29]

10^9 to 10^{10} M^{-1}s^{-1}, indicating a nearly diffusion-controlled association of the alcohol, whereas the dissociation rate constant k_1^- decreases upon increasing alcohol chain length (see Table 3.2).[163,166,172–174] The values of the rate constant k_1^- determined

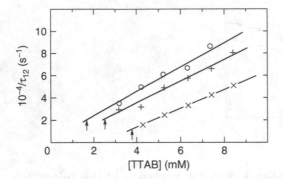

Figure 3.13 Variation of the relaxation time for the TTAB exchange with the surfactant concentration in water ($\times$); water + 0.2 M butanol (+); and water + 0.1 M pentanol (O) at 5°C. Reproduced from Reference 166 with permission of Elsevier/Academic Press.

using Aniansson theory[29] or Hall phenomenological theory[27] can be significantly different[163,173] (see Table 3.2). This, however, does not affect the qualitative conclusions concerning the exchange of the cosurfactant.

The reciprocal of the relaxation time τ_{12} characterizing the surfactant exchange varied linearly with the surfactant concentration at constant alcohol concentration (see Figure 3.13). The cosurfactant was found to have relatively little effect on the exchange rate constants k^+ and k^- of the surfactant.[166,172–174] Some examples are given in Table 3.3.

2. Studies of the Formation/Breakdown of the Mixed Cosurfactant/Surfactant Micelles

The length of the cosurfactant alkyl chain (carbon number m_1) has a very strong effect on the relaxation time τ_2 for the micelle formation/breakdown. Indeed, the longer the cosurfactant (alcohol), the more it partitions in the micelles and the more it affects the micelle size distribution curve. Some representative results are shown in Figure 3.14 for alcohol additions to CTAC.[179] Similar results were reported for alcohol additions to tetradecyl and hexadecylpyridinium

TABLE 3.3 Values of the Rate Constants for the Surfactant Exchange in the Presence of Alcohol[a]

Surfactant	Alcohol	$10^{-6} k_1^-$ (s^{-1})	$10^{-8} k_1^+$ $(M^{-1}s^{-1})$	Reference
DeTAB	none	300	45	173
DeTAB	propanol	140	22	173
DeTAB	butanol	110	17	173
DeTAB	pentanol	95	15	173
TTAB	none	2.4	7	166
TTAB	0.2 M butanol	1.2	5	166
TTAB	0.1 M pentanol	0.76	5	166

[a] Interpretation of the kinetic data according to Aniansson theory for mixed micellization[29]

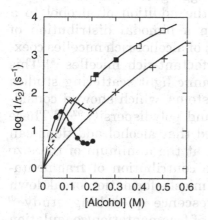

Figure 3.14 Effect of alcohol on the relaxation time for the micelle formation/breakdown in 0.3 M CTAC solution: (+) butanol; (☐) pentanol; (×) hexanol; and (●) heptanol. Reproduced from Reference 179 with permission of the American Chemical Society.

chlorides[179] (see Figure 5.2, Chapter 5) and SDS.[177] Short-chain alcohols up to pentanol induce a monotonic increase of $1/\tau_2$. This increase is very large and can reach a factor of 10^4 (see Figure 5.2, Chapter 5). Longer chain alcohols show a more complex behavior, with $1/\tau_2$ going through a maximum at an alcohol concentration, which decreases when the alcohol chain length is increased. Nevertheless, the primary effect of the alcohol, at low alcohol concentration, is an increase of $1/\tau_2$. The relaxation behavior was related to the solubilization site of the alcohol. Short-chain alcohols at any concentration and long-chain alcohols at low concentration are solubilized close at the micelle surface and labilize the micelles. At higher concentration long-chain alcohols are solubilized in the micelle core and render the micelles less labile. They act like oils (see Chapter 5, Section IV.B). The changes of $1/\tau_2$ reflect at least qualitatively the changes of micelle lifetime. Thus the initial effect of the cosurfactant is to increase the micelle lability.

The explanation of results such as those in Figure 3.14 is rather complicated because alcohols are known to effect all micellar properties: cmc, micelle aggregation number and ionization degree, etc.[183] Besides, the addition of alcohol to a micellar solution may result in a bimodal distribution of aggregate size, with a population of alcohol-rich micelles coexisting with a population of surfactant-rich micelles.[184] This possibility is supported by dynamic light scattering studies performed on closely related systems, which showed considerable changes of micelle size and polydispersity.[92,179] These observations combined suggested that alcohol additions can greatly affect the aggregates A_r at the minimum of the size distribution curve and also the contribution of fragmentation/coagulation reactions to the micelle formation/breakdown process. A time-resolved fluorescence quenching study[180] directly revealed the occurrence of fragmentation/coagulation reactions at a microsecond time scale, that is, much faster than in the chemical relaxation studies.[179] It was suggested that chemical relaxation methods probe fragmentation/coagulation reactions between submicellar aggregates whereas time-resolved fluorescence quenching reveals reactions between a micelle proper and a submicellar aggregate.[179] The labilizing effect of short-chain alcohols may be of importance when using surfactant solutions for solubilizing water-insoluble compounds or when performing chemical reactions using micelles as microreactors.

Section IV in Chapter 5 provides further details on the dynamics of cosurfactant/surfactant mixed micelles.

V. DYNAMICS OF SOLUBILIZED SYSTEMS

Solubilized systems refer to surfactant solutions in which the micelles have solubilized compounds (solubilizates) that are generally poorly soluble in water. Most studies address the rate constants for the exchange of the solubilizate. A few studies examine the effect of solubilizate other than cosurfactants on the surfactant exchange and micelle formation/breakdown.

A. Rate Constants for the Exchange of the Solubilizate

Time-resolved luminescence quenching,[185–190] flash photolysis,[191–195] time-resolved optical spectroscopy,[196] and electron spin resonance[197–199] have been extensively used to measure the exit rate constants k_S^- of many solubilizates from micelles. The determination of the rate constant k_S^+ for the association (entry) of the solubilizate into micelles requires the additional measurement of the solubilizate binding constant to the micelles. A number of values of k_S^+ and k_S^- for conventional solubilizates are listed in Table 3.4. The last five entries in this table give data for more special solubilizates.[200–204] The trends in the values of k_S^+ and k_S^- are qualitatively similar to those for the exchange of surfactants.

Table 3.4 shows that the association (entry) of aromatic solubilizates in micelles is very close to being diffusion-controlled, as indicated by the entries 1, 2, and 7–16 for which k_S^+ is in the 10^9–10^{10} $M^{-1}s^{-1}$ range. The k_S^+ values are somewhat smaller for compounds having an alkyl chain as in entries 18–20 but remain close to those for a diffusion-controlled process. The difference may reflect the detail of the solubilization process: aromatics are solubilized close or at the micelle surface, in the palisade layer, whereas compounds with an alkyl chain must have this alkyl chain penetrating in the micelle core.

Nevertheless, values of k_S^+ both much larger and much smaller than for a diffusion-controlled process have been reported. Thus Moroi et al.[205] reported for naphthalene and pyrene values of k_S^+ in the range of 10^{12} and 10^{13} $M^{-1}s^{-1}$, respectively. These values are in strong disagreement with reported

TABLE 3.4 Values of k_S^+ and k_S^- for Selected Compounds at 25°C

Entry	Compound	Surfactant[a]	Method[b]	k_S^+ ($M^{-1}s^{-1}$)	k_S^- (s^{-1})	Reference
1	Naphthalene	SDS	TRL	1.9×10^{10}	1.3×10^6	185
2	1-Bromonaphthalene	SDS	TRLQ	4×10^{10}	2.5×10^4	186
3	1-Bromonaphthalene	SDS	TRLQ		4×10^4	187
4	1-Chloronaphthalene	SDS	TRLQ		4.3×10^4	188
5	1-Chloronaphthalene	CTAB	TRLQ		1.9×10^3	188
6	Anthracene	CTAB	FP		2×10^2	191
7	Biphenyle	CTAB	TRLQ		1.2×10^5	186
8	m-Dicyanobenzene	SDS	TRLQ	10^{10}	6×10^6	189
9	Methylene iodide	CTAB	TRFQ	2.5×10^{10}	9.5×10^6	191
10	Ethyliodide	SDS	TRLQ	2×10^{10}	5×10^6	190
11	Acetophenone	SDS	FP	2.6×10^{10}	7.8×10^6	192
12	Propiophenone	SDS	FP	1.4×10^{10}	3×10^6	192
13	Benzophenone	SDS	FP	5.2×10^{10}	2×10^6	192
14	Xanthone	SDeS	FP	2.4×10^{10}	2.7×10^6	193
15	Xanthone	SDS	FP	1.2×10^{10}	1.6×10^6	193
16	Xanthone	SDS/0.3 M NaCl	FP	1.5×10^{10}	2.1×10^6	193
17	Xanthone	STS	FP	7.2×10^9	1.1×10^6	193
18	Xanthone	CTAC	FP	9×10^8	6×10^5	193
19	cis 1,3-Pentadiene	SDS	FP	1.2×10^9	8.9×10^6	194
20	trans 1,3-Pentadiene	SDS	FP	9.5×10^8	6.9×10^6	194
21	1,3-Hexadiene	SDS	FP		2.3×10^6	194
22	1,3-Octadiene	SDS	FP	8.3×10^8	1.3×10^5	194

TABLE 3.4 Values of k_S^+ and k_S^- for Selected Compounds at 25°C (continued)

No.	Compound	Surfactant	Method	k_S^+	k_S^-	Ref.
22	1,3-Cyclooctadiene	SDS	FP		3.5×10^5	194
23	Acetone	SDS, CTAB	FP	$>10^{10}$	$1\text{-}4 \times 10^8$	195
24	Benzyl radical	SDeS	TROS		2.7×10^6	196
25	Benzyl radical	SDS	TROS		1.8×10^6	196
26	Benzyl radical	STS	TROS		1.2×10^6	196
27	Toluene	SDS	Solubility		1.4×10^6	186
28	t-Butyl-(1,1-dimethylpentyl)-nitroxide	SOS	ESR		7.5×10^5	197
29	t-Butyl-(1,1-dimethylpentyl)-nitroxide	SDeS	ESR		5.9×10^5	197
30	t-Butyl-(1,1-dimethylpentyl)-nitroxide	SDS	ERS		3.7×10^5	197
31	t-Butyl-(1,1-dimethylpentyl)-nitroxide	TDS	ESR		2.2×10^5	197
32	t-Butyl-(1,1-dimethylpropyl)-nitroxide	SDS	ESR		1.1×10^6	198
33	t-Butyl-(1,1-dimethylbutyl)-nitroxide	SDS	ESR		3.1×10^5	198
34	Di-t-butylnitroxide	SDS	ESR		3.5×10^5	199
35	Safranine$^+$, Cl$^-$	SDS	TRFQ	1.5×10^9	1×10^5	200
36	Erythrosin B^{2-}	SDS	TRFQ	9.6×10^8	2.5×10^2	201
37	Rose bengal^{2-}	CTAB	FP	4×10^7	$2\text{-}5 \times 10^5$	202
38	Molecular oxygen	SDS	FP, ESR		$< 5 \times 10^7$	203
39	Muonium	SDS, CTAB, SDS, DTAB, $C_{12}EO_3$	TRFQ	1.3×10^{10}	5×10^8	204

a SOS, SDeS, SDS, STS: sodium octyl, decyl, dodecyl and tetradecyl sulfates; DTAB, CTAB: dodecyl and hexadecyl trimethylammonium bromides; $C_{12}EO_3$: triethyleneglycol monododecylether;
b TRL: time-resolved luminescence; TRLQ: time-resolved luminescence quenching; FP: flash photolysis; TROS: time-resolved optical spectroscopy.

ones for the same compounds.[185,186] The difference may result from the different surfactant used (dodecylammonium trifluoroacetate instead of SDS), but it may also be associated with the solubilization model used for the quantitative analysis of the data. The entries 35 and 37 show k_s^+ values lower than for a diffusion-controlled process. These lower values are the result of electrostatic repulsion between the dye and the micelles that slows down the dye incorporation.

Table 3.4 shows that the exit rate constant varies very much with the hydrophobicity of the solubilizate whether it is aromatic (compare entries 1, 2, and 5; entries 6 and 24) or aliphatic (see entries 30, 32, and 33) or is aromatic with an increasing aliphatic moiety (see entries 10, 11, and 23). The results indicate that an additional methylene group to the aliphatic part results in a decrease of k_s^- by a factor close to 3, as in the case of surfactants. For a given solubilizate, k_s^- decreases as the micelle radius increases (see entries 13, 14, and 16 for xanthone; entries 24–26 for the benzyl radical; entries 28–31 for the ESR probe). This effect has been accounted for theoretically by Almgren.[157] Entries 1–3 show a strong decrease of k_s^- upon attaching a bromine or chlorine atom to naphthalene, again as for surfactants. In many of the studies performed by TRLQ or FP, the measured exit rate constant referred to the triplet state of the solubilizate or of a radical arising from the solubilizate. However, it has been verified that the values of the partition coefficient of the solubilizate in the triplet state and in the ground state are equal.[185] It is thus likely that the values of the exchange rate constants are also the same. Besides the exit rate constant for toluene (entry 27) and the benzyl radical (entry 25) from SDS, micelles are nearly equal.[196]

The values of k_s^+ and k_s^- have been shown to depend significantly on the isomerism of the solubilizate. Thus, k_s^- values of 2.8×10^6, 0.9×10^6 and 0.2×10^6 s^{-1} have been reported for the exit of the 2-, 3-, and 4-nitrophenol, respectively, from SDS micelles.[206]

A detailed analysis of the fluorescence decay of 2-ethylnaphthalene in micelles of cationic surfactants has shown that the probe constantly exchange between the micelle core and the head group region with a characteristic time of 5–10 ns.[207]

The exit rate constants for cyclohexane and *p*-xylene from the micelles present in the cubic phase of the sodium octanoate/octane/water system have been measured by combining TRLQ and NMR self-diffusion data on the system.[137] The value found for *p*-xylene compares well with that determined on the basis of solubility measurements.[186]

The entry/exit rate constants of alkylpyridinium ions into/from hexadecyltrimethylammonium chloride and acetate (CTAC and CTAAc) micelles have been determined by TRLQ.[208] The nature of the surfactant counterion was found to have only a slight effect on the values of the rate constants. The effect of the chain length of the alkylpyridinium was important, and the results showed the usual decrease of k_S^- by a factor close to 3 per additional methylene group. The rate constant k_S^+ was around 10^9 $M^{-1}s^{-1}$ irrespective of the counterion and the alkyl chain length.

The above results suggest that the order of magnitude of the exit rate constant of a solubilizate from a given surfactant micelle can be estimated from the partition coefficient or the binding constant $K = k_S^+ / k_S^-$ of the solubilizate to the micelles taking for k_S^+ a value of, say, 3×10^9 $M^{-1}s^{-1}$. This method that replaces complex kinetic measurements by solubility measurements was first proposed by Almgren et al.[186] and shown to yield correct estimates of k_S^-.

The solubilizates considered thus far were of relatively small size. Some interesting results have started to accumulate for solubilizates that are extremely large and very hydrophobic. In this case the micelle dynamics can limit the rate of intermicellar exchange of the solubilizate. We alluded to this when discussing the occurrence of fragmentation/coagulation processes at the end of Section III.A. The pyrene-labeled diheptadecyl triglyceride probe is so hydrophobic that it cannot be exchanged between micelles via the aqueous phase. It is transported from one micelle to another attached to a small premicellar aggregate that breaks away from one micelle and fuses with another one.[93,94] These previous stopped-flow studies have been extended to other nonionic surfactants, the Synperionic A7 and A50, using the same probe in addition to the octyl and dodecylpyrene.[209] The reported values of the incorporation rate constants do not depend on the probe and are in the range 10^5 to 10^6 $M^{-1}s^{-1}$. These values are in the same range as those obtained from TRFQ. A rigorous theory of the use of the stopped-flow method for studies such as those just discussed has been reported.[210]

Several years earlier in a little-quoted paper it was suggested that pyrene migrates between SDS micelles in the aqueous phase under the form "clumps" of pyrene with associated monomeric surfactants.[211] These clumps can be considered as premicellar aggregates. Supporting these ideas, evidence has been reported that the redistribution of dodecylpyrene between SDS micelles takes many hours at room temperature.[212]

Thus the literature has considered only two limiting cases of solubilizates, those of low to medium hydrophobicity and those of extremely high hydrophobicity. For the first ones, the association is close to diffusion-controlled. For the second one, the association is completely controlled by the dynamics of the micelles. No exchange may take place even at the time scale of months (the case of the pyrene diheptadecyltriglyceride probe in SDS solutions[93,94]) if fragmentation/coagulation processes do not occur in the system. Note that such processes may not occur spontaneously but may take place in the presence of pyrene. It would be interesting to investigate probes of strong hydrophobicity with exit rate constants comparable to the rate of fragmentation. This can probably be best done by using a series of alkylpyrenes.

B. Effect of Solubilizates on the Micelle Dynamics

There have been few investigations of the effect of aromatic or alkyl solubilizates on the micellar dynamics. Alkanes have been found to have almost no effect on the value of the relaxation time τ_1 for the surfactant exchange in micellar solutions of sodium heptylsulfate and hexylammonium chloride.[108] Likewise, cyclohexane has very little effect on the exchange in micellar solutions of sodium octylsulfate.[213] In these studies the amount of solubilizate was relatively small. In contrast, a complex variation of $1/\tau_1$ with the concentration of benzene, toluene, and *p*-xylene was reported, with the presence of a maximum and sometimes also of a minimum in the representative plots.[213] It is tempting to relate these variations to the known solubilization sites of these compounds. Alkanes are solubilized in the micelle core and would only slightly affect the exit/entry of the surfactant. The aromatic compounds are solubilized first in the palisade layer (surface layer of the micelles) and then in the micelle core once this region is saturated. The effect of aromatics on τ_1 resembles that of alcohols on τ_2 (see Section III.C.2).

The effect of oils on the micelle formation/breakdown has been essentially performed on mixed alcohol + surfactant micelles in order to be able to solubilize a sufficient amount of oil in the system.[92,179,180] These studies are reviewed in detail in Chapter 5, Section IV.B. Here, it will only be stated that additions of alkanes to mixed alcohol + surfactant micelles always result in a large decrease of $1/\tau_2$, an effect opposite to that of alcohol additions to pure surfactant solutions (see Chapter 5, Figure 5.3). This effect was attributed to the formation of an oil core and a decrease of the micelle polydispersity that prevented the occurrence of fragmentation/coagulation processes. Additions of toluene resulted in a plot of $1/\tau_2$ against the toluene concentration showing a maximum. The initial increase of $1/\tau_2$ was attributed to a surface solubilization of toluene at low concentration. At higher concentration toluene is solubilized in the core and has an effect similar to that of alkanes. The maximum was not observed with butylbenzene.[179] This interpretation was supported by TRFQ and dynamic light scattering studies of similar systems.[92,180]

VI. DYNAMICS OF POLYMER/SURFACTANT SYSTEMS

A. Reminder on Polymer/Surfactant Systems

Surfactants can bind to water-soluble polymers.[214–216] In many instances the binding occurs only when the surfactant concentration is larger than a certain value referred to as critical aggregation concentration, or cac, and when the molecular weight of the polymer is larger than a minimum value corresponding to a degree of polymerization above, say, 20.[216] The cac is generally lower than the cmc. The polymer/surfactant interaction is cooperative, and the surfactant binds to the polymer under the form of aggregates that can be referred to as *polymer-bound micelles*. The capacity of a polymer to bind a surfactant is restricted. Generally the investigations show that above a certain concentration, often referred to as C_2, free micelles appear in the polymer/surfactant system. In the early studies and in some recent ones as well, the concentration C_2 was taken as that where the polymer is *saturated* by the bound surfactant. However, it was later shown that the polymer can still bind a small amount of additional surfactant at $C > C_2$.[217,218] The stoichiometry of the binding (amount of bound surfactant per gram of polymer) has been reported for a number of systems.

The characteristic of the interaction between a surfactant and a polymer (occurrence and stoichiometry of the binding, value of the cac, value of the surfactant aggregation number, and value of the ionization degree of the bound aggregates, for instance) depend very much on the nature of both the surfactant and polymer.[214–216] Thus hydrophilic polymers such as PEO (poly(ethylene oxide)) or PVP (poly(vinylpyrrolidone)) clearly interact with anionic surfactants. However, their interaction with nonionic or cationic surfactants is very weak or absent. Strongly hydrophilic polyelectrolytes interact strongly with oppositely charged surfactants but only weakly or not at all with nonionic surfactants or surfactants having a like charge. A progressive increase of the polymer hydrophobicity results in dramatic changes in these behaviors.[216] The interaction becomes progressively less cooperative, and a sufficiently hydrophobic but still water-soluble polymer (hydrophobically modified polymer or polysoap) will bind any type of surfactant irrespective of its charge and nature.

Polymers and surfactants are simultaneously present in many formulations used for cosmetics, paints, food, detergents, mineral processing, and enhanced oil recovery, to cite but a few. Indeed, the polymer is frequently introduced for controlling the viscosity of the formulation or the process in which the formulation is used. An understanding of the mechanism of interaction between polymer and surfactant can thus be important in the basic and applied aspects of these systems. Given the close relationship that exists between dynamics of surfactant solutions and their properties that are important in applications such as foaming, wetting, emulsification, solubilization, detergency, and thin film stability,[219,220] it is not surprising that the dynamics of surfactant aggregates bound to water soluble polymers has been much investigated.

Most of these investigations were performed using chemical relaxation methods, ultrasonic absorption relaxation for the exchange process and T-jump, p-jump, or stopped-flow for the micelle formation/breakdown. More recently, NMR methods have started to be used for the same purpose.

At the outset it must be pointed out that several of the reported investigations involved polymer/surfactant systems that contained free and polymer-bound micelles (surfactant concentration above C_2). The interpretation of the data is then difficult, if at all possible, as the relaxation signals are then complex weighted averages of the signals from the different micellar species. Another difficulty is that the aggregation number of the polymer-bound micelles depends much on the composition of the system and vary fairly rapidly with the concentration of both the polymer and surfactant. A correct analysis of the kinetic data for polymer/surfactant systems thus requires fairly accurate values of the stoichiometry of the binding and of the aggregation number of the bound micelles. Unfortunately, in all reported studies one or both sets of data were not available.

B. Surfactant Exchange from Polymer-Bound Micelles

The first study mainly involved mixed solutions of PVP or poly(vinyl alcohol) and sodium octylsulfate.[221] No information on the binding stoichiometry and on the aggregation number of the bound aggregates was reported. The relaxation time for the exchange process was obtained from ultrasonic absorption relaxation. It was not affected much by the presence of PVP in the range where free micelles were present in the system. The authors concluded that polymer addition had little effect on the kinetics of the surfactant exchange. Nevertheless, a relaxation process was observed at concentration below the cmc that is less clearly seen in the absence of PVP.

The next study involved the PVP/sodium hexadecylsulfate system and reported both kinetic and equilibrium binding data.[222] The much longer-chain surfactant used permitted easier and more accurate measurements of the relaxation time τ_1 by means of p-jump than in Reference 221. The main relaxation observed above the cac of the system was attributed to the exchange process. In some of the reported experiments, the surfactant concentration exceeded C_2 but the occurrence of free micelles in the system did not show on the plot of $1/\tau_1$ vs a (see Equation 3.9). From the analysis of the results the authors concluded that the phenomenological theory for the kinetics of the surfactant exchange[223] provided a better fit of their results than Aniansson et al. treatment.[24,25] This conclusion was based on the fact that the phenomenological theory gave a single plot for the kinetic data obtained using two different PVP concentrations (see Figure 3.15) whereas two distinct plots were obtained when plotting $1/\tau_1$ vs a as suggested by Aniansson et al. treatment. The slope of the plot in Figure 3.15 yields $k^-/N = 1.8 \times 10^3$ s^{-1} and, thus, $k^- \approx 8 \times 10^4$ s^{-1}. This value is relatively close to that reported for sodium hexadecylsulfate in aqueous solution.[25]

The next studies involved kinetic and binding studies and used PVP or PEO and various alkylsulfate surfactants.[224,225] The value of k^-/N was found to be independent of the polymer concentration and molecular weight and to be rather close to those in the absence of polymer[225] when taking into account the smaller value of N for the polymer-bound micelles, as compared to free micelles. The association of a surfactant with a polymer-bound micelle was found to be diffusion-controlled.[224,225]

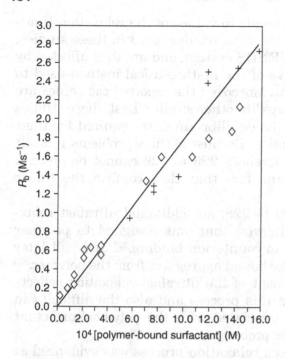

Figure 3.15 Sodium hexadecylsulfate/PVP system: variation of the backward rate R_b with the amount of polymer-bound surfactant at 35°C and PVP concentration of 0.01 wt% (◊) and 0.025 wt% (+). Reproduced from Reference 222 with permission of the Royal Society of Chemistry.

Subsequent ultrasonic relaxation studies[226–229] also involved the PVP/SDS system. The reported results are very different from those on the same system given in Reference 225, in the sense that in the binding range the value of $1/\tau_1$ is claimed to be nearly independent of the surfactant concentration. This is apparently true for the data in References 226 and 227 (PVP/SDS systems) but certainly not for the results reported in Reference 228 for several other polymer/surfactant systems. There it is clearly seen that $1/\tau_1$ is nearly constant over a very short range of concentration but increases linearly with C in the binding range determined from ultrasonic velocity measurements,[228] as predicted by Equation 3.9. Besides in Reference 226 to Reference 228, the range where $1/\tau_1$ is nearly constant always corresponds to values of the relaxation times that are clearly below the time window used in these studies, particularly for the PVP/SDS system, and are thus affected by large errors irrespective of the mathematical method used to extract the value of $1/\tau_1$. Moreover, the reported cac values are much larger than reported in other studies. Last, these studies did not report any of the equilibrium data required for their quantitative interpretation. Because of these problems, it is felt that the results in References 226 to 229 cannot be given a meaning other than the fact that they confirm the linear increase of $1/\tau_1$ with C.

In References 226 to 228, an additional ultrafast relaxation process was observed that was assigned to polymer stretching,[226] later on to counterion binding,[227] and still later to the detachment of the bound aggregates from the polymer.[228] This changing assignment of the ultrafast relaxation reflects the lack of data about this process and also the difficulty in separating processes associated with the polymer/surfactant interactions from other processes.

In several studies a relaxation process was evidenced at concentration below the cac.[221,226–228] It has been attributed to the binding of isolated surfactants on the polymer, occurring before the cooperative binding stage, a behavior similar to the binding of surfactant to surfaces. However, a similar process has been reported to occur also in surfactant solutions in the absence of polymer, although less pronounced.[221] It may therefore not be associated with polymer/surfactant interactions. More work on this process is required before any definite statement can be made about its origin.

A fast exchange of the surfactant (sodium alkylbenzene-sulfonate) between water and micelles bound to hydrophobically modified poly(acrylamide) with different substitution degrees was inferred from ^{1}H NMR studies.[230] More recently, ^{19}F NMR has been used to study the dynamics of the exchange of SDS or cesium perfluorooctanoate between the aqueous phase and micelles bound to a hydrophobically modified poly(sodium acrylate) bearing $C_7F_{15}CH_2$ pendent groups.[231] The results showed that the exchange of both surfactants was fast on the NMR time scale, that is, with characteristic times well below 1 ms. The investigated polymer is able to form small hydrophobic microdomains through the self-association of several pendent alkyl chains. The time characterizing the exchange of a pendent chain between the aqueous environment and the environment provided by the hydrophobic microdomains was found to be of a few microseconds, a value consistent with those reported for regular micelles.[231]

In conclusion, the results reported in Reference 224 and Reference 225 are the most valuable ones for understanding the kinetics of the surfactant exchange process involving polymer-bound micelles. These results clearly show that the kinetics is little affected by the binding. The polymer simply shifts the cmc to a lower value, i.e., the cac. This results in lower values of $1/\tau_1$, as if the surfactant alkyl chain was longer and its cmc lower. This explains why the $1/\tau_1$ versus C plot shows no peculiarity when the surfactant concentration becomes larger than C_2. This is the same behavior as that observed when plotting the $1/\tau_1$ versus C data for a series of surfactants of increasing chain length (see Section III.A). All data points also nearly fall on a single straight line. Nevertheless, more studies should be performed on systems where τ_1 can be measured accurately at as many concentrations of polymer and surfactant as possible. This requirement suggests the use of long-chain surfactants and of the p-jump or T-jump techniques. Measurements of micelle aggregation numbers and equilibrium binding studies should be performed on the same systems in order to permit a quantitative interpretation of the data.

C.   Kinetics of Formation/Breakdown of Polymer-Bound Micelles

Most studies dealt with the kinetics of micelle formation/breakdown in systems containing PVP or PEO and sodium alkylsulfates and were performed using jump methods (T-jump or p-jump).[232-238] In all of these studies, the presence of the polymer was found to strongly decrease the value of the relaxation time τ_2 associated with the micelle formation/breakdown. At the outset, it must be pointed out that most of these studies were performed in conditions where the system may have included free and polymer-bound micelles. Even then, the polymer was often found to have a large effect on the value of τ_2. This is illustrated in Figure 3.16 for the SDS/PEO system.[234] The $1/\tau_2$ vs. PEO concentration plots show a break while the relaxation amplitude vs. PEO plots show a minimum at a certain polymer concentration, in experiments performed at constant surfactant concentration. The composition of the systems at the break or minimum has been shown to correspond to that where the polymer is nearly saturated by surfactant (concentration C_2 discussed earlier). At above the break concentration, the larger the polymer concentration, the larger the increase of

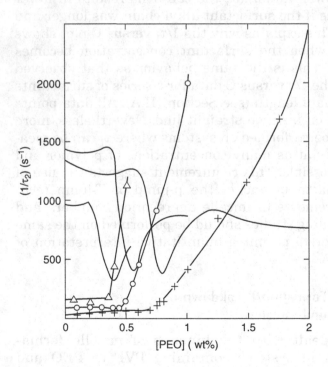

Figure 3.16 SDS/PEO system: influence of the PEO (M_w = 10,000) concentration on the value of $1/\tau_2$ and on the amplitude of the relaxation (lines without symbols) in experiments performed at constant SDS concentration of 35 mM (Δ); 45 mM (O); and 70 mM (+). Reproduced from Reference 234 with permission of the American Chemical Society.

$1/\tau_2$. Similar results were reported for the SDS/PVP system.[232] These variations are probably related to the decrease of the surfactant aggregation number upon increasing polymer concentration and the corresponding changes in the micelle size distribution curve when the surfactant aggregates on the polymer.[239] The submicellar aggregates at the minimum of the distribution curve may be stabilized when forming on the polymer.

D. Miscellaneous Studies

The study of the lithium perfluorononanoate/PVP system by ^{19}F NMR showed separate resonance lines corresponding to free and polymer-bound micelles at 12°C.[240] These lines coalesced at 27°C indicating a time of exchange between free and bound micelles of the order of 0.1 ms.

There has been some preliminary chemical relaxation investigations of the kinetics of protein/surfactant systems.[232,241] Both studies involved the bovine serumalbumine/SDS system and used the p-jump with conductivity detection. The introduction of the protein in the micellar solution of SDS was found to result in an increase of $1/\tau_2$, a result similar to that obtained with regular polymers. This effect probably also reflects changes in the micelle size distribution curve. The surfactant aggregates bound to proteins are also smaller than in the absence of protein.

Narita et al.[242] investigated the kinetics of binding of a surfactant to a weakly cross-linked gel of poly(2-acrylamido-2-methylpropanesulfonic acid). The penetration of the surfactant into the gel was shown to be extremely slow, stretching over hours.

VII. DYNAMICS OF VARIOUS PROCESSES OCCURRING IN MICELLAR SYSTEMS

A. Counterion Exchange Kinetics

Counterions bound to the micelles are only electrostatically attracted by the oppositely charged micelle surface.[243,244] For instance, alkali metal counterions bind to micelles with a negligible loss of hydration water.[245,246] This also appears to be the case, to a large extent, for divalent counterions, since not even an outer-sphere complex is formed by a bound divalent counterion and a head group at the micelle surface.[247] As a result, the exchange between free and bound counterions is always extremely rapid, as systematically observed in NMR studies.[243]

Most quantitative studies of counterion exchange have been performed using TRFQ.[248-255] All these studies used counterions that can quench the fluorescence of a micelle solubilized probe, most often pyrene. Table 3.5 lists some values of the rate constants k_C^+ and k_C^- for the association and dissociation of a counterion to/from a micelle. Large errors may affect the listed values as they are often obtained as the difference between two large numbers. Nevertheless, it is seen that the association is close to being diffusion-controlled. The dissociation rate constant clearly decreases as the ion charge is increased, but for a given ionic charge the value depends relatively little on the nature of the counterion. This of course reflects the fact that the counterions are not complexed by surfactant head groups. In agreement with the large values of k_C^-, a relaxation process with a characteristic time of about 50 ns has been seen in solutions of cetylpyridinium iodide using the E-field jump method and assigned to iodide dissociation from micelles.[256]

The dissociation rate constant of all quenching counterions has been found to increase slightly with the surfactant concentration and more with the concentration of an added salt that does not quench the fluorescence.[248-255] This effect was first attributed to the occurrence of intermicellar collisions, during which some quenching counterions are transferred from one micelle to another by a hopping mechanism. Later Alonso and Quina[254] explained this effect using the pseudophase ion exchange model, which assumes that the counterion dissociation rate constant is determined by the potential at the micelle surface. They reported an equation that related k_C^- to the concentration of nonquenching counterion in the aqueous phase. An example of fit of variation of k_C^- with the counterion concentration is shown in Figure 3.17. This new interpretation appears to provide a more reasonable

TABLE 3.5   Exchange Rate Constants for Counterions in Different
Micellar Systems from TRFQ Studies[a]

Surfactant/probe	Counterion	$10^{-9} k_C^+$ ($M^{-1}s^{-1}$)	$10^6 k_C^-$ (s^{-1})	Reference
DTAC/pyrene	I⁻	50	2.4	248
CTAC/pyrene	I⁻		0.25	250
CTAC/methylpyrene	I⁻	26	3.6	250
SDS/pyrene	Tl⁺	29	7.6	251
SDS/pyrene	Ag⁺	<16	8.0	251
SDS/pyrene	Cs⁺	20	2	251
SDS/pyrene	Cu²⁺	2.9	0.5	252
SDS/methylpyrene	Cu²⁺		0.2	253
SDS/pyrene	Cu²⁺		0.1	255
SDS/pyrene	Ni²⁺	1	0.1	251
SDS/pyrene	Co²⁺	1	0.1	251
SDS/pyrene	Eu³⁺	< 0.1	< 1	251
SDS/pyrene	Cr³⁺	< 0.1	< 1	251

[a] The listed values of k_C^- correspond to low surfactant concentration or ionic strength.

explanation of the overall results than the hopping mechanism. Indeed, some values of the calculated rates of collision
between micelles with counterions hopping were extremely
large, in the 10^8 to 10^9 $M^{-1}s^{-1}$ range, and did not appear reasonable for the ionic micelles used.

Although their study does not deal with counterion
exchange, Hedin and Furo[257] recently confirmed that counterions diffuse very rapidly on the micelle surface.

B. Micellar Kinetics at the Approach of a Critical Point

Poly(oxyethylene) monoalkylether nonionic surfactants show
clouding and phase separation when the temperature is
increased.[258] It is now agreed that the clouding is mainly
associated with large fluctuations of composition that occur
at the approach of a critical point.[258] Micelle growth also
contributes but to a much lesser extent.[259] Two studies used
the T-jump method with light-scattering detection to investigate the effect of the approach of the critical point on micellar

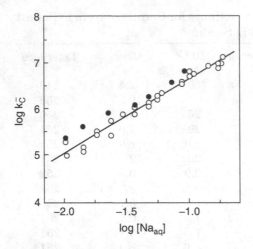

Figure 3.17 Variation of the rate constant for Cu^{2+} escape from SDS micelles with the concentration of sodium ions in the aqueous phase from Reference 253 (•) and Reference 254 (O). The solid line going through the data from Reference 253 is a fit based on the pseudophase ion exchange model. Reproduced from Reference 254 with permission of the American Chemical Society.

kinetics.[260,261] The two studies agreed that the method used can distinguish between critical fluctuations and micellar kinetics. Indeed, composition fluctuations give rise to rapid relaxation effects with wavelength-dependent time constants. In contrast, the time constants associated with micellar kinetics are independent of the wavelength.[261] An ultrasonic absorption study[114] also permitted a clear distinction between relaxations due to composition fluctuations and to surfactant exchange in solutions of poly(oxyethylene) monoalkylether nonionic surfactants.

C.   Micellar Kinetics in the Presence of Cyclodextrins

Cyclodextrins are known to form various types of complexes with surfactants. Verrall et al.[262–264] investigated the effect of complexation on the surfactant exchange kinetics, using the ultrasonic relaxation method. The results revealed multiple relaxation both in the presence and absence of micelles, indicating that the molecular dynamics of the cyclodextrin/surfactant complex was detected in the ultrasonic time scale used. For systems containing micelles, the authors separated the contribution resulting from the dynamics of the cyclodextrin/surfactant complex from that resulting from the monomer/micelle exchange. The analysis of the contribution arising from the surfactant exchange by means of Aniansson and Wall theory led the authors to conclude that the kinetics of this process was not affected by the cyclodextrin/surfactant complex.

D.   Dynamic Surface Tension and Micelle Formation/Breakdown Kinetics

The dynamic surface tension (DST) at long time $\gamma(t)_{t\to\infty}$ for surfactant solutions at concentrations above the cmc has been related to the relaxation time for micelle breakdown τ_2 according to[265]

$$\gamma(t)_{t\to\infty} = \gamma_{eq} + \frac{RT\Gamma^2}{2Ct}\left(\frac{\tau_2}{D}\right)^{1/2} \qquad (3.26)$$

In Equation 3.26, Γ is the equilibrium surface excess, C the bulk concentration, t the time, and D the surfactant monomer diffusion coefficient. Eastoe et al. have measured the time dependence of the DST and the relaxation time τ_2 for solutions of many surfactants: nonionic, dimeric, and zwitterionic.[266,267] In all instances the fitting of the data to Equation 3.26 with the experimentally determined value of τ_2 was poor. The authors concluded that the micelle dissociation may have an effect on the measured DST only if the concentration of monomeric surfactant in the subsurface diffusion layer is limiting or when the micelle lifetimes are extremely long. No surfactant for which this last condition is fulfilled was evidenced by the authors. They also concluded that the rapid dissociation of monomers from micelles present in the subsurface was not likely to limit the surfactant adsorption and thus the DST.

VIII. RELEVANCE OF MICELLAR KINETICS TO PRACTICAL ASPECTS OF SURFACTANT SOLUTIONS

In a series of recent studies summarized in two review papers.[219,220] Shah et al. have shown that whereas the relaxation time for micelle formation/breakdown of SDS micelles goes through a maximum at a concentration around 200 mM, many properties involving liquid/gas, liquid/liquid, and solid/liquid phenomena go through a maximum or a minimum at the same SDS concentration (see Figure 3.18). These experimental results are certainly sound. However, the model used to explain the maximum of τ_2 in terms of increased micellar stability upon decreasing intermicellar distance and vice versa is open to criticism. The SDS micelles grow moderately with concentration in a very continuous fashion in the 0.02 to 0.8 M range, with no discontinuity or peculiarity in the aggregation number versus concentration plot.[268] They grow because the packing parameter[34] of SDS is slightly larger than 1/3 or become so as the concentration is increased. The maximum in τ_2 reflects the fact that the process of fragmentation/coagulation starts contributing to the micelle formation/breakdown, thereby shortening the micelle lifetime. This occurs because the micelles become progressively more polydisperse and the intermicellar repulsion is reduced by the increased ionic strength of the system. With the occurrence of this process monomers become more rapidly replaced if consumed by some process involving them in any of the properties represented in Figure 3.18. It is then easily accepted that this property may go through a maximum or minimum as the surfactant concentration is increased.

A recent theory of the rate of solubilization into micellar solutions involves the dynamics of the solubilizate exchange between the micelles and the continuous phase.[269]

IX. CONCLUSIONS AND SUMMARY

Our understanding of the dynamics of micellar solutions (surfactant and counterion exchange, micelle formation/break-

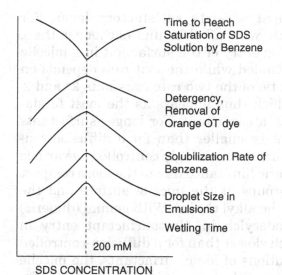

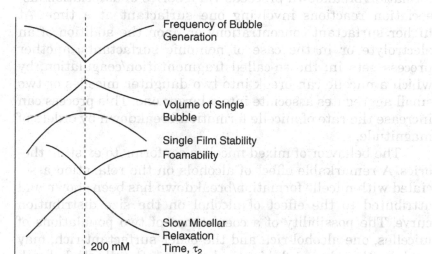

Figure 3.18 Liquid/gas phenomena (*bottom*) and liquid/liquid and liquid/solid phenomena (*top*) exhibiting a maximum or a minimum at an SDS concentration of 200 mM. Reproduced from Reference 219 with permission of Elsevier/Academic Press.

down) has now reached a fairly satisfactory level. For conventional surfactants with alkyl chains not longer than, say, 14 carbon atoms, the entry of the surfactant in a micelle is nearly diffusion-controlled while the exit rate depends on its chain length. The ratio of the two rate constants k^+ and k^- is equal to the cmc, which thus appears as the most fundamental characteristic of a surfactant. For longer surfactants the entry rate constant is smaller than for a diffusion-controlled process and becomes increasingly controlled by various factors, among them steric hindrance due to the more compact packing of the head groups at the micelle surface and the increased bulkiness of the alkyl chains. With gemini (dimeric) surfactants with a dodecylchain, the surfactant entry in micelles is already much slower than for a diffusion-controlled process. For dilute solutions of ionic surfactants, the micelle formation/breakdown proceeds via a series of association/dissociation reactions involving one surfactant at a time. At higher surfactant concentrations or upon the addition of an electrolyte or in the case of nonionic surfactants, another process sets in: the so-called fragmentation/coagulation, by which a micelle can break into two daughter micelles or two small aggregates associate into a larger one. This process can increase the rate of micelle formation/breakdown by orders of magnitude.

The behavior of mixed micelles conforms to existing theories. A remarkable effect of alcohols on the relaxation associated with micelle formation/breakdown has been shown and attributed to the effect of alcohol on the size distribution curve. The possibility of a coexistence of two populations of micelles, one alcohol-rich and the other surfactant-rich, may explain the observed kinetic behavior. Small alcohol-rich micelles may play the role of carrier between surfactant-rich micelles.

The exchange of a solubilizate between micelles and bulk phase has been shown to present the same characteristics as the surfactant exchange.

The dynamics of polymer-bound surfactant aggregates has not been sufficiently investigated. Nevertheless, it appears that the surfactant exchange is little affected by the binding while the micelle formation/breakdown can be considerably modified.

Last, the importance of micellar kinetics in various technological processes that all use surfactants has been clearly demonstrated.

In the course of this chapter, areas where additional studies appear to be required have been pointed out. However, the present trend is to move toward more complicated systems such as microemulsions (see Chapter 5), vesicles (see Chapter 6) and lyotropic mesophases (see Chapter 7).

REFERENCES

1. Eigen, M. *Pure Appl. Chem.* 1963, 6, 97.

2. Jaycock, M.J., Ottewill, R.H. *IVth Inter. Congress on Surface Active Substances* (Brussels, 1964), Section B, paper 8.

3. Nakagawa, T., Inoue, H. *IVth Inter. Congress on Surface Active Substances* (Brussels, 1964), Section B, paper 11.

4. Mijnlieff, P.F., Ditmarsch, R. *Nature* 1965, 208, 889.

5. Czeniarwski, M. *Rocz. Chem.* 1965, 39, 1469.

6. Kresheck, G.C., Hamori, E., Davenport, G., Scheraga, H.A. *J. Am. Chem. Soc.* 1966, 88, 246.

7. Bennion, B.C., Tong, L.K., Holmes L.P., Eyring, E.M. *J. Phys. Chem.* 1969, 73, 3288.

8. Bennion, B.C., Eyring, E.M. *J. Colloid Interface Sci.* 1970, 32, 286.

9. Lang, J., Eyring, E.M. *J. Polym. Sci. A-2* 1972, 10, 89.

10. Lang, J., Auborn, J.J., Eyring, E.M. *J. Colloid Interface Sci.* 1972, 41, 484.

11. Takeda, K., Yasunaga, T. *J. Colloid Interface Sci.* 1972, 40, 127.

12. Takeda, K., Yasunaga, T. *J. Colloid Interface Sci.* 1973, 45, 406.

13. Janjic, T., Hoffmann, H. *Zeit. Phys. Chem. N. F.* 1973, 86, 322.

14. Zana, R., Lang, J. *C. R. Acad. Sci. (France) ser. C* 1968, 266, 1347.

15. Graber, E., Lang, J., Zana, R. *Koll. Z. Z. Polym.* 1970, 238, 470.

16. Graber, E., Zana, R. *Koll. Z. Z. Polym.* 1970, 238, 479.

17. Rassing, J., Sams, P.J., Wyn-Jones, E. *J. Chem. Soc. Faraday Trans. 2* 1973, 69, 180.

18. Yasunaga, T., Oguri, H., Miura, M. *J. Colloid Interface Sci.* 1967, 23, 352.

19. Yasunaga, T., Fujii, S., Miura, M. *J. Colloid Interface Sci.* 1969, 30, 399.

20. Zana, R. In *Chemical and Biological Applications of Relaxation Spectrometry*, Wyn-Jones, E., Ed., D. Reidel Publ. Co., Dordrecht, Holland, 1975, p. 133.

21. Folger, R., Hoffmann, H., Ulbricht, W. *Ber. Bunsenges. Phys. Chem.* 1974, 78, 986.

22. Muller, N. In *Reaction Kinetics in Micelles*, Cordes, E., Ed., Plenum Press, New York, 1973, p. 1.

23. Nakagawa, T. *Colloid Polym. Sci.* 1974, 252, 56.

24. Aniansson, E.A.G., Wall, S.N. *J. Phys. Chem.* 1974, 78, 1024 and 1975, 79, 857.

25. Aniansson, E.A.G., Wall, S.N., Almgren, M., Hoffmann, H., Kielmann, I., Ulbricht, W., Zana, R., Lang, J., Tondre, C. *J. Phys. Chem.* 1976, 80, 905.

26. Lessner, E., Teubner, M., Kahlweit, M. *J. Phys. Chem.* 1981, 85, 1529.

27. Hall, D.G. *J. Chem. Soc. Faraday Trans. 2* 1981, 77, 1973.

28. Lessner, E., Teubner, M., Kahlweit, M. *J. Phys. Chem.* 1981, 85, 3167.

29. Aniansson, E.A.G. In *Techniques and Applications of Fast Reactions in Solution*, Gettins, W.J., Wyn-Jones, E., Eds., D. Reidel Publ. Co., Dordrecht, Holland, 1979, p. 249.

30. Wall, S., Elvingson, C. *J. Phys. Chem.* 1985, 89, 2695.

31. Hall, D.G. *J. Chem. Soc. Faraday Trans. 2* 1987, 83, 967.

32. Tanford, C. *J. Phys. Chem.* 1974, 78, 2469.

33. Ruckenstein, E., Nagarajan, R. *J. Phys. Chem.* 1975, 79, 2622.

34. Israelachvili, J.N., Mitchell, D.J., Ninham, B.W. *J. Chem. Soc. Faraday Trans. 2* 1976, 72, 1525.

35. Porschke, D., Eggers, F. *Eur. J. Biochem* 1972, 26, 4901.

36. Garland, F., Patel, R. *J. Phys. Chem.* 1974, 78, 848.

37. Robinson, B., Seelig-Löffler, A., Schwarz, G. *J. Chem. Soc. Faraday Trans. 1* 1975, 71, 815.

38. Almgren, M., Aniansson, E.A.G., Holmaker, K. *Chem. Phys.* 1977, 19, 1.

39. Wall, S.N., Aniansson, E.A.G. *J. Phys. Chem.* 1980, 84, 727.

40. Aniansson, E.A.G. In *Techniques and Applications of Fast Reactions in Solution*, Gettins, W.J., Wyn-Jones, E., Eds., D. Reidel, Dordrecht, Holland, 1979, p. 249. Also in *Aggregation Processes in Solution*, Wyn-Jones, E., Gormally, J., Eds., Elsevier, New York, 1983.

41. Kegeles, G. *J. Colloid Interface Sci.* 1980, 73, 274 and *Arch. Biochem. Biophys.* 1980, 200, 279.

42. Aniansson, E.A.G., Wall, S.N. *J. Colloid Interface Sci.* 1980, 78, 567 and *J. Phys. Chem.* 1980, 84, 727.

43. Kahlweit, M., Teubner, M. *Adv. Colloid Interface Sci.* 1980, 13, 1.

44. Kahlweit, M. *Pure Appl. Chem.* 1981, 53, 2069 and *J. Colloid Interface Sci.* 1982, 90, 92.

45. Hall, D.G. *Colloids Surf.* 1982, 4, 367.

46. Aniansson, E.A.G. *Ber. Bunsenges. Phys. Chem.* 1978, 82, 981 and *J. Phys. Chem.* 1978, 82, 2805.

47. Aniansson, E.A.G. *Prog. Colloid Polym. Sci.* 1985, 70, 2.

48. Zana, R. In *Surfactant Solutions. New Methods of Investigation*, Zana, R., Ed., Plenum Press, New York, 1987, Chapter 9, p. 463.

49. Elvingson, C. *J. Phys. Chem.* 1987, 91, 1455.

50. Kahlweit, M. In *Physics of Amphiphiles, Micelles, Vesicles and Microemulsions*, Degiorgio, V., Corti, M., Eds., North-Holland, Amsterdam, 1985.

51. Coleen, A. *J. Phys. Chem.* 1974, 78, 1676.

52. Hall, D.G. In *Organized Solutions. Surfactants in Science and Technology*, Friberg, S., Bothorel, P., Eds., M. Dekker Inc, New York, 1992, p. 11.

53. Waton, G. *J. Phys. Chem.* 1997, 101, 9727.

54. Turner, M.S., Cates, M.E. *J. Phys. France* 1990, 51, 307.

55. Teubner, M. *J. Phys. Chem.* 1979, 83, 2917.

56. Yiv, S., Zana, R. *Can. J. Chem.* 1980, 58, 1780.

57. Chan, S.K., Hermann, U., Ostner, W., Kahlweit, M. *Ber. Bunsenges. Phys. Chem.* 1977, 81, 60.

58. Chan, S.K., Hermann, U., Ostner, W., Kahlweit, M. *Ber. Bunsenges. Phys. Chem.* 1977, 81, 396.

59. Aniansson, E.A.G., Wall, S.N. *Ber. Bunsenges. Phys. Chem.* 1977, 81, 1293.

60. Chan, S.K., Hermann, U., Ostner, W., Kahlweit, M. *Ber. Bunsenges. Phys. Chem.* 1977, 81, 1294.

61. Teubner, M., Diekmann, S., Kahlweit, M. *Ber. Bunsenges. Phys. Chem.* 1978, 82, 1278.

62. Baumgardt, K., Klar, G., Strey, R. *Ber. Bunsenges. Phys. Chem.* 1979, 83, 1222.

63. Pakusch, A., Strey, R. *Ber. Bunsenges. Phys. Chem.* 1980, 84, 1163.

64. Baumgardt, K., Klar, G., Strey, R. *Ber. Bunsenges. Phys. Chem.* 1982, 86, 912.

65. Diekmann, S. *Ber. Bunsenges. Phys. Chem.* 1979, 83, 528.

66. Elvingson, C., Wall, S. *J. Phys. Chem.* 1986, 90, 5250.

67. Sams, P., Wyn-Jones, E., Rassing, J. *Chem. Phys. Lett.* 1972, 13, 233.

68. Inoue, T., Tashiro, R., Shimozawa, R. *Bull. Chem. Soc. Jap.* 1981, 54, 971.

69. Inoue, T., Tashiro, R., Shibuya, Y., Shimozawa, R. *J. Phys. Chem.* 1978, 82, 2037.

70. von Gottberg, F., Smith, K.A., Hatton, T. A. *J. Chem. Phys.* 1998, 105, 2232.

71. Elvingson, C., Wall, S. *J. Colloid Interface Sci.* 1988, 121, 414.

72. Hoffmann, H., Nagel, R., Platz, G., Ulbricht, W. *Colloid Polym. Sci.* 1976, 254, 812.

73. Hoffmann, H., Lang, R., Pavlovic, D., Ulbricht, W. *Croatica Chem. Acta* 1979, 52, 87.

74. Baumüller, W., Hoffmann, H., Ulbricht, W., Tondre, C., Zana, R. *J. Colloid Interface Sci.* 1978, 64, 418.

75. Lang, J. *J. Phys. Chem.* 1982, 86, 992.

76. Bauernschmitt, D., Hoffmann, H., Platz, G. *Ber. Bunsenges. Phys. Chem.* 1981, 85, 203.

77. Hermann, C.-U., Kahlweit, M. *J. Phys. Chem.* 1980, 84, 1536.

78. Hoffmann, H., Kielman, H.S., Pavlovic, D., Platz, G., Ulbricht, W. *J. Colloid Interface Sci.* 1981, 80, 237.

79. Hoffmann, H., Ulbricht, W. *Z. Phys. Chem. NF* 1977, 106, 167.

80. Ulbricht, W., Zana, R. *Colloids Surf. A.* 2001, 185, 487.

81. Inoue, T., Ikeuchi, M., Kuroda, T., Shimozawa, R. *Bull. Chem. Soc. Jap.* 1981, 54, 2613.

82. Hoffmann, H. *Prog. Colloid Polym. Sci.* 1978, 65, 140.

83. Hoffmann, H. *Ber. Bunsenges. Phys. Chem.* 1978, 82, 988.

84. Lang, J., Zana, R. In *Surfactant Solutions. New Methods of Investigation*, Zana, R., Ed., M. Dekker Inc., New York, 1987, Chapter 8.

85. Kato, S., Nomura, H., Zielinski, R., Ikeda, S. *J. Colloid Interface Sci.* 1991, 146, 53.

86. Nomura, H., Koda, S., Matsuoka, T., Hiyama, T., Shibata, R., Kato, S. *J. Colloid Interface Sci.* 2000, 230, 22.

87. Hermann, C.-U., Kahlweit, M. *Ber. Bunsenges. Phys. Chem.* 1973, 77, 1119.

88. Lang, J., Tondre, C., Zana, R., Bauer, R., Hoffmann, H., Ulbricht, W. *J. Phys. Chem.* 1975, 79, 276.

89. Tondre, C., Lang, J., Zana, R. *J. Phys Chem.* 1975, 52, 372.

90. Tondre, C., Zana, R. *J. Colloid Interface Sci.* 1978, 66, 544.

91. Malliaris, A., Lang, J., Zana, R. *J. Phys. Chem.* 1986, 90, 655.

92. Lang, J., Zana, R., Candau, S. J. *Ann. Chim.* (*Italy*) 1987, 77, 103.

93. Rharbi, Y., Winnik, M.A., Hahn Jr. K.G. *Langmuir* 1999, 15, 4697.

94. Rharbi, Y., Winnik, M.A. *J. Phys. Chem.* B 2003, 107, 1491.

95. Rassing, J., Sams, P.J., Wyn-Jones, E. *J. Chem. Soc. Faraday Trans. 2* 1974, 70, 1247.

96. Kato, S., Nomura, H., Honda, H., Zielinski, R., Ikeda, S. *J. Phys. Chem.* 1988, 92, 2305.

97. Frindi, M., Michels, B., Levy, H., Zana, R. *Langmuir* 1994, 10, 1140.

98. Frindi, M., Michels, B., Zana, R. *J. Phys. Chem.* 1992, 96, 6095.

99. Frindi, M., Michels, B., Zana, R. *J. Phys. Chem.* 1992, 96, 8137.

100. Frindi, M., Michels, B., Zana, R. *J. Phys. Chem.* 1994, 98, 6607.

101. Wan-Badhi, W.A., Palepu, R., Bloor, D.M., Hall, D.G., Wyn-Jones, E. *J. Phys. Chem.* 1991, 95, 6642.

102. Wan-Badhi, W.A., Lukas, T., Bloor, D.M., Wyn-Jones, E. *J. Colloid Interface Sci.* 1995, 169, 462.

103. Kato, S., Harada, H., Nakashima, H, Nomura, H. *J. Colloid Interface Sci.* 1992, *150*, 305.

104. Gettins, J., Jobling, P., Walsh, M., Wyn-Jones, E. *J. Chem. Soc. Faraday Trans. 2* 1980, 76, 794.

105. Zana, R., Yiv, S., Kale, K. *J. Colloid Interface Sci.* 1980, 77, 456.

106. Inoue, T., Shibuya, Y., Shimozawa, R. *J. Colloid Interface Sci.* 1978, 65, 370.

107. Adair, D., Reinsborough, V., Plavac, N., Valleau, J. *Can. J. Chem.* 1974, 52, 429.

108. Adair, D., Reinsborough, V., Zamora, S. *Adv. Mol. Relaxation Proc.* 1977, 11, 63 and references therein.

109. Takeda, K. *J. Sci. Hiroshima Univ.* A 1976, 40, 87.

110. Lianos, P., Lang, J. *J. Colloid Interface Sci.* 1983, 96, 222.

111. Lee, M.K. *Daehan Hwahak Hwoejee* 1973, 17, 72 and 1976, 20, 193.

112. Telgmann, T., Kaatze, U. *J. Phys. Chem.* 1997, 101, 7758 and 1997, 101, 7766.

113. Kato, S., Harada, S., Sahara, H. *J. Phys. Chem.* 1995, 99, 12570.

114. Telgmann, T., Kaatze, U. *Langmuir* 2002, 18, 3068.

115. Hoffmann, H., Platz, G., Rehage, H., Reizlein, K., Ulbricht, W. *Makromol. Chem* 1981, 182, 451.

116. Hoffmann, H., Tagesson, B., Ulbricht, W. *Ber. Bunsenges. Phys. Chem.* 1979, 83, 148.

117. Hoffmann, H., Platz, G., Ulbricht, W. *J. Phys. Chem.* 1981, 85, 1418.

118. Fox, K.K. *Trans. Faraday Soc.* 1971, 67, 2802.

119. Schmidt, D., Gähwiller, Ch., Von Planta, C. *J. Colloid Interface Sci.* 1981, 83, 191.

120. Almgren, M., Grieser, F., Thomas, J.K. *J. Chem. Soc. Faraday Trans. 1* 1979, 75, 1674.

121. Henglein, A., Proske, Th. *J. Am. Chem. Soc.* 1978, 100, 3706.

122. Bolt, J., Turro, N.J. *J. Phys. Chem.* 1981, 85, 4029.

123. Bossev, D.P., Matsumoto, M., Nakahara, M. *J. Phys. Chem. B* 1999, 103, 8251.

124. Guo, W., Brown, T.A., Fung, B.M. *J. Phys. Chem.* 1991, 95, 1829 and private communication.

125. Guo, W., Fung, B.M., O'Rear, E.A *J. Phys. Chem.* 1992, 96, 10068.

126. Kondo, Y., Miyazawa, H., Sakai, H., Abe, M., Yoshino, N. *J. Am. Chem. Soc.* 2002, 124, 6516.

127. Cantu, L., Corti, M., Degiorgio, V. *J. Phys. Chem.* 1991, 95, 5981.

128. Cantu, L., Corti, M., Del Favero, E., Muller, E., Raudino, A., Sonnino, S. *Langmuir* 1999, 15, 4975.

129. Fung, B.M., Mamrosh, D.L., O'Rear, E.A., Frech, C.B., Afzal, J. *J. Phys. Chem.* 1988, 92, 4405.

130. Takisawa, N., Thomason, M., Bloor, D.M., Wyn-Jones, E. *J. Colloid Interface Sci.* 1993, 157, 77.

131. Yamashita, T., Yano, H., Harada, S., Yasunaga, T. *Bull. Chem. Soc. Jap.* 1982, 55, 3403.

132. Jones, P., Wyn-Jones, E., Tiddy, G. *J. Chem. Soc. Faraday Trans. 1* 1987, 83, 2735.

133. Thomason, M.A., Bloor, D.M., Wyn-Jones, E. *J. Phys. Chem.* 1991, 95, 6017.

134. Ross, L., Barclay, C., Balcom, B.J., Forrest, B.J. *J. Am. Chem. Soc.* 1986, 108, 761.

135. Causon, D., Gettins, J., Gormally, J., Greenwood, R., Natarajan, N., Wyn-Jones, E. *J. Chem. Soc. Faraday Trans. 2* 1981, 77, 143.

136. Wan-Badhi, W., Mwakibete, H., Bloor, D.M., Palepu, R., Wyn-Jones, E. *J. Phys. Chem.* 1992, 96, 918.

137. Söderman, O., Johansson, L. *J. Colloid Interface Sci.* 1996, 179, 570.

138. Hakansson, B., Hansson, P., Regev, O., Söderman, O. *Langmuir* 1998, 14, 5730.

139. Hoffmann, H., Tagesson, B. *Z. Physik. Chem. NF* 1978, 110, 113.

140. Eriksson, J., Lundgren, S., Henriksson, U. *J. Chem. Soc. Faraday Trans. 2* 1985, 81, 833.

141. Kaneshina, S., Tanaka, M., Tomida, T., Matuura, R. *J. Colloid Interface Sci.* 1974, 48, 450.

142. Inoue, T., Kawaguchi, T., Tashiro, R., Shimozawa, R. *J. Colloid Interface Sci.* 1982, 87, 572.

143. Kern, F., Lequeux, F., Zana, R., Candau, S.J. *Langmuir* 1994, 10, 1714.

144. Oda, R., Narayanan, J., Hassan, P.A., Manohar, C., Salkar, R.A., Kern, F., Candau, S.J. *Langmuir* 1998, 14, 4364.

145. Michels, B., Waton, G. *J. Phys. Chem. B* 2000, 104, 228.

146. Candau, S.J., Merikhi, F., Waton, G., Lemaréchal, P. *J. Phys. France.* 1990, 51, 977.

147. Oda, R., Weber, V., Lindner, P., Pine, D.J., Mendes, E., Schosseler, F. *Langmuir* 2000, 16, 4859.

148. Oelschlager, C., Waton, G., Buhler, E., Candau, S.J., Cates, M.E. *Langmuir* 2002, 18, 3076.

149. Oelschlager, C., Waton, G., Candau, S.J., Cates, M.E. *Langmuir* 2002, 18, 7265.

150. Berret, J.-F., Lerouge, S., Decruppe, J.-P. *Langmuir* 2002, 18, 7279.

151. Barentin, C., Liu, A.H. *Europhys. Lett.* 2001, 55, 432.

152. Oda, R., Panizza, P., Schmutz, M., Lequeux, F. *Langmuir* 1997, 13, 6407.

153. Cates, M.E., Candau, S.J. *Europhys. Lett.* 2001, 55, 887.

154. Sams, P.J., Rassing, J., Wyn-Jones, E. *Adv. Mol. Relax. Proc.* 1975, 6, 255.

155. Hall, D.G., Jobling, P.L., Wyn-Jones, E., Rassing, J.E. *J. Chem. Soc. Faraday Trans.* 2 1977, 73, 1582.

156. Malliaris, A., Lang, J., Zana, R. *J. Chem. Soc. Faraday Trans. 1* 1986, 82, 109.

157. Almgren, M. *J. Am. Chem. Soc.* 1980, 102, 7882.

158. Binana-Limbelé, W., Zana, R., Platone, E. *J. Colloid Interface Sci.* 1988, 124, 647.

159. Muto, Y., Yoda, K., Yoshida, N., Esumi, K. Meguro, K., Binana-Limbelé, W., Zana, R. *J. Colloid Interface Sci.* 1989, 130, 165.

160. Patist, A., Chhabra, V., Pagidipati, R., Shah, R, Shah, D.O. *Langmuir* 1997, 13, 432.

161. Bauernschmitt, D., Hoffmann, H. *Makromol. Chem.* 1981, 181, 2365.

162. Cantu, L., Corti, M., Salina, P. *J. Phys. Chem.* 1991, 95, 5981.

163. Verrall, R.E. *Chem. Soc. Rev.* 1995, 24, 135.

164. Gettins, J., Hall, D.G., Jobling, P.L., Rassing, J.E., Wyn-Jones, E. *J. Chem. Soc. Faraday Trans.* 2 1977, 74, 1957.

165. Yiv, S., Zana, R. *J. Colloid Interface Sci.* 1978, 65, 286.

166. Yiv, S., Zana, R., Ulbricht, W., Hoffmann, H. *J. Colloid Interface Sci.* 1981, 80, 224.

167. Gormally, J., Sztuba, B., Wyn-Jones, E., Hall, D.G. *J. Chem. Soc. Faraday Trans. 2* 1985, 81, 395.

168. Rao, N.P., Verrall, R.E. *J. Phys. Chem.* 1982, 86, 4777.

169. Kato, S., Jobe, D., Rao, N.P., Ho, C.H., Verrall, R.E. *J. Phys. Chem.* 1986, 90, 4167.

170. Smith, P., Gould, C., Kelly, G., Bloor, D.M., Wyn-Jones, E. In *Reactions in Compartmentalized Liquids*, Knoche, W., Schumaker, R., Eds., Springer-Verlag, Berlin, 1989, p. 83.

171. Wan-Badhi, W., Bloor, D.M., Wyn-Jones, E. *Langmuir* 1994, 10, 2219.

172. Kelly, G., Takisawa, N., Bloor, D.M., Hall, D.G., Wyn-Jones, E. *J. Chem. Soc. Faraday Trans. 1* 1989, *85*, 4321.

173. Jobe, D., Verrall, R.E., Skalski, B., Aicart, E. *J. Phys. Chem.* 1992, 96, 6811.

174. Aicart, E., Jobe, D., Skalski, B., Verrall, R.E. *J. Phys. Chem.* 1992, 96, 2348.

175. Verrall, R.E., Jobe, D., Aicart, E. *J. Mol. Liquids* 1995, 65/66, 195.

176. Uehara, H. *J. Sci. Hiroshima Univ. A* 1976, 40, 305.

177. Inoue, T., Tashiro, R., Shimozawa, R., Matuura, R. *Mem. Fac. Sci. Kyushu Univ. C* 1979, 11, 251.

178. Oh, S.-G., Shah, D.O. *Langmuir* 1991, 7, 1318.

179. Lang, J., Zana, R. *J. Phys. Chem.* 1986, 90, 5258.

180. Malliaris, A., Lang, J., Sturm, J., Zana, R. *J. Phys. Chem.* 1987, 91, 1475.

181. Michels, B., Waton, G. *J. Phys. Chem.* 2003, 107, 1133.

182. Patist, A., Axelberg, T., Shah, D.O. *J. Colloid Interface Sci.* 1998, 208, 259.

183. Zana, R. *Adv. Colloid Interface Sci.* 1995, 57, 1.

184. Lang, J. *J. Phys. Chem.* 1990, 94, 3734.

185. Kowalczyk, A.A., Vecer, J., Hodgson, B.W., Keene, J.P., Dale, R.E. *Langmuir* 1996, 12, 4358.

186. Almgren, M., Grieser, F., Thomas, J.K. *J. Am. Chem. Soc.* 1979, 101, 279.

187. Gläsle, K., Klein, U.K., Hauser, M. *J. Mol. Struct.* 1982, 84, 353.

188. Turro, N.J., Aikawa, M. *J. Am. Chem. Soc.* 1980, *102*, 4866.

189. Croonen, Y., Gelade, E., Van der Zegel, M., Van der Auwerauer, M., Vanderdriessche, H., De Schryver, F.C., Almgren, M. *J. Phys. Chem.* 1983, 87, 1426.

190. Lofroth, J.-E., Almgren, M. *J. Phys. Chem.* 1982, 86, 1636.

191. Infelta, P.P., Grätzel, M., Thomas, J.K. *J. Phys. Chem.* 1974, 78, 190.

192. Scaiano, J.C., Selwyn, J.C. *Can. J. Chem.* 1981, 59, 2368.

193. Mohtat, N., Cozens, F., Scaiano, J.C. *J. Phys. Chem. B* 1998, 102, 7557.

194. Selwyn, J.C., Scaiano, J.C. *Can. J. Chem.* 1981, 59, 663.

195. Leigh, W., Scaiano, J.C. *J. Am. Chem. Soc.* 1983, 105, 5652.

196. Turro, N.J., Zimmt, M., Lei, X., Gould, I., Nitsche, K., Cha, Y. *J. Phys. Chem.* 1987, 91, 4544.

197. Nakagawa, T., Jizomoto, H. *Colloid Polym. Sci.* 1974, 252, 482.

198. Nakagawa, T., Jizomoto, H. *Colloid Polym. Sci.* 1979, 257, 502.

199. Atherton, N.M., Strach, S.J. *J. Chem. Soc. Faraday Trans. 2* 1972, 68, 374.

200. Gehlen, M. *J. Phys. Chem.* 1995, 99, 4181.

201. Flamigni, L. *J. Phys. Chem.* 1992, 96, 3331.

202. Seret, A., Van de Voorst, A. *J. Phys. Chem.* 1990, 94, 5293.

203. Turro, N.J., Aikawa, M., Yekta, A. *Chem. Phys. Lett.* 1979, 64, 473.

204. Venkateswaran, K., Barnabas, M., Ng, B., Walker, D. *Can. J. Chem.* 1988, 66, 1979.

205. Moroi, Y., Morisue, T., Matsuto, H., Yonemura, H., Humphrey-Baker, R., Grätzel, M. *J. Chem. Soc. Faraday Trans.* 1997, 93, 3545.

206. Senz, A., Gsponer, H. *J. Colloid Interface Sci.* 1997, 195, 94.

207. Cang, H., Brace, D.D., Fayer, M.D. *J. Phys. Chem. B* 2001, 105, 10007.

208. Ranganathan, R., Okano, L., Yihwa, C., Alonso, E., Quina, F.H. *J. Phys. Chem. B* 1999, 103, 1977.

209. Rharbi, Y., Bechthold, N., Landfester, K., Salzman, A., Winnik, M.A. *Langmuir* 2003, 19, 10.

210. Hilczer, M., Barzykin, A.V., Tachiya, M. *Langmuir* 2001, 17, 4196.

211. Hunter, T.F., Szczepanski, A. *J. Phys. Chem.* 1984, 88, 1231.

212. Bohne, C., Konuk, R., Scaiano, J.C. *Chem. Phys. Lett.* 1988, 152, 156.

213. Jobe, D.J., Reinsborough, V., White, P. *Can. J. Chem.* 1982, 60, 279.

214. Goddard, D.E., Ananthapadmanabhan, K.P., Eds. *Interactions of Surfactants with Polymers and Proteins.* CRC Press, Boca Raton, FL, 1993.

215. Kwak, J.C.T., Ed. *Polymer-Surfactant Systems*, Surfactant Sci. Ser. No 77, M. Dekker Inc., New York, 1998.

216. Zana, R. In *Structure-Performance Relationships in Surfactants*, Esumi, K., Ueno, M., Eds. 2nd edition, M. Dekker Inc, New York, 2003, p. 547.

217. Bloor, D.M., Mkakibete, H., Wyn-Jones, E. *J. Colloid Interface Sci.* 1996, 178, 334.

218. Ghoreishi, S., Fox, G., Bloor, D.M., Holtzwarth J.F., Wyn-Jones, E. *Langmuir* 1999, 15, 5474.

219. Patist, A., Kanicky, J.R., Shukla, P.K., Shah, D.O. *J. Colloid Interface Sci.* 2002, 245, 1.

220. Patist, A., Oh, S.G., Leung, R., Shaha, D.O. *Colloids Surf. A* 2001, 176, 3.

221. Gettins, J., Gould, C., Hall, D.G., Jobling, P.L., Rassing, J.E., Wyn-Jones, E. *J. Chem. Soc. Faraday Trans. 2* 1980, 76, 1535.

222. Painter, D.M., Bloor, D.M., Takisawa, N., Hall, D.G., Wyn-Jones, E. *J. Chem. Soc. Faraday Trans. 1*, 1988, 84, 2087.

223. Hall, D.G., Gormally, J., Wyn-Jones, E. *J. Chem. Soc. Faraday Trans. 1*, 1983, 79, 645.

224. Takisawa, N., Brown, P., Bloor, D.M., Hall, D.G., Wyn-Jones, E. *J. Chem. Soc. Faraday Trans. 1*, 1989, 85, 2099.

225. Wan-Badhi, W.A., Wan-Yunus, W.M.Z., Bloor, D.M., Hall, D.G., Wyn-Jones, E. *J. Chem. Soc. Faraday Trans.* 1993, 89, 2737.

226. D'Aprano, A., La Mesa, C., Persi, L. *Langmuir* 1997, 13, 5876.

227. Persi, L., La Mesa, C., D'Aprano, A. *Ber. Bunsenges. Phys. Chem.* 1997, 101, 1949.

228. La Mesa, C., Persi, L., D'Aprano, A. *Ber. Bunsenges. Phys. Chem.* 1998, 102, 1459.

229. La Mesa, C. *Colloids Surf. A* 1999, 160, 37.

230. Effing, J., McLennan, I.,Van Oss, N., Kwak, J.C.T. *J. Phys. Chem.* 1994, 98, 12397.

231. Furo, I., Iliopoulos, I. *Langmuir* 2001, 17, 8049.

232. Bloor, D.M., Wyn-Jones, E. *J. Chem. Soc. Faraday Trans. 2*, 1982, 78, 657.

233. Bloor, D.M., Knoche, W., Wyn-Jones, E. In *Techniques and Applications of Fast Reactions in Solution*, Gettins, W.J., Wyn-Jones, E., Eds., D. Reidel Publ. Co., Dordrecht, Holland, 1979, p. 265.

234. Tondre, C. *J. Phys. Chem.* 1985, 89, 5101.

235. Leung, R., Shah, D.O. *J. Colloid Interface Sci.* 1986, 113, 484.

236. Gharibi, H., Rafati, A.A. *Langmuir* 1998, 14, 2191.

237. Dhara, D., Shah, D.O. *J. Phys. Chem. B* 2001, 105, 7133.

238. Dhara, D., Shah, D.O. *Langmuir* 2001, 17, 7233.

239. Zana, R., Lianos, P., Lang, J. *J. Phys. Chem.* 1985, 89, 41.

240. Segre, A.L., Proietti, N., Sesta, B., D'Aprano, A., Amato, M.E. *J. Phys. Chem. B* 1998, 102, 10248.

241. Takeda, K. *Bull. Chem. Soc. Jpn.* 1982, 55, 2547.

242. Narita, T., Gong, J.P., Osada, Y. *Macromol. Rapid Commun.* 1997, 18, 853, *J. Phys. Chem. B* 1998, 102, 4566.

243. Lindman, B., Wennerström, H. *Topics Current Chem.* 1980, 87, 1.

244. Chevalier, Y., Zemb, T. *Rep. Prog. Phys.* 1990, 53, 279.

245. Mukerjee, P. *J. Phys. Chem.* 1962, 66, 1733.

246. Kale, K.M., Zana, R. *J. Colloid Interface Sci.* 1977, 61, 312.

247. Oakes, J. *J. Chem. Soc. Faraday Trans. 2*, 1973, 69, 1321.

248. Grieser, F. *Chem. Phys. Lett.* 1981, 83, 59.

249. Malliaris, A., Lang, J., Zana, R. *J. Chem. Soc. Faraday Trans. 1*, 1986, 82, 109.

250. Roelants, E., Geladé, E., Van der Auweraer, M., Croonen, Y., De Schryver, F.C. *J. Colloid Interface Sci.* 1983, 96, 288.

251. Dederen, J.C., Van der Auweraer, M., De Schryver, F.C. *Chem. Phys. Lett.* 1979, 68, 451, *J. Phys. Chem.* 1981, 85, 1198.

252. Grieser, F., Tausch-Treml, R. *J. Am. Chem. Soc.* 1980, 102, 7258.

253. Almgren, M., Linse, P., Van der Auweraer, M., De Schryver, F.C., Geladé, E., Croonen, Y. *J. Phys. Chem.* 1984, 88, 289.

254. Alonso, E.O., Quina, F.H. *Langmuir* 1995, 11, 2459.

255. Almgren, M., Gunnarson, G., Linse, P. *Chem. Phys. Lett.* 1982, 85, 541.

256. Grunhagen, H.H. *J. Colloid Interface Sci.* 1975, 53, 282.

257. Hedin, N., Furo, I. *J. Phys. Chem. B* 1999, 103, 9640.

258. Degiorgio, V., Piazza, R., Corti, M., Minero, C. *J. Chem. Phys.* 1985, 82, 1025.

259. Zana, R., Weill, C. *J. Phys. Lett.* 1985, 46, L-953.

260. Platz, G. *Ber. Bunsenges. Phys. Chem.* 1981, 85, 1155.

261. Strey, R., Pakusch, A. In *Surfactants in Solution*, Mittal, K., Bothorel, P., Eds., Plenum Press, New York, 1986, vol. 4, p. 465.

262. Jobe, D., Verrall, R.E., Junquera, E., Aicart, E. *J. Phys. Chem.* 1993, 97, 1243.

263. Jobe, D., Verrall, R.E., Junquera, E., Aicart, E. *J. Phys. Chem.* 1994, 98, 10814.

264. Jobe, D., Verrall, R.E., Junquera, E., Aicart, E. *J. Colloid Interface Sci.* 1997, 189, 294.

265. Makievski, A.V., Fainerman, V.B. *Kolloid Zh.* (English translation) 1992, 54, 890.

266. Eastoe, J., Dalton, J.S., Rogueda, P., Crooks, E.R., Pitt, A.R., Simister, E.A. *J. Colloid Interface Sci.* 1997, 188, 423.

267. Eastoe, J., Dalton, J.S., Heenan, R.K. *Langmuir* 1998, 14, 5719.

268. Zana, R., Lang, J. Unpublished results show that the micelle aggregation number of SDS micelles increases nearly linearly from about 65 at 0.02 mM to 140 at 0.8 M.

269. Sailaja, D., Suhasini, K., Kumar, S., Gandhi, K. *Langmuir* 2003, 19, 4014.

4

Dynamics in Micellar Solutions of Amphiphilic Block Copolymers

RAOUL ZANA

CONTENTS

I. INTRODUCTION

Amphiphilic copolymers are obtained by the copolymerization
of two (or more) monomers. They can be synthesized with an
enormous variety of structures of the block type, graft type,
comb type, alternating type, or random type.[1] Within each
type of structure, the chemical diversity of the monomers that
can be used to generate copolymers is very large and
restricted only by the ability of the polymer chemist to syn-
thesize the monomers and perform the polymerization
required for obtaining the desired copolymer with the desired
structure. This diversity, together with the possible changes
of composition, molecular weight, and architecture of the
copolymers, underlines the large number of parameters that

are going to determine the equilibrium as well as the dynamic properties of amphiphilic copolymers in solution.

This chapter deals mainly with block copolymers. Block copolymers are made up of two or more polymeric blocks. The most common block copolymers are of the diblock $A_{N_A}B_{N_B}$ and triblock $A_{N_A}B_{N_B}A_{N_A}$ or $B_{N_B}A_{N_A}B_{N_B}$ types. N_A and N_B are the degrees of polymerization of the blocks of monomers A and B. Block copolymers are *intrinsically* amphiphilic whenever A and B are of sufficiently differing nature because of the large values of N_A and N_B, usually tens to hundreds of units and more. Consider for instance the poly(styrene)-poly(isobutylene) copolymer, where the two blocks are hydrophobic, in the usual sense of this word. Nevertheless, this copolymer forms micelles in the organic (hydrophobic) solvent cyclohexane which is a "good" solvent for the poly(isobutylene) block but a "bad" solvent for the poly(styrene) block at temperatures above 34.5°C, the so-called *theta* temperature of the poly(styrene)/cyclohexane system. This example can be generalized, and it may be stated that block copolymers greatly extend the concept of amphiphilicity with respect to small amphiphilic molecules. For the latter, the difference in nature between A and B must be more pronounced. Another important fact is that small amphiphilic molecules are generally surface active, while a block copolymer characterized by large values of N_A and N_B may display little surface activity. This behavior is observed when the isolated copolymer molecule (referred to as *unimer*) can form a monomolecular micelle where the block with the least affinity for the solvent forms a compact coil.[2-6] In this conformation the contacts between solvophobic repeating units and solvent are largely prevented. The coiling of the block with the most affinity for the solvent may further help in preventing such contacts. Monomolecular micelles can form, however low the copolymer concentration. In this conformation the unimer may not adsorb at interfaces, being in a state of lower energy with respect to the adsorbed state. Indeed, upon adsorption the copolymer must undergo a conformational reorganization that can be costly in free energy. Amphiphilic block copolymers that form micelles in aqueous solution but that do not affect the surface tension of water have been reported.[7]

Throughout this chapter the A block is supposed to be solvophilic (hydrophilic) and the B block solvophobic (hydrophobic). The A and B blocks are also referred to as *soluble* and *insoluble* blocks, respectively. Diblock copolymers $A_{N_A} B_{N_B}$ are the equivalent of conventional surfactants with one alkyl chain and one head group. Triblock copolymers $B_{N_B} A_{N_A} B_{N_B}$ are the equivalent of *gemini* (dimeric) surfactants; triblock copolymers $A_{N_A} B_{N_B} A_{N_A}$ the equivalent of *bolaform* surfactants, and multiblock copolymers $A_{N_A} B_{N_B} A_{N_A} B_{N_B}$ ...... the equivalent of *oligomeric* surfactants.

In a selective solvent of one of the blocks, copolymers can form micelles that contain from one to a very large number of copolymer molecules. The micelle *core* is made up of insoluble blocks and is surrounded by a *corona* made up of soluble blocks. As for surfactants the concentration above which micelles form is referred to as critical micellization concentration (cmc). The cmc of amphiphilic copolymers is often not easy to determine because it can be so small as to be beyond the range accessible with the available methods and/or stretch over a not-so-narrow range of concentration owing to the polydispersity in composition and molecular weight of the copolymer (see below).[6] As for surfactants, the cmc and micelle aggregation number of amphiphilic copolymers are expected to depend on the nature and length of the blocks and on the quality of the solvent. The solvent can be water, an organic solvent, a water/organic solvent mixture, or a mixture of two organic solvents. Fluorescence probing methods where the fluorescent probe can be extrinsic (pyrene in the case of aqueous copolymer solutions) or intrinsic, that is, attached to the copolymer molecule, and scattering methods (light, x-ray, neutron) have been much used to detect the formation of micelles in block copolymer solutions.[1,8]

Similarly to surfactant micelles, amphiphilic block copolymer micelles are not frozen objects, at least when the copolymer molecular weight is relatively low. The same processes as those discussed in Chapter 3 — that is, exchange of surfactants between micelles and bulk phase and micelle formation/breakdown — also occur in micellar solutions of copolymers. In view of the structure of amphiphilic copolymers, it is readily realized that these processes are going to be considerably more complex and also probably much slower than with surfactants. In fact, it is shown later in this chapter that copolymers often give rise to kinetically frozen systems.

For copolymers in aqueous solutions, the most widely used hydrophilic block is the poly(ethylene oxide), referred to as $EO_{N(EO)}$. Other hydrophilic blocks are poly(sodium acrylate), poly(methacrylic acid), poly(sodium methacrylate), and poly(vinylpyridine). Commonly used hydrophobic blocks are the poly(propylene oxide) referred to as $PO_{N(PO)}$, poly(1,2-butylene oxide), and poly(styrene). The triblock copolymers poly(ethylene oxide)-poly(propylene oxide)-poly(ethylene oxide) $EO_{N(EO)}PO_{N(PO)}EO_{N(EO)}$ have been by far the most investigated, for both their equilibrium and dynamic properties, although they are not the simplest ones from the structural viewpoint.[1] These copolymers are relatively easily prepared with a wide range of composition and molecular weight and are commercially available.[1,9] They may be obtained from companies manufacturing them (ICI, Serva, BASF) under various trademarks: Synperionic®, Poloxamer®, Pluronic®. They have found many applications and uses.[9]

Some basic differences that exist between surfactants and amphiphilic block copolymers must be pointed out. First, the length of a surfactant alkyl chain or of its head group, if it is a poly(ethylene oxide) group, is much shorter than the length of the blocks making up a copolymer. Whereas the alkyl chain of a surfactant is short and nearly linear, in copolymers both blocks are sufficiently long to adopt a more or less coiled conformation. In the nonmicellar state, this may result in the formation of a unimolecular or unimer micelle as mentioned earlier. Besides, in the core of a plurimolecular micelle the chains of the insoluble block are *entangled*. These two effects will have a significant impact on the micellar dynamics. A second and extremely important difference concerns the nature of the insoluble block. In surfactants this moiety is an alkyl or alkylaryl hydrocarbon chain or a mixed hydrocarbon/fluorocarbon chain. These chains are totally hydrophobic, over their whole length. In copolymers the hydrophobic (solvophobic) monomer often contains atoms or groups of atoms, such as ether or ester oxygen atoms or amino groups that are partly hydrophilic (solvophilic). As a result the hydrophobic (solvophobic) character of such moieties is less clearly defined because these atoms or groups of atoms may very well carry water (solvent) within the copolymer micelle core. A third difference concerns the purity of the samples. Surfactants can be obtained in an extremely pure form if one is willing to spend the time required by the purification procedure. In contrast, block copolymers can never be obtained with a comparable degree of purity. They often contain homopolymers that affect the determination of basic properties such as cmc and aggregation number.[10,11] If the homopolymers are eliminated, there still remains the polydispersity in block length. This introduces some variability that is difficult to control, even in copolymers prepared by identical methods of synthesis. This polydispersity can be extremely large.

This chapter is organized as follows. This introduction is followed by a section that reviews the available theories and simulations for the dynamics of block copolymer micelles. Section III reviews historical and general aspects of the dynamics of block copolymer micelles. Sections IV and V review experimental results for the $EO_{N(EO)}PO_{N(PO)}EO_{N(EO)}$ triblock copolymers in aqueous solution and for other block copolymers in hydro-organic or organic solution, respectively. Section VI deals with the dynamics of the exchange of solubilizates between block copolymer micelles and bulk phase. Section VII reviews dynamics in solutions of associating polymers. Section VIII reviews the dynamics of miscellaneous processes involving copolymers. Section IX gives some general conclusions.

II. THEORETICAL ASPECTS OF THE DYNAMICS OF MICELLES OF BLOCK COPOLYMERS

The first experimental studies of the dynamics of block copolymer micelles all invoked the theory of Aniansson and Wall (AW).[12–14] and its extensions by Kahlweit et al. (K)[15–18] that were developed for micellar solutions of surfactants. Later on, theories specific to amphiphilic block copolymers were developed that drew much from AW and K theories (referred to as AWK theory below), at least for the mechanisms by which surfactants are exchanged between micelles and by which micelles form and break down. In this section the main features of AWK theory are first recalled. Analytical treatments[19,20] and Monte Carlo simulations[21,22] specific to amphiphilic block copolymers are presented next.

A. AWK Theory for the Dynamics of Surfactant Micelles and Main Conclusions Drawn from Results for Surfactant Solutions

The central assumption in the AW treatment[12-14] is that micelles form and break down via a series of stepwise association/dissociation reactions (4.1) where one surfactant S at a time associates to aggregate S_{s-1} or dissociates from aggregate S_s (s and s-1 are the numbers of surfactant making up aggregates S_s and S_{s-1}). These reactions are also referred to as entry/exit or insertion/expulsion[19] reactions:

$$S_{s-1} + S \underset{k^-}{\overset{k^+}{\rightleftarrows}} S_s \qquad (4.1)$$

In AW treatment, reactions (4.2) of fusion/fission of aggregates are excluded on ground that such reactions are unlikely because of the low concentration of aggregates S_i and S_j. Reactions (4.2) are also referred to as fragmentation/coagulation or merger/splitting reactions.[22]

$$S_i + S_j \rightleftarrows S_{i+j} \qquad (4.2)$$

AW theory predicts that a micellar solution responds to a rapid perturbation with two time constants or relaxation times, in agreement with a large number of experimental observations (see Chapter 3, Section III.A). The short relaxation time τ_1 corresponds to the fast process by which the aggregates adjust to the perturbation by gaining or losing a small number of surfactants via exchange reactions (4.1), but their number remains constant during this process. At the end of the fast process, the system is in a state of pseudo-equilibrium because oligomers and micelles are not yet in equilibrium with each other. This equilibration takes place via a slower process, with a relaxation time τ_2, during which the concentration of the aggregates varies, i.e., micelles form or break down. The rate of this process is determined by the concentration $[S_r]$ of the aggregates S_r at the minimum of the micelle size distribution curve that occurs at s = r (see Chapter 3, Figure 3.1). The quantity $1/k^-[S_r]$ was shown to be equivalent to a resistance opposing the flow of matter between the range of oligomers and the range of micelle proper. In this range the rate constants k^+ and k^- are assumed to be independent of the aggregation number s. AW theory applies to dilute micellar solutions of nonionic surfactants at concentration close to the cmc, i.e., to systems where the micelles are nearly spherical and monodisperse. The expression of τ_1 is given by Equation 4.3 (see Chapter 3, Sections II.A and II.B).[13,14] It has been used to analyze kinetic data for micellar solutions of block copolymers.

$$1/\tau_1 = (k^-/\sigma^2) + (k^-/N)X \qquad (4.3)$$

In Equation 4.3 N is the average micelle aggregation number, that is, the average number of surfactants making up a micelle. σ characterizes the micelle polydispersity (σ^2 is the variance or second moment of the aggregation number distribution function $[S_s] = f(s)$). X is the reduced surfactant concentration $(C - cmc)/cmc$, where C is the total surfactant concentration. Thus $1/\tau_1$ should increase linearly with C provided that all other quantities in Equation 4.3 are independent of C. From such plots, one can obtain k^- and σ^2, provided that N is known. The *average residence time of a given surfactant in the micelle* is $T_R = N/k^-$ whereas $1/k^-$ is the average time between two successive exchange reactions. The entry rate constant k^+ can be obtained from[14]

$$k^+ \cong k^-/cmc \qquad (4.4)$$

Aniansson[14] also derived Equation 4.5:

$$k^- = N(D/l_b^2)\exp[-\Delta G°/kT] \qquad (4.5)$$

In Equation 4.5 D is the diffusion coefficient of a free surfactant, l_b is a molecular distance close to 0.13 nm, and

$\Delta G°$ is the activation energy for the transfer of a surfactant from the micellar to the free state. $\Delta G°$ is close to $\Delta G°_{tr}$, free energy of transfer of a surfactant from a micelle proper to the aqueous phase. $\Delta G°_{tr}$ varies linearly with the alkyl chain carbon number m and can be expressed as

$$\Delta G°_{tr} = \alpha m + \beta \text{ with } \alpha \approx 1.1 \, kT. \qquad (4.6)$$

AW theory assumes that at low surfactant concentration micelles form and break down via a series of stepwise reactions (4.1) that involve one surfactant at a time. This gives rise to a peculiar dependence of the associated relaxation time τ_2 on C, with $1/\tau_2$ decreasing on increasing C (see Chapter 3, Section III). At high C and/or ionic strength for ionic surfactants Kahlweit et al. postulated that reactions (4.2) of fusion/fission start contributing to the micelle formation/breakdown.[15–18] Indeed, repulsion between ionic micelles is screened out and the micelles are polydisperse. This makes reactions (4.2) more likely to occur. Since the initial state (free surfactant) and final state (micellized surfactant) are identical in both reactions (4.1) and (4.2), the micelle formation/breakdown remains characterized by a single relaxation time τ_2. The expression of τ_2 was derived by Kahlweit et al.[15-18]

The conclusions drawn from the analysis of the chemical relaxation data for surfactant solutions are detailed in Chapter 3, Section III. The three main conclusions of these studies are recalled. First, the value of the entry rate constant k$^+$ of a surfactant in a micelle is large, in the range of 10^9 M^{-1}s^{-1}, that is close to the value for diffusion–controlled processes, irrespective of the nature and alkyl chain length of the surfactant. However, surfactants with very long alkyl chains, with 16 or more carbon atoms, and dimeric (gemini) surfactants are characterized by values of k$^+$ smaller than for a diffusion-controlled process. Second, the value of k$^-$ is very sensitive to the surfactant chain length and decreases by a factor of about 3 per additional methylene group in this chain. The variation of k$^-$ with the alkyl chain carbon number m reproduces that of the cmc and is well predicted by Equations 4.5 and 4.6. The free energy increment per methylene group inferred from the variations of the cmc and of k$^-$ with m are nearly equal. Third, at low concentration of ionic surfactant the micelle formation/breakdown proceeds by a series of stepwise reactions (4.1), as postulated by Aniansson and Wall.[12,13] At higher surfactant concentration and/or ionic strength fragmentation/coagulation reactions (4.2) set in and rapidly become the main process by which micelles form and break down. This results in a variation of $1/\tau_2$ with C that is V-shaped as predicted by AWK theory (see Chapter 3, Figure 3.7). For nonionic surfactants, $1/\tau_2$ increases with C in the whole range of concentration. This suggests that fragmentation/coagulation predominates at all concentrations owing to the weak intermicellar repulsion in these systems.

B. Analytical Treatments of the Dynamics of Block Copolymer Micelles

The first theoretical treatment is that of Halperin and Alexander.[19] This theory considers block copolymers $A_{N_A}B_{N_B}$ solubilized in a highly selective solvent of block A_{N_A} in which block B_{N_B} is insoluble. The copolymer gives rise to spherical micelles with a core made up of B_{N_B} blocks and a corona (shell) made up of A_{N_A} blocks. Two limiting cases are considered: that of micelles with thin dense corona ($N_A \ll N_B$) and that of starlike micelles with stretched corona ($N_A \gg N_B$). The core is assumed to be at a temperature above the glass transition temperature (melted state). Based on a qualitative argument (steric repulsion between coronas of two approaching block copolymer micelles) and a semi-quantitative argument that considers the energy of the "activated state" in micelle formation/breakdown, the authors conclude that block copolymer micelles form and break down probably via the AW mechanism, that is, by a series of stepwise association/dissociation reactions (4.1). The theory assumes a fast perturbation of very small amplitude of the system (relaxation condition). The authors picture the expulsion of a copolymer from a micelle as a two-stage process. First, the insoluble block progressively leaves the core, forming what the authors call a spherical "bud" (collapsed state, as postulated in a unimolecular micelle). In a second stage that takes place once the insoluble block is completely out of the core and the budding process completed, the expulsion of the entire copolymer from the corona occurs. The authors then make use of the Kramers theory of rate constants in the high viscosity limit and properly adapted to block copolymers to obtain the expressions of the exit rate constant k^- and of the relaxation time τ_1 for the exchange process in the case of solutions of diblock copolymers. The expressions of k^- are given by Equation 4.7 for micelles with a dense and thin corona ($N_A \ll N_B$) and Equation 4.8 for starlike micelles ($N_A \gg N_B$), respectively:

$$k^- \sim N_B^{-2}\exp(-N_B^{2/3}\gamma a^2/kT) \tag{4.7}$$

$$k^- \sim N_B^{-2/25}N_A^{-9/5}\exp(-N_B^{2/3}\gamma a^2/kT) \tag{4.8}$$

In Equations 4.7 and 4.8, a is a typical monomer size and γ is a surface tension in which are lumped all pair interactions. As noted by the authors, τ_1 depends on the lengths (polymerization degrees) of the soluble and insoluble blocks for the starlike micelles but does not depend on the length of the soluble block for micelles with a thin corona. The comparison of the exponential factors in Equation 4.5 and in Equations 4.7 and 4.8 shows that α is the equivalent of γa^2 and that the variation of k^- with the chain length of the insoluble block (m or N_B) is slower for block copolymers than for surfactants. The $N_B^{2/3}$ factor in Equations 4.7 and 4.8 reflects the assumed collapsed conformation of the insoluble block of the copolymer. The interaction of the insoluble block with the solvent and the soluble block occurs at the surface of the spheroid formed by the collapsed insoluble block in the unimer state. It is thus proportional to the surface of this block that scales as $N_B^{2/3}$.

The above theory was extended by Dormidontova[20] to larger deviations from equilibrium and as to include the time for the disentanglement of the insoluble block from the micelle core as well as micelle fusion/fission processes. The insoluble (hydrophobic) block in free copolymers is assumed to be in a collapsed state as in Reference 19. The author first derived scaling laws for the energetic barriers to the different processes: unimer insertion/expulsion and micelle fusion/fission. The Kramers theory of rate constants in the high viscosity limit was then used to obtain expressions for the rate constants of these processes. The expressions were further used to solve numerically the kinetic equations corresponding to the initial micellization. When only unimer exchange occurred (no fusion/fission), the initial micellization was found to start by fast unimer association that yielded oligomers (especially dimers but also some trimers and tetramers). This fast process rapidly exhausted the supply of unimer because unimer expulsion from existing micelles was very slow. As a result, the association process froze and during a significant period of time the system remained in a state of quasi-equilibrium until the release of unimer by micelles became significant. The association process then resumed. The increase of the aggregation number of micelles with the largest concentration was gradual, going successively through dimers, trimers, tetramers, i-mers until the true equilibrium was reached.

A very different scenario emerged in the case of simultaneous unimer exchange and fusion/fission. No quasi-equilibrium was observed and association proceeded quickly. Odd i-mers never prevailed and dimers turned into tetramers, then hexamers, octamers, and so on. At the beginning of the micellization process, the micelle fusion mechanism was practically the only active one, until unimer expulsion from micelles became active. It is only toward the end of the process that unimer exchange became significant, in conjunction with micelle fusion, particularly between unequal micelles.

C. Dynamic Monte Carlo Simulations

Dynamic simulations revealed that micellization occurs in two steps with different time scales.[21] First the free chains equilibrate rapidly with aggregates of all sizes. Then in a slower step, the aggregates equilibrate with one another. This behavior was demonstrated by arbitrarily selecting one chain as a "tracer" and monitoring the number of chains in the aggregate to which the tracer belongs.[21]

The method was later extended to identify the mechanisms by which chain exchange occurs and study their relative contributions to the overall dynamics of block copolymer micelles.[22] The simulations were run on a cubic lattice and involved block copolymers $A_{N_A}B_{N_B}$ of low molecular weight. The pairwise interaction parameters for species in the lattice were taken as $e = e_{AB} = e_{BSolvent} > 0$ and all other pair interaction parameters set equal to zero. The authors noted that e is proportional to the Flory-Huggins interaction parameter, χ. It is also related to the quantity α in Equation 4.6 and γa^2 in Equations 4.7 and 4.8. The dynamic simulations permitted the authors to follow each chain during a long time span and to reveal three types of exchange mechanisms: (a) chain insertion/expulsion (entry/exit); (b) micellar merger/spitting (fragmentation/coagulation); and (c) micellar spanning. In process (c) a chain can transfer from one micellar core to another without ever being a free chain when the exteriors of the core of the two micelles are separated by a distance shorter than the length of the fully extended insoluble B block. The numbers of transitions per unit time (transition rates) were obtained by counting the number of transitions of each type that occurred during a specified number of Monte Carlo steps. Figure 4.1 (top) shows that chain insertion/expulsion is predominant at low concentration but that micelle merger/splitting governs the dynamics at high volume fraction, whereas process (c) contributes little throughout. Figure 4.1 (bottom) shows the pronounced effect of the pair interaction parameter on the kinetics: a larger e value makes chain expulsion slower. No transition appears to occur at values of $e > 0.5$. At lower values of e the contributions of processes (a) and (b) are first very close and much larger than that of process (c). At still lower e values process (b) becomes predominant, process (c) remaining unimportant. The simulations showed that the effect of an increase of e is not compensated by a decrease of N_B that leaves the product eN_B constant. A kinetic scheme was postulated and the eigenvalues ($\propto$ reciprocal relaxation times) calculated. In all cases three nonzero eigenvalues were found on two times scales differing by about one order of magnitude. The two smaller eigenvalues are quite close and are mainly due to micelle merger/splitting while the larger eigenvalue originates from insertion/expulsion. The difference decreases with an increase of the copolymer volume fraction

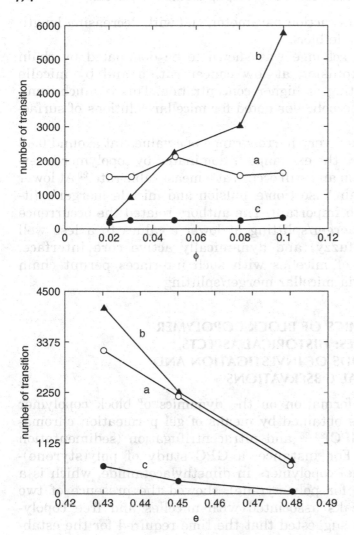

Figure 4.1 Variation of the number of transition via chain insertion/expulsion (a); micelle merger/splitting (b); and micelle spanning (c) with the copolymer volume fraction ϕ (*top*; $N_A = N_B = 10$; $e = 0.45$) and the pairwise interaction parameter e (*bottom*; $N_A = N_B = 10$; $\phi = 0.05$). Number of Monte Carlo steps: 35,482. Reproduced from Reference 22 with permission of the American Chemical Society.

and of the interaction parameter, and with decreasing length of the insoluble block.

Chain exchange was shown to be dominated by chain insertion/expulsion at low concentration and by micelle merger/splitting at higher concentration. This is much reminiscent of the behavior noted for micellar solutions of surfactants.

There is a very narrow range of e-values (at around 0.48 ± 0.02) where the exchange is dominated by copolymer insertion/expulsion and still occurs at a measurable rate.[22] At lower e values chain insertion/expulsion and micelle merger/splitting are both important. The authors related the occurrence of micelle merger/splitting at lower e-values to a less well organized (fuzzy) and dynamically active core interface. Encounters of micelles with such interfaces permit chain exchanges via micellar merger/splitting.

III. DYNAMICS OF BLOCK COPOLYMER MICELLES: HISTORICAL ASPECTS, METHODS OF INVESTIGATION AND GENERAL OBSERVATIONS

The first information on the dynamics of block copolymer micelles was obtained by means of gel permeation chromatography (GPC)[23–28] and ultracentrifugation (sedimentation velocity).[29] For instance, a GPC study of poly(styrene)-poly(isoprene) copolymers in dimethylacetamide, which is a bad solvent for poly(styrene), showed the presence of two separate peaks associated with micelles and free copolymers.[23] This suggested that the time required for the establishment of the unimer/micelle equilibrium was long with respect to the time required for the experiments, that is, hours.[23] A similar observation made in a GPC study of a $EO_{N(EO)}PO_{N(PO)}EO_{N(EO)}$ copolymer was interpreted as indicating a residence time of the copolymer in the micelles of several hours and an extremely long micelle lifetime.[25] As noted by Tuzar and Kratochvil.[30] such conclusions must be considered with caution because GPC and ultracentrifugation are indirect methods and the interpretation of the results in terms of kinetics of the micelle/unimer equilibrium is often controversial. Positive adsorption of the unimer on the column can result in a complete misinterpretation of the results. For instance, a GPC study of poly(ethylene-*stat*-butene)-poly(styrene) copolymer in dioxane/heptane mixtures first concluded that the time for micelle kinetics was in the range of 10 min to hours.[26] Later on it was realized that the unimer adsorbed positively on the material filling the GPC column.[27] The results were reanalyzed and the time required for micelle formation was calculated to be 4.5 ms.[27] In the GPC studies in Reference 28 and probably also in Reference 25, the mobile phase did not contain copolymer. This can cause serious problems.

The first reliable results concerning the dynamics of block copolymer micelles in aqueous solutions were obtained by means of essentially three techniques: ultrasonic absorption relaxation for studying processes with characteristic times between 1 ns and 1 μs;[31] temperature-jump with light-scattering detection that permits the study of processes with relaxation time between 10 μs and 1 s[31,32]; and stopped-flow or simple mixing with light scattering[33] or fluorescence detection[30] that can be used for the study of reactions with half-life between 10 ms and hours (see Chapter 2, Section II). Some fluorescence studies made use of the nonradiative energy transfer (NRET).[30,34] In these studies the block copolymer was labeled with either an energy donor D or an energy acceptor A. Two micellar solutions of the same block copolymer at the same concentration, one of the A-labeled copolymer, the other of the D-labeled copolymer, were mixed and the fluorescence emission of the donor and/or that of the acceptor measured as a function of time. The time dependence of the intensity reflects the progressive hybridization of the two differently labeled micellar populations. However, such studies do not provide a direct measure of the rate constant of exit of a copolymer from the micelles.[34] Nevertheless, they give much information on the time scale of the exchange process and on the effect of various parameters on the rate of this process. Besides, the NRET fluorescence technique can be used for copolymer solutions in organic solvents as well as in water.[30,34] Other fluorescence studies simply mixed a micellar solution of the unlabeled copolymer and a unimer solution of the same copolymer but labeled.[35] More recently the kinetics of block copolymer micelles has been followed by small angle neutron scattering (SANS) after mixing two copolymer micellar solutions differing only by the degree of deuteration of the copolymer.[36] The kinetics of block copolymer micelles has also been investigated by means of static and dynamic light scattering in conjunction with temperature quench.[37] Such studies were possible because the rate of unimer exchange between micelles was very slow.

These studies revealed an enormous difference in behavior between micelles of poly(ethylene oxide)-poly(propylene oxide)-poly(ethylene oxide) copolymers in aqueous solution and of other copolymers in solution in water or in organic solvent. For the former, the micellar kinetics can be quite fast, often in the subsecond time scale, even for micelle formation/breakdown. For the other copolymers, the kinetics is often much slower, stretching over hours or possibly days, even for the exchange process. Kinetically frozen systems are often encountered when water or water/organic solvent mixtures are used as solvent. To a large extent, these differences arise mainly because of the nature of the insoluble block in the two types of copolymers. Differences in molecular weight also play an important role.

IV. DYNAMICS OF MICELLES OF POLY(ETHYLENE OXIDE)-POLY(PROPYLENE OXIDE)-POLY(ETHYLENE OXIDE) BLOCK COPOLYMERS IN AQUEOUS SOLUTION

A. Qualitative Features

The dynamics of micelles of $EO_{N(EO)}PO_{N(PO)}EO_{N(EO)}$ block copolymers in aqueous solution has been investigated by means of ultrasonic absorption relaxation,[31,38–40] T-jump,[31,32,41–46] and stopped-flow[43,47] with light-scattering detection. The reported results have given rise to discussions and controversies, and the situation in the literature may look quite confusing. For this reason, it was felt useful to first critically review the main qualitative features of the reported results before going into quantitative aspects.

The first study was performed by Hecht and Hoffmann[32] by means of T-jump with light-scattering detection of the relaxation. It concerned the copolymers F127, P123, and F88 (see molecular weights, M_w, and EO contents listed in Table 4.1). The relaxation of the system resulted in an increase of intensity of scattered light, as expected in view of the fact that the cmc of $EO_{N(EO)}PO_{N(PO)}EO_{N(EO)}$ copolymers decreases and the amount of micellized copolymer and micelle molecular weight increases upon increasing temperature.[9] A single relaxation process was observed and assigned to the copolymer exchange. However, even though the authors used small perturbations, the relaxation signals were not fully single exponential. This may be due to the polydispersity in composition of the samples that were "used as received." The average relaxation time fell in the 0.1 to 10 ms range. A much slower relaxation was observed at concentrations below the cmc[32] and a similar observation reported for another copolymer.[31] This slow process disappeared at a temperature above which all the copolymer was micellized and it was attributed to premicellar aggregation.[32] Another and perhaps more likely explanation is that this process is associated with the presence of hydrophobic impurities, mainly a diblock copolymer, that aggregate below the cmc or cmT of the triblock copolymer (cmT = critical micellization temperature: temperature above which micelles form in a solution of block copolymer of a given concentration C; at the cmT, cmc = C). The presence of such impurities has been shown in several studies, and they give rise to an anomalous light scattering below the cmc or cmT.[10,11] Above the cmc or cmT the impurities are dispersed among triblock copolymer micelles and the relaxation associated with their aggregates disappear. Nevertheless, these impurities can affect the kinetics of the copolymer exchange process.[42]

Two main relaxation processes were evidenced by Michels et al.[31] in a study of solutions of the block copolymers L64 and PF80 (see characteristics in Table 4.1). The ultrasonic absorption relaxation method revealed for L64 solutions a main relaxation process with a relaxation time in the 0.1 to 1 μs range. Faster processes of much smaller amplitude were detected and attributed to conformational changes or hydra-

TABLE 4.1 Characteristics of the Block Copolymers and Values of the Exchange Rate Constants

Copolymer[a]	M_W[a] (g/mol)	wt% PO[b]	T[c] (°C)	Cmc[a] (mM)	N[d]	k^+ ($M^{-1}s^{-1}$)	k^- (s^{-1})	Reference
L64 $EO_{13}PO_{30}EO_{13}$ (laser T-jump)	2920	60	40	0.8[e]	40[e]	2.4×10^8	2×10^5	40
L64 $EO_{13}PO_{30}EO_{13}$ (ultrasonic relaxation)	2920	60	40	0.69[e]	80[e]	6×10^9	4×10^6	31
P84 $EO_{19}PO_{43}EO_{19}$	4200	60	27	3.7	25	10^8	4×10^5	39
P85 $EO_{26}PO_{40}EO_{26}$	4600	50	37.7	0.22	25	1.5×10^8	3×10^4	41
P123 $EO_{19}PO_{69}EO_{19}$	5750	70	21.4	0.174	25	1.1×10^7	1.9×10^3	32
P104 $EO_{27}PO_{61}EO_{27}$	5900	60	24.2	0.7	25	3×10^6	2×10^3	41
F88 $EO_{103}PO_{39}EO_{10}$	11400	20	40	0.55	25	2.9×10^7	1.6×10^4	32
F127 $EO_{100}PO_{65}EO_{100}$	12600	30	21.1	1.9	25	2.4×10^6	4.6×10^3	32
F108 $EO_{132}PO_{50}EO_{132}$	14600	20	32	0.3	25	3×10^6	9×10^2	41

[a] From Reference 48.

[b] From Reference 9.

[c] Temperature at which the τ_1 values are reported.

[d] The value N = 25 was chosen somewhat arbitrary but it is within a factor of 2 of the values reported at temperature close to the cmc T.[32]

[e] Values of the cmc and N used in the calculations.

tion of the copolymer. The concentration dependence of the relaxation amplitude of the main relaxation process did not conform to what would be expected for a conformational equilibrium, contrary to what was postulated in an earlier study.[38] Besides, the comparison of the results for the L64 and PF80 copolymers indicated that the composition fluctuations occurring at the approach of the critical point of L64 cannot be responsible for the observed relaxation. Indeed, the phase diagram of copolymer PF80 shows no critical point, but PF80 solutions show the tail of an ultrasonic relaxation process. This led the authors to assign the main ultrasonic relaxation process to copolymer exchange.

The study of L64 solutions by T-jump with light-scattering detection[31] revealed two relaxation processes. Examples of relaxation signal are given in Figure 4.2. The increase of intensity of scattered light associated with the faster process yields a relaxation time τ_1 that is dependent on the scattering

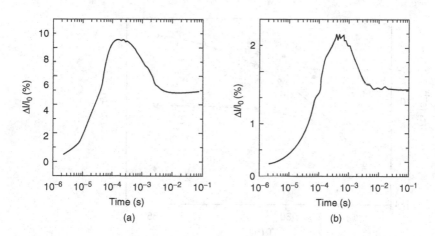

Figure 4.2 Time dependence of the relative change of intensity of the light scattered by a 1.5 wt% L64 solution at a final temperature of 40°C after a T-jump of 0.3°C. (A) scattering angle 90°; relaxation times: 30 ± 10 μs and 1.40 ± 0.15 ms; (B) scattering angle 20°; relaxation times: 170 ± 60 μs and 1.37 ± 0.15 ms. Reproduced from Reference 31 with permission of the American Chemical Society.

angle. A similar behavior was reported for another low molecular weight copolymer, the P84 (see Figure 4.3 and Table 4.1).[39] The scattering angle dependence was explained on the basis of the fact that in T-jump relaxation with light-scattering detection, the shortest relaxation time that can be measured is not the temperature rise time (that can be made very short by increasing the electrolyte content of the solution) but the time characterizing the fluctuations of concentration of the scattering particles. These fluctuations are affected by the copolymer exchange.[31] This explanation permitted the authors to account nearly quantitatively for the dependence of the short relaxation time on the scattering angle.[31] Note that τ_1 was found to depend on the scattering angle only when its value was below 100 µs (see Figure 4.3) and thus comparable to the time characterizing the fluctuations of micelle concentration. This occurs for copolymers with relatively low molecular weight and at relatively low copolymer concentration.

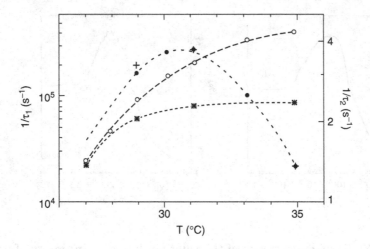

Figure 4.3 Temperature dependence of the relaxation times τ_1 for the fast process (○, *) and τ_2 for the slow process (•, +) for a 2 wt% P84 solution at scattering angle 90° (○, •) and 20° (*,+). The angle dependence is observed only for the fast process and at small values of τ_1. Reproduced from Reference 39 with permission of Academic Press/Elsevier.

The second (slow) relaxation process seen by T-jump is characterized by a decrease of light intensity (see Figure 4.2) and a relaxation time independent of the scattering angle for L64[31] and P84[39] (see Figure 4.3). The slow relaxation was attributed to copolymer micelle formation/breakdown. Indeed, immediately after the T-jump the micelles absorb free copolymer molecules in order to permit the required diminution of cmc. This process is much faster than the formation of additional micelles by stepwise association of free copolymers and occurs with very little variation of the number of micelles. Then in a much slower process the larger micelles progressively adjust their size to the new equilibrium conditions and equilibrate with the oligomers. As for surfactants, this process involves a change of micelle number. The slow relaxation was not observed for the high molecular weight copolymer PF80,[31] as in the Hecht and Hoffmann study.[32]

Relaxation signals very similar to those represented in Figure 4.2 were reported in another study of several copolymers using a laser T-jump apparatus with light-scattering detection.[41] The authors did not investigate the angle dependence of the fast process and they directly identified the time constant of the fast relaxation to the relaxation time of the copolymer exchange.[41] The slow process was attributed to the micelle formation/breakdown. The authors also noted that this slow relaxation was not always observed, given that its amplitude was very small in a fairly large range of temperature and concentration. In subsequent studies they reported on the effect of impurities on the relaxation behavior of L64 solutions[42] (see Figure 4.4) and criticized[43,44] the earlier assignment by Michels et al.[31] of the fast ultrasonic relaxation process to copolymer exchange. They assigned this process to the relaxation of composition fluctuations in spite of the results concerning the copolymer PF80 that shows no critical point.[31] This point demands further investigation. Besides, on the basis of a variation of the amplitude of the relaxation signal from negative to positive (see Figure 4.4, bottom) they claimed that a third relaxation process occurred in L64 solutions at a temperature above 45°C. This process was assigned to micelle clustering into larger aggregates[43,44] Stopped-flow

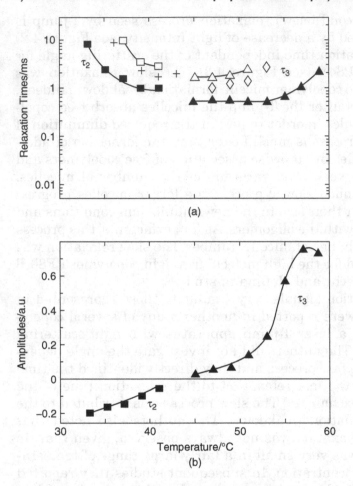

Figure 4.4 Temperature dependence of the relaxation time (a) and relaxation amplitude (b) of the second and postulated third relaxation processes measured for a 2.5 wt% L64 solution. The effect of impurities can be seen by comparing the results for the relaxation times for a purified L64 (■, ▲); an industrial L64 (□, △) and a mixture of L64 and L61 (0.25 wt%) (+, ◇). Adapted and reproduced from Reference 42 with permission of the American Chemical Society.

experiments that used very large perturbations apparently confirmed the existence of this third process. However, the time constants measured in T-jump and stopped-flow experiments were quite different.[47]

This postulated third relaxation process was shown to have no real existence.[45] Theoretical calculations of the effect of temperature on the amplitude of the relaxation associated with the micelle formation/breakdown showed that this amplitude must go from negative to positive as the temperature is increased, as experimentally observed, without the need to invoke an additional relaxation process.[45,46] This directly results from the fact that the cmc of $EO_{N(EO)}PO_{N(PO)}EO_{N(EO)}$ copolymers decreases rapidly upon increasing temperature and that the number of micelles remains nearly constant during the exchange process. Besides, new experimental results led to the same conclusion.[45] Figure 4.4 shows the relaxation time and amplitude data on the basis of which the existence of the third relaxation process was postulated.[42,43] The laser T-jump technique used in these studies was apparently not sensitive enough for measurements in the temperature interval separating the ranges of existence of the second relaxation (micelle formation/breakdown) and of the postulated third relaxation (micelle clustering), where the relaxation amplitude is small. Such measurements were possible with the Joule-effect T-jump apparatus used in Reference 45. The results represented in Figure 4.5 clearly show that there is continuity between the results below and above the zero amplitude temperature that correspond to the second process and the postulated third process, respectively. These results demonstrate that the second and third relaxation processes arise from a single and the same process, the micelle formation/breakdown.

In conclusion, the two main relaxation processes detected in micellar solutions of triblock $EO_{N(EO)}PO_{N(PO)}EO_{N(EO)}$ copolymers are associated with copolymer exchange (fast process) and micelle formation/breakdown (slow process), as for micellar solutions of surfactants. The slow process is often not detected because of its small amplitude in T-jump experiments with light-scattering detection. The fast process can be convoluted by the relaxation of the fluctuations of micelle concentration through diffusion when its relaxation time is very short (case of low molecular weight polymers). When this process is slow enough, it can be conveniently investigated by T-jump.

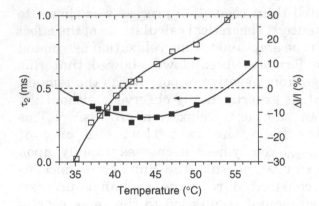

Figure 4.5 Temperature dependence of the relaxation time (■) and relaxation amplitude (□), expressed as relative change ΔI/I of the scattered intensity I, for the micelle formation/breakdown in a 5 wt% L64 solution in the presence of 50 mM KCl. The results show the complete continuity of these variations as the relaxation amplitude goes from negative to positive. Reproduced from Reference 45 with permission of the American Chemical Society.

B. Quantitative Features

1. Exchange Rate Constants

The determination of the exchange rate constants k^+ and k^- makes use of Equation 4.3 and Equation 4.4 and requires experimental values of the relaxation time τ_1 as a function of the copolymer concentration C. References 32, 40, and 42 report such data that concern the copolymers L64, P123, F88 and F127. Figure 4.6 illustrates the variations of $1/\tau_1$ with C for the three last copolymers.[32] The plots are not as linear as for conventional surfactants owing to the fact that the relaxation signals were not fully single exponential.[32] Nevertheless these results were used to extract the values of k^+ and k^- listed in Table 4.1, assuming $N = 25$.[32] Values of τ_1 have been also reported for other copolymers at a given temperature[31] or in the form of plots of τ_1 versus $T - cmT$ at a given concentration.[39,41] These data have been used to extract the values of the rate constants, using for this purpose the cmT and cmc values reported for the copolymers[48] and assuming $N = 25$[32]

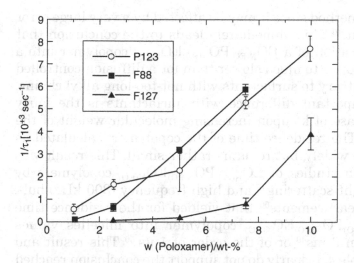

Figure 4.6 Variations of the relaxation time for the exchange process with the copolymer concentration at the temperature of 21.1, 21.4 and 40.0°C for F127, P123, and F88, respectively, where most of the copolymer is in the micellar state. Reproduced from Reference 32 with permission of Elsevier.

and $\sigma^2/N \approx 1$, as was often found for surfactants (see Chapter 3, Section III.B). The calculations made use of the largest values reported for τ_1, which correspond to the lowest values of C and of $T - cmT$. This procedure minimized possible errors on τ_1 values arising from the convolution of the relaxation signal by the temperature rise time and also the time characteristic of the fluctuations of micelle concentration (see above). It should be stressed that the errors on the values of k^+ and k^- are probably quite large. For instance, for P85 the values of k^+ calculated from the results reported at 32 and 37.7°C in Reference 41 differed by a factor 7. Nevertheless, as expected, the largest value of k^+ was found for the higher temperature. Also the values of k^+ and k^- listed in Table 4.1 do not refer to the same temperature. This means that the discussion of the results that follows is only of semi-quantitative or qualitative character.

The values of k^+ are all in the range between 2.4×10^6 and 2.4×10^8 $M^{-1}s^{-1}$, if we eliminate the value found by the ultrasonic relaxation method since it may be affected by a very large error or irrelevant.[40] This immediately leads to the conclusion that the incorporation of a $EO_{N(EO)}PO_{N(PO)}EO_{N(EO)}$ copolymer into a micelle is slower to much slower than for a diffusion-controlled process, contrary to surfactants with not-too-long alkyl chains. Another important difference with surfactants is the fairly rapid decrease of k^+ upon increasing molecular weight of the copolymer. The residence time of the copolymer (calculated as N/k^-) varies widely, but remains rather small. This result was confirmed in studies of $EO_{N(EO)}PO_{N(PO)}EO_{N(EO)}$ copolymers by dynamic light scattering[49] and high frequency (300 kHz) bulk modulus measurements[50] that yielded for the residence time of a $EO_{N(EO)}PO_{N(PO)}EO_{N(EO)}$ copolymer into micelles values shorter than 3 ms[49] or of the order of 1 μs.[50] This result and those in Table 4.1 clearly do not support the conclusion reached in a GPC study that the residence time of the copolymer $EO_{99}PO_{65}EO_{99}$ in micelles is of the order of hours.[25]

The results in Table 4.1 show that k^+ and k^- decrease when the length of the PO block is increased at constant length of the EO block — compare the results for P85 and P104 for which $N(EO) = 26$; or the results for F88 and F127 for which $N(EO)$ is around 100. Figure 4.7 shows the variations of k^+ and k^- with the number of PO units, $N(PO)$. Straight lines can be drawn through the data points corresponding to k^+ and k^-. Recall that for surfactants the variation of log k^- with the number of methylene groups in the surfactant alkyl chain yields the free energy of transfer per methylene group from micelle to water, $\Delta G°_{tr}(CH_2)$. Similarly, the log k^- versus $N(PO)$ plot in Figure 4.7 yields the value of the free energy of transfer per propylene oxide unit from micelle to water, $\Delta G°_{tr}(PO) = (0.12 \pm 0.03) \, kT$. A similar value was obtained from the log k^+ versus $N(PO)$ plot in Figure 4.7. However, Equations 4.7 and 4.8 suggest that for copolymers log k^- should be plotted against $N(PO)^{2/3}$, when neglecting the prefactor in these equations. This plot (see Figure 4.8) yields $\Delta G°_{tr}(PO) = 0.65 \, kT$. The situation is more complicated if one wants to take into account the prefactor in Equations 4.7 and 4.8, which apply to micelles with a dense core and thin corona ($N_A \ll N_B$) and to starlike micelles ($N_A \gg N_B$), respectively.

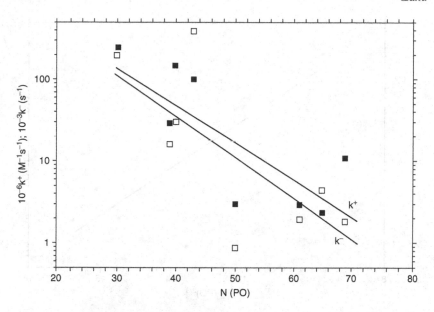

Figure 4.7 Variations of the exit rate constant k^- ($\square$, see text) and the entry rate constant k^+ ($\blacksquare$) with the number of propylene oxide units for triblock $EO_{N(EO)}PO_{N(PO)}EO_{N(EO)}$ copolymers. The solid lines are the best fit of the data.

Indeed, the investigated copolymers satisfy neither one nor the other condition since the values of N_A and N_B are all within one order of magnitude. An attempt has been made to fit the data to an expression of k^- with the prefactor $N(PO)^{-53/75}N(EO)^{-0.9}$ that is a geometric average of the prefactors in Equations 4.7 and 4.8. The experimental values of $k^-_{cor} = N(PO)^{53/75}N(EO)^{0.9}k^-$ have been plotted against $N(PO)$ or $N(PO)^{2/3}$. The quality of the plots was similar to those in Figure 4.7 and Figure 4.8 and their slopes yielded values of $\Delta G°_{tr}(PO)$ close to those given above. The value $\Delta G°_{tr}(PO) = 0.3\ kT$ has been reported.[51] Besides, the reported log cmc versus $N(PO)$ plot for $EO_{N(EO)}PO_{N(PO)}EO_{N(EO)}$ copolymers yielded the value $\Delta G°_{tr}(PO) = 0.16kT$.[52] The value of $\Delta G°_{tr}(PO)$ obtained from Figure 4.7 is thus in the right range, but not that from Figure 4.8, however inaccurate both values are. Recall that the dependence of log k^- on $N(PO)^{2/3}$ is based on

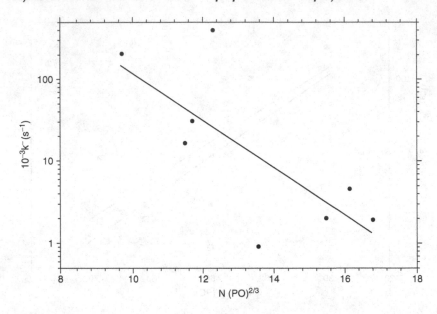

Figure 4.8 Variation of the entry rate constant k^- with $N(PO)^{2/3}$ as suggested by Halperin and Alexander theory[19] for triblock $EO_{N(EO)}PO_{N(PO)}EO_{N(EO)}$ copolymers. The solid line is a best fit of the data.

the assumption of a compact conformation of the PO block in the free state. It thus appears that such may not be not the case for $EO_{N(EO)}PO_{N(PO)}EO_{N(EO)}$ copolymers.

The results in Table 4.1 reveal no clear trend for the effect of the length of the hydrophilic poly(ethylene oxide) block on the values of k^+ and k^-. For instance, the results for P84, P85, and F88 for which N(PO) is close to 40 show a decrease of k^+ and k^- by factors of about 3 and 25, respectively, as N(EO) is increased from 19 to 103. However, the comparison of the results for the copolymers P123, P104, and F127 that have nearly the same N(PO) value of 65 ± 4 shows that the increase of N(EO) from 19 to 100 brings about a decrease of k^+ by a factor of about 4 and an increase of k^- by a factor of about 2. The decrease of k^+ consistently observed upon increasing length of the hydrophilic block may simply reflect the increased steric hindrance experienced by the hydrophobic block when it enters a micelle with a thicker corona.

The results reviewed here concerned aqueous solutions. Only one study reported on the kinetics of copolymer exchange in reverse micellar solutions of a $EO_{N(EO)}PO_{N(PO)}EO_{N(EO)}$ copolymer in an organic solvent. It concerns the L64 copolymer (see Table 4.1) in solution in xylene, in the presence of water.[53] The authors reported the lower bound values $k^+ = 6.3 \times 10^5$ $M^{-1}s^{-1}$ and $k^- = 1.9 \times 10^4$ s^{-1}. These values appear to be much too small in view of the rather large cmc value of L64 in xylene, about 33 mM. More work remains to be done to achieve a better understanding of the dynamics of $EO_{N(EO)}PO_{N(PO)}EO_{N(EO)}$ micelles in organic solvents.

Note that the theories and simulations reviewed in Section II concern diblock copolymers while the experimental results discussed above refer to triblock copolymers. It is not clear how this difference can affect the above discussion. However, diblock poly(ethylene oxide)-poly(propylene oxide) copolymers have a lower cmc than triblock copolymers of the same composition and molecular weight.[52] Thus, they will probably be characterized by a slower micellar dynamics. Also the theories and simulations of the dynamics of block copolymer micelles assume that in free copolymers the insoluble block is in a collapsed state.[19,20,22] This may not be the case for the $EO_{N(EO)}PO_{N(PO)}EO_{N(EO)}$ copolymers due to the weak hydrophobic character of the poly(propylene oxide) block.

2.   Micelle Formation/Breakdown

The relaxation time τ_2 associated with the formation/breakdown of micelles of $EO_{N(EO)}PO_{N(PO)}EO_{N(EO)}$ copolymers (i.e., the slow relaxation process detected by T-jump) has often been found to go through a minimum upon increasing temperature.[39,42,43,45] This is probably due to the fact that the cmc of these copolymers is a very rapidly decreasing function of temperature while the micelle aggregation number increases with temperature. Owing to this effect, it was not possible to obtain the activation energy of the micelle formation/breakdown process.

More importantly, at a given temperature, $1/\tau_2$ has been found to increase with the copolymer concentration C for L64,[31,43] P84[39], and P85[39] (see Figure 4.9). This behavior has been interpreted as indicating that the micelle formation/breakdown occurs via fragmentation/coagulation reactions (2).[31,39,43] Such an assignment would support the predictions of Dormidontova's theoretical treatment[20] and of the dynamic Monte Carlo simulations.[22] It should be recalled that in Aniansson and Wall theory a special type of distribution of the micelle aggregation number can result in an increase of $1/\tau_2$ with C.[14] However, this behavior is obtained for unrealistic values of σ^2/N (see Reference 14, Figure 6, plot 6) and this possibility can therefore be discarded. The occurrence of fragmentation/coagulation as the dominant process in micelle formation/breakdown probably results from the fact that, in the absence of electrical charges, the intermicellar interaction is too weakly repulsive to prevent collisions between $EO_{N(EO)}PO_{N(PO)}EO_{N(EO)}$ micelles. Recall that in solutions of the nonionic poly(ethylene glycol) monoalkylether

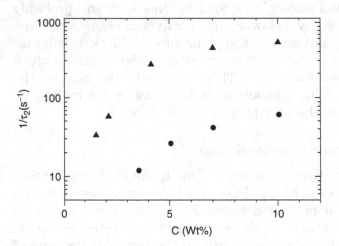

Figure 4.9 Variation of the relaxation time for micelle formation/breakdown with the concentration of the P84 (•) and P85 (▲) copolymers at 36°C. Reproduced from Reference 39 with permission of Elsevier.

surfactants, micelle formation/breakdown also occurs via frag-
mentation/coagulation reactions (see Chapter 3, Section III.C).

C.   Conclusions

The exchange of copolymer between micelles in solutions of
$EO_{N(EO)}PO_{N(PO)}EO_{N(EO)}$ copolymers is slower than for a diffusion-controlled process. Besides, the rate constants for copolymer exit and entry decrease as the length of the hydrophobic
poly(propylene oxide) block increases. Nevertheless, the
exchange process remains relatively fast, with rate constants
for the association increasing from 2×10^6 to 2×10^8 $M^{-1}s^{-1}$ as
the copolymer molecular weight decreases from about 15,000
to 3,000. This behavior reflects the low free energy cost for
transferring a propylene oxide unit from the micelle core to
water, 0.16 to 0.3 kT, as compared to 1.1 kT for the transfer
of a methylene group. The micelle formation/breakdown
appears to take place via fragmentation/coagulation reactions
under the experimental conditions used in the reviewed studies that all concerned copolymers of relatively low molecular
weights. The results in Figure 4.7 suggest that a considerable
slowing down of the micellar kinetics may be observed when
using triblock copolymers with a much larger poly(propylene
oxide) block and when going from a triblock to a diblock
copolymer of the same molecular weight and composition.

V.   DYNAMICS OF MICELLES OF COPOLYMERS
OTHER THAN $EO_{N(EO)}PO_{N(PO)}EO_{N(EO)}$ IN AQUEOUS
SOLUTION AND IN ORGANIC SOLVENT

A.   General Considerations

The results in the preceding section showed that low molecular
weight $EO_{N(EO)}PO_{N(PO)}EO_{N(EO)}$ copolymers are characterized by
relatively labile micelles. Two questions arise immediately:
What about other copolymers with a more hydrophobic block?
What about copolymers in other solvents?

Some part of the answer to the first question can be
found by considering the hydrophobicity of the repeating
units. Propylene oxide is the least hydrophobic repeating
unit, with a reported free energy of transfer from the micelle

core to water $\Delta G°_{tr}$(PO) of about 0.16 to $0.3kT$.[51,52] Other repeating units in the order of increasing hydrophobicity are: lactide < methylene < oxybutylene < styrene $\approx$ styrene oxide.[52] The reported cmc values permitted us to obtain the values $\Delta G°_{tr}$(butylene oxide) = $1.5kT$[52] and $\Delta G°_{tr}$(lactide) = $0.7kT$.[54] The ε-caprolactone repeating unit is reported to be characterized by a $\Delta G°_{tr}$ value intermediate between those of styrene and butylene oxide.[52] A precise value of $\Delta G°_{tr}$ could not be determined for styrene and styrene oxide but appears to be at least twice larger than for butylene oxide, that is, above $3kT$. Some results indicate that in the free state poly(styrene)-poly(ethylene oxide) diblock copolymers form unimolecular (unimer) micelles.[55,56] The same appears to occur for poly(butylene oxide)-poly(ethylene oxide) copolymers at a higher degree of polymerization of the butylene oxide block.[55,56] This last result is an important one because it will have a strong bearing on the micellar dynamics. The values of $\Delta G°_{tr}$ of other repeat units (methyl methacrylate, for instance) that have been much used in copolymers are unfortunately not available.

Turning to the second question, it is obvious that the effect of the solvent can be extremely important. Results for the conventional surfactant cetyltrimethylammonium bromide (CTAB) provide a good example of this effect on the equilibrium and dynamic properties of surfactant micelles. The reported cmc and micelle aggregation number of CTAB are respectively about 0.9 mM and 100 in water[57] and 100 mM and less than 10 in formamide.[58] The residence time of a cetyltrimethylammonium ion in its micelle can be estimated to be of the order of 0.1 ms in water and 0.1 μs in formamide (see Chapter 3, Section III). Block copolymers are expected to show comparable changes.

B. Dynamics of Copolymer Micelles in Aqueous and Hydro-Organic Solution

Many studies of aqueous micellar solutions of block copolymers reported that the exchange equilibrium was extremely slow and often frozen, with no exchange occurring on the time scale of weeks and even months. This situation has been encountered with poly(styrene)-poly(methacrylic acid),[59,60] poly(methyl methacrylate)-poly(acrylic acid) when the hydrophobic poly(methyl methacrylate) block contains 70 repeating units,[61] poly(1,2-butadiene)-poly(ethylene oxide),[62] poly(styrene)-poly(acrylic acid),[63] poly(styrene)-poly(ethylene oxide) at room temperature when the styrene block contains more than 10 repeating units,[64,65] and poly(t-butylacrylate)-poly(2-vinylpyridine),[66] for instance. Note that most of these studies involved copolymers of relatively low molecular weights, say below 20,000. It is most likely that the copolymer exchange is frozen for all block copolymers that are insoluble in water and for which aqueous micellar solutions are prepared by first solubilizing the copolymer in an organic solvent in which the block copolymer is molecularly dispersed and replacing the organic solvent by dialysis against water. The mechanism of micelle formation and of rapid freezing of the copolymer exchange during this procedure has been well described.[63] The freezing occurs over a fairly narrow range of composition, a result that supports the conclusion of dynamic Monte Carlo simulations[22] (see Section II.C). Indeed, these simulations predicted that the micelle/monomer exchange equilibrium occurs predominantly via reactions (4.1) only for a very narrow range of value of the pair interaction parameter e, and therefore also for a narrow range of solvent composition.[22]

Most of the above-mentioned studies were performed at a temperature that was below the glass transition temperature of the hydrophobic block, poly(styrene), or poly(methyl methacrylate). This led the authors to attribute the freezing of the copolymer exchange to a very strong hindrance of the motion of the hydrophobic chains in the glassy micellar core.[34,64,67,68] However, two recent studies[62,69] clearly showed that the absence of exchange kinetics or its great slowness may also have a thermodynamic origin. The effect then arises from the large difference in free energy of the block copolymer in the free and micellar states. In one study, Rager et al.[69] showed that additions of the cosolvent methanol or dioxane to aqueous solutions of poly(methyl methacrylate)-poly(acrylic acid) diblock copolymers promoted the micelle formation/breakdown. However, these additions left the rate of chain exchange practically unaffected, while the glass transition temperature was probably much affected. The authors concluded that the block copolymer micelles must remain very stable even near their disintegration limit and that the range of solvent composition in which exist dynamic micelles must be very small. This last point is again in agreement with the theoretical prediction of dynamic Monte Carlo simulations.[22] The addition of *p*-xylene, which is a good solvent for poly(methyl methacrylate) and thus significantly affects its glass transition temperature, was also found to have little effect on the kinetics of chain exchange.[69] More complicated results were obtained upon the addition of dimethyladipate that promotes the copolymer association but they led to the same conclusion.[69] The second study involved poly(butadiene)-poly(ethylene oxide) copolymers.[62] The glass transition temperature of poly(butadiene) is $-12°C$, well below the temperature at which the experiments were performed. The authors investigated by small angle neutron scattering (SANS) solutions of a mixture of two copolymers of differing molecular weight and composition. They showed that the SANS pattern for a solution prepared by simply mixing solutions of the two copolymers and equilibrated for 8 days was identical to the pattern that could be calculated from the spectra of solutions of the isolated copolymers. Thus no copolymer exchange took place after 8 days. The SANS pattern was very different from that of an aqueous solution of an intimate mixture of the same two copolymers prepared by solubilizing the copolymer mixture in chloroform and evaporating the solvent. The investigated micelles are in a frozen state because of the hydrophobicity of the poly(butadiene) block.

Nevertheless, some aqueous and hydro-organic block copolymer solutions showed a slow copolymer exchange and values of k⁻ were reported. These studies used labeled copolymers and the exchange was detected by NRET. Thus copolymer exchange occurred in solutions of poly(styrene)-poly(ethylene oxide) copolymers at 60°C with a rate constant k⁻ of about 10^{-5} s⁻¹.[65] At room temperature the same copolymers showed no exchange even after several days. Similar observations were reported for the diblock copolymers poly(styrene)-poly(sodium methacrylate) and poly(*t*-butyl-styrene)-poly(sodium methacrylate).[70] The copolymer exchange was frozen at room temperature but was detected at 60°C (time scale: 1 hour). In both studies this behavior was attributed to the fact that the glass transition temperature of the polystyrene block, even though of relatively low molecular weight, was below room temperature. At 60°C the value of k^- for the poly(styrene)-poly(sodium methacrylate) was much larger, in the range of 10^{-3} s^{-1}, than that reported for the poly(styrene)-poly(ethylene oxide) copolymer,[65] probably because this last study used a larger styrene block. Also, the value of k^- was about three times larger for the poly(styrene) than for the poly(*t*-butylstyrene) copolymer (1.85×10^{-3} s^{-1} against 0.51×10^{-3} s^{-1}), as expected on the basis of the weaker hydrophobic character of the former.[70] The copolymer exchange kinetics was investigated in solutions of di- and triblock copolymers of (dimethylamino)alkyl methacrylate and sodium methacrylate of differing molecular weight, composition, alkyl chain length, and architecture (see Table 4.2 and Figure 4.10).[71,72] The cmc values are all within a factor of 5 while the k^- values are all within a factor of 17 (see Table 4.2). These differences are small in view of the drastic changes in the different parameters characterizing the copolymers when going from one copolymer to the other and particularly when considering the two copolymers SC495 and SC 704, which have reversed architectures. A possible explanation for these results is that all the investigated block copolymers form unimolecular micelles in the free state. In such a case differences in molecular weight, structure, composition, and architecture between the different copolymers have much less impact on the cmc and k^- values. Indeed, unimolecular micelles have a core made up of the collapsed (dimethylamino)alkyl methacrylate block and a more or less extended corona of sodium methacrylate. The contacts between the hydrophobic block and the solvent are thus largely eliminated. Both the cmc and k^- values will vary as N_B^{δ} (N_B = number of repeating units of the hydrophobic block) with $\delta < 1$. For instance, the cmc of block copolymers

TABLE 4.2   Characteristics of the Sodium Methacrylate-
(Dimethylamino)alkyl Methacrylate Block Copolymers
Investigated in References 71, 72

Polymer	M_n^x, M_n^y	x/y	cmc (µM)	$10^3 \times k^-(s^{-1})$
(Dimethylamino)ethyl Methacrylate-Sodium Methacrylate				
SC184	3600, 7000	23/49	11.2	2.43
SC240	2400, 2900	15/20	2.8	1.48
SC495	3500, 3600	23/25	10.8	0.50
SC704	1800, 7000	11/49	14.2	1.28
(Dimethylamino)propyl Methacrylate-Sodium Methacrylate				
SC367	2600, 4300	15/30	3.8	1.35
[2-(Dimethylamino)-1-methyl]ethyl Methacrylate-Sodium Methacrylate				
SC32B	3500, 5300	20/37	2.8	0.32
SC31B	1500, 5500	9/39	2.8	0.13

M_n^x and M_n^y are the molecular weights of the (dimethylamino)alkyl methacrylate and sodium methacrylate blocks, respectively, and x and y their
degrees of polymerization (see Figure 4.10).

has been predicted to vary as $N_B^{-1/3}$,[6] while the rate constant
k^- has been predicted to vary as $N_B^{-2/3}$.[19] The copolymer
exchange kinetics was also observed for poly(methyl methacrylate)–poly(acrylic acid) copolymer when the degree of
polymerization of the methyl methacrylate block was
N(MMA) = 20 or 40, in the time scale of 25 hours. The
exchange was frozen even over weeks for N(MMA) = 70.[61,69]
In this study additions of NaCl were found to result in a
considerable slowing down of the exchange.

Tian et al.[68] used sedimentation velocity measurements
to study the rate of micelle hybridization when two micellar
solutions containing micelles made up from two different di-
or triblock copolymers of styrene and methacrylic acid are
mixed. Dioxane/water mixtures were used as solvents. In contrast to results reported in References 71 and 72, the exchange
kinetics was found to depend very strongly on the composition, molecular weight, and architecture of the copolymer and
also on the quality of the solvent (thermodynamic control).

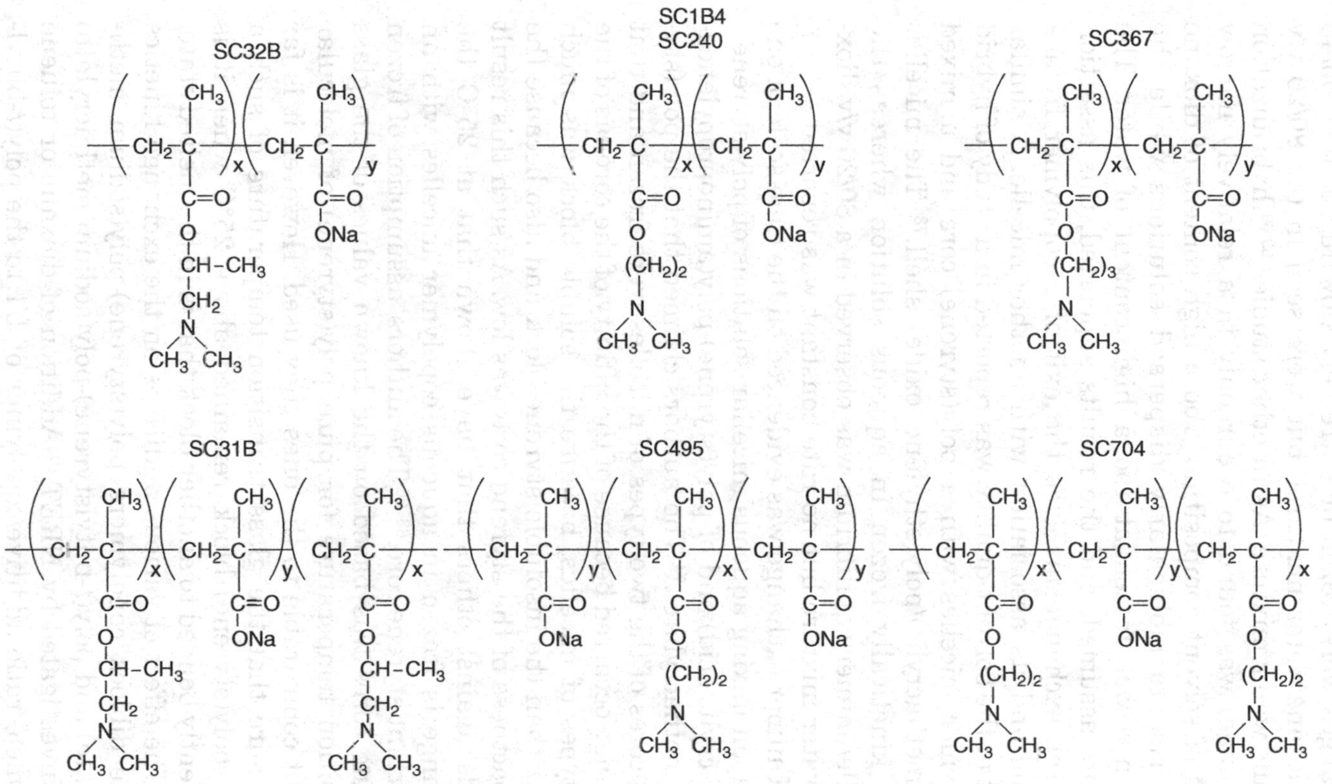

Figure 4.10 Structures and abbreviations of the block copolymers investigated in References 71 and 72. Reproduced from Reference 72 with permission of the American Chemical Society.

Exchanges were frozen or extremely slow in 70/30 or 55/45 v/v dioxane/water mixtures but were seen in the 80/20 v/v dioxane/water mixture. As in other studies,[63,69] hybridization (exchange) was found to occur only in a relatively narrow range of solvent composition. Too a high content of dioxane gave rise to molecularly dispersed solutions while the exchange was frozen at too a high content of water. The authors assumed, and the results supported, this assertion that the exchange involves the exit of a copolymer from a micelle and its association with another micelle. A similar effect of the solvent quality was reported in a study of hybrid copolymer micelles with a poly(styrene) core and a mixed poly(methacrylic)/poly(ethylene oxide) shell.[73] The micelles were kinetically frozen in aqueous solution whereas the micelle/unimer equilibrium was observed in a 80/20 v/v dioxane/water mixture but no rate constant was reported.

Unimer exchange was evidenced on the time scale of one day upon mixing aqueous micellar solutions of poly(styrene)-poly(acrylic acid) and of poly(styrene)-poly(aminopropyleneglycol methacrylate).[74] The authors claimed that the poly(styrene) cores of the two types of micelles were frozen and that exchange occurred because of the affinity of the coronas of the two types of micelles, because the soluble block was much longer than the insoluble styrene block and also because the compactness of the styrene core was low. As such this result stands against others that have shown that at 25°C the exchange is frozen in aqueous copolymer micelles with an important styrene core.[63–65] The authors' assumption of frozen styrene cores was based on the known values of the glass transition temperature for pure poly(styrene) of molecular weight comparable to the ones they used. However, it is far from sure that the glass transition temperature of such a short poly(styrene) block remained above 25°C when it is covalently bonded to another block that is in the melted state.

The effect of various additives on the exchange kinetics of the diblock copolymers poly(styrene)-poly(sodium methacrylate) and poly(t-butylstyrene)-poly(sodium methacrylate) was investigated by NRET.[70] Addition of dioxane or toluene in a mole ratio additive/copolymer of 1/1 to the poly(t-butylstyrene)-poly(sodium methacrylate) copolymer resulted in an acceleration of the exchange kinetics and a value of the rate constant k⁻ at room temperature close to (dioxane addition) or about three times larger (toluene addition) than at 60°C in the absence of additive. Surfactants (anionic sodium dodecyl-sulfate and nonionic Triton X100) were also found to strongly accelerate the copolymer exchange. These effects were discussed in terms of the Flory-Huggins interaction parameters between the additive and micelle core.

In conclusion, the exchange kinetics of copolymers other than $EO_{N(EO)}PO_{N(PO)}EO_{N(EO)}$ in aqueous and hydro-organic solution is often frozen and, when detectable, quite slow even for copolymers with a relatively short insoluble (hydrophobic) block. The exchange process can be frozen either for thermodynamic reasons or because the micelle core is in the glassy state (strong steric hindrance for chain motion). The reported values of the copolymer exit rate constant k⁻ are extremely small, many orders of magnitude smaller than for surfactants. This was expected because the insoluble blocks are always much larger in copolymers than in surfactants. Some studies reported k⁻ values that are very dependent on the insoluble block molecular weight, architecture, and composition of the block copolymer as well as on the quality of the solvent. Other studies showed only little dependence. These seemingly contradictory behaviors indicate that the exchange kinetics is system-dependent to a great degree. An important result is that when the quality of the solvent is modified by the addition of a cosolvent the range of solvent composition in which the exchange equilibrium occurs is rather limited. Too good a solvent gives rise to molecularly dispersed solutions whereas a decrease of the solvent quality rapidly leads to frozen systems. This is in agreement with the results of simulations of copolymer exchange kinetics. The literature does not report on the kinetics of the micelle formation/breakdown process. In view of the slowness of the copolymer exchange, it is likely that this process is so slow that it cannot be detected on the time scale of laboratory experiments (weeks or months).

C. Dynamics of Copolymer Micelles in Organic Solutions

The use of organic solvents permits a fine-tuning of the rate of copolymer exchange and micelle formation/breakdown in block copolymer solutions. Indeed, the quality of the solvent can be adjusted in a more progressive manner than in aqueous/organic solvent mixtures. If the block copolymer does not contain a polyelectrolyte block, it is in principle easier to find an organic solvent, pure or mixed, in which the exchange is detectable. Nevertheless, there have been reports of frozen exchanges even for block copolymer solutions in organic solvents. The solutions of poly(isoprene)-poly(methyl methacrylate) in acetonitrile, which is a modest solvent for the poly(methyl methacrylate) block, are an example.[75] The degree of polymerization of the poly(isoprene) block was relatively large, but the experiments were performed above its glass temperature. Other examples are the solutions of poly(ethylene-*alt*-propylene)-poly(ethylene oxide) in dimethylformamide at room temperature[36]; of poly(styrene)-poly(isoprene) and poly(styrene)-poly(ethylene-*co*-propylene) in decane at 21°C[76]; and of poly(styrene)-poly(acrylic acid) in aqueous dioxane with more than 11% water.[77] In these investigations the molecular weight of the insoluble block was generally fairly large.

NRET has often been used to investigate the exchange dynamics in micellar solutions of block copolymers in organic solvent.[33–35,78–81] However, as discussed by Wang et al.,[34] this method measures relaxation times that are related in a complex and still unclear manner to the exchange rate constants k^+ and k^- for the copolymer exchange. Nevertheless, it permits the study of the effect of various parameters on the copolymer exchange rate. Liu[79] has discussed the experimental conditions in which NRET experiments must be performed in order to yield values of the exchange rate constants. Other methods such as GPC,[26,27] stopped-flow,[33] sedimentation velocity,[76] static and dynamic light scattering,[37,82,83] small angle neutron scattering,[36] and electron microscopy[84,85] have also been used to study the dynamics of block copolymer micelles. The main results are summarized as follows:

1.   In most instances, regardless of whether the perturbation applied to the system was small (system close to equilibrium)[34–36,78,80,81] or very large[33,37,82,83] and irrespective of the method used, the block copolymer solution gave rise to relaxation signals that were not single exponential. These signals could be fitted by a double exponential function, thus yielding two relaxation times. It has been noted that "the two exponential terms are used as a reasonable approximation of a probably continuous relaxation spectrum of relaxation times."[33,78] Such a spectrum may arise from the unavoidable polydispersity in composition and molecular weight of the copolymer samples. Recall that the relaxation signals reported in a T-jump study of $EO_{N(EO)}PO_{N(PO)}EO_{N(EO)}$ copolymers were also not single exponential.[32] The values of the two relaxation times were often within a factor of 10, and in several instances the authors based their discussion on an average relaxation time.[34,78,80,81] One study reported a single relaxation time but the relaxation signals clearly show systematic deviations that reveal the presence of a slower relaxation process or the existence of a spectrum of relaxation times.[35] Nevertheless, some authors argued that the existence of two relaxation times may indicate that in addition to the copolymer exchange via entry/exit reactions (4.1), exchange may occur via a still unidentified second mechanism.[34,36]

2.   The relaxation process observed by NRET,[34,35,78,80,81] SANS,[36] and sedimentation velocity[76] (systems close to equilibrium) has been attributed to the copolymer exchange. The results showed that the copolymer exchange rate is extremely sensitive to the solvent quality and to the temperature. This is illustrated in Figure 4.11 for solutions of poly(styrene)-poly(hydrogenated isoprene) in dioxane/heptane mixtures in which micelles form with a poly(styrene) core.[81] A

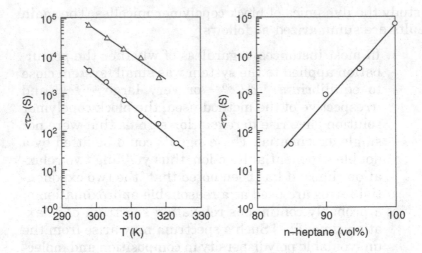

Figure 4.11 Variations of the average relaxation time associated with the copolymer exchange in solutions of poly(styrene)-poly(hydrogenated isoprene) (Kraton G-1701) in dioxane/heptane mixtures. Left: effect of temperature at heptane volume fraction 100% (○) and 95% (△). Right: effect of the heptane volume fraction at 298 K. Reproduced from Reference 81 with permission of Elsevier.

similar effect of the solvent quality was noted for the copolymer exchange rate when increasing the water content in solutions of poly(styrene)-poly(ethylene oxide) in methanol/water mixtures.[34] At 30°C the average relaxation time was increased by a factor close to 30 when the water content of the system was increased from 5 to 10 volume %. An extreme sensitivity of the exchange rate to temperature was noted in a study of mixtures poly(styrene)-poly(ethylene-*co*-propylene) and poly(styrene)-poly(isoprene) in decane.[76] The exchange was frozen at 20°C and was very fast at 35°C. It could be investigated at the intermediate temperature of 29°C.

3. A fairly comprehensive study was reported for the copolymer exchange rate in micellar solutions of poly(styrene)-poly(2-cinnamoylethyl methacrylate) (PS-PCEMA) in tetrahydrofuran/cyclopentane

(THF/CP) where micelles form with a PCEMA core.[80] The exchange rate was little affected by the polymer concentration but increased significantly with increasing temperature and content of THF, a plasticizer of the PCEMA core. The exchange rate decreased as the length of the PCEMA block increased. Similar results were reported for other copolymers. Surprisingly, the exchange rate also increased with the length of the soluble PS block. The authors explained this result by the fact that as the PS block in the corona becomes longer, repulsion between neighboring PS chains become larger, that may result in an increased uptake of THF by the PCEMA core. In turn, this may increase the copolymer exit rate.

4. Values of the copolymer exchange rate constants k^+ and k^- were reported only for solutions of poly(methyl methacrylate)-poly(methacrylic acid) in ethylacetate/methanol mixtures where micelles form with a poly(methyl methacrylate) core when the ethyl acetate volume fraction is above 80%.[35] The rate constants k^+ and k^- were found to be independent of the copolymer concentration and to be around 2×10^3 M^{-1}s^{-1} and 5×10^{-4} s^{-1}, respectively, in the 90/10 (v/v) ethyl acetate/methanol mixture. Both rate constants were increased by a factor of about 2 when the methanol volume fraction was increased to 15%. The values of k^+ are much lower than for a diffusion-controlled process, by several orders of magnitude. Thus there is significant barrier opposing the entry of the copolymer into the micelle, probably arising from the poly(methacrylic acid) corona that must be crossed by the insoluble block.

5. An interesting result was reported in a sedimentation study of the hybridization of micelles of poly(styrene)-poly(isoprene) (PS-PI) and poly(styrene)-poly(ethylene-co-propylene) (PS-PEP) in decane that is a poor solvent of the poly(styrene) block.[76] The hybridization takes place by the exit of unimers from one micelle

and incorporation into another micelle. The authors showed that the transport of unimers from the PS-PI micelles into PS-PEP micelles is much faster than the reverse process. A similar result was reported for solutions of poly(styrene)-poly(methacrylic acid) in water/dioxane mixtures.[68]

6. The reported values of the relaxation time associated with the copolymer exchange are often quite large, hours to days. However, the hybridization of micelles of poly(styrene)-poly(4-vinylpyridine) in solution in toluene, where micelles form with a poly(4-vinylpyridine) core, took place in less than 10 s.[86] This result may look surprising, but it can be probably explained in terms of the quality of the solvent toluene for the "insoluble" poly(4-vinylpyridine) block. If the solvent is not very bad, it may penetrate to a more or less greater extent in the core, thereby greatly facilitating the exchange.

7. Very large perturbations were used in a study by stopped-flow with light-scattering detection of micellar solutions of poly(styrene)-poly(hydrogenated isoprene) in dioxane/heptane mixtures in which micelles form with a poly(styrene) core at high dioxane content.[33] Very short relaxation times, in the 0.1 s range, were measured when forming micelles by a 1/1 v/v mixing of a unimer solution in a dioxane/heptane mixture (60/40 v/v) with dioxane and when dissociating micelles by a 1/1 v/v mixing of a micellar solution in dioxane with heptane. The authors assumed that the fast micelle dissociation was due to a rapid influx of heptane into the micelle core and a subsequent rapid separation of the copolymers by diffusion. The average relaxation time for the association process was proposed to mainly reflect the growth of associated species with low aggregation number into the final micelles. The authors noted that such experiments cannot be directly compared with experiments performed in a near-equilibrium situation. Such experiments were performed using NRET and

labeled polymers and yielded much longer relaxation times, in the range of 100 s.[78]

8. Studies of the dynamics of micelle formation/breakdown that used very large perturbations were also performed on poly(α-methylstyrene)-poly(vinylphenethyl alcohol) in benzyl alcohol where micelles form with a poly(α-methylstyrene) core.[37,82] The processes were very slow, on the time scale of half an hour to tens and hundreds of hours. Thus they could be followed by static and dynamic light-scattering measurements of the apparent molecular weight, radius of gyration, and hydrodynamic radius of the micelles, $M_{w,app}$, $R_{g,app}$, and $R_{h,app}$, respectively. The micelle formation and breakdown were induced by temperature variations from the unimer to the micellar range (temperature decrease) and vice versa (temperature increase). A study that examined the micelle formation induced by a quenching from 60°C (unimer state) to 35°C evidenced two well-separated relaxation times.[37] The results were interpreted following the Aniansson and Wall model, taking into account the information revealed by the simultaneous variations measured for $M_{w,app}$, $R_{g,app}$, and $R_{h,app}$. The fast process was assumed to correspond to the self-association of unimers into micelles with an aggregation number close to but not equal to that of the equilibrium micelles. The number of micelles increased until the unimer concentration reached a value around the cmc. Then a slow micelle growth to the equilibrium size took place by micelle formation/breakdown via unimer entry/exit reactions (1). Reactions of fusion/fission (2) were excluded. Figure 4.12 schematically represents the various stages of micelle formation, according to this model. The main difference with AW mechanism is that one starts with a system that contains no copolymer micelles.

In a second study of the same system, smaller temperature variations were used to study micelle formation and breakdown.[82] The effect of the amplitude

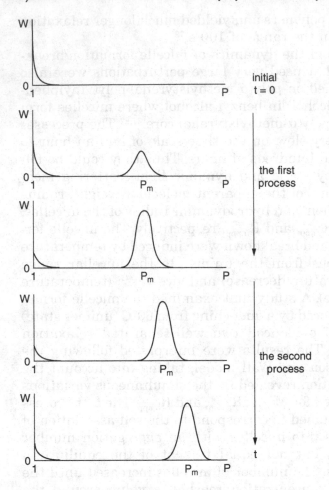

Figure 4.12 Schematic representation of the shape of the micelle size distribution curve (w is the fraction of micelle of aggregation number P; t is the time) during the first and second relaxation processes (see text). Reproduced from Reference 37 with permission of the American Chemical Society.

of the quench of temperature on the micelle formation was twofold. First, only the slowest of the two relaxation processes evidenced in Reference 37 was observed because at the starting temperature (40 or

45°C) the solution already contained micelles. Second, the measured relaxation times were larger. In the micelle breakdown experiments a temperature increase was applied to a solution equilibrated at 35°C, which contained micelles. The final temperature was also in the micelle range. This temperature jump gave rise to variations of $M_{w,app}$ and $R_{g,app}$ going through a minimum as time elapsed, on the time scale of up to 200 hours. These variations, together with the effect of surfactant concentration on the relaxations, indicated that the number of micelles and their aggregation number first decreased (micelle decomposition), then increased (micelle recovery). As noted by the authors, this process is quite complex but fits in AW mechanism. The effect of the amplitude of the perturbation is clearly seen in Figure 4.13 that represents the variations of the relaxation times with the copolymer concentration.

In conclusion, the reviewed studies of the dynamics of micelles of block copolymers other than $EO_{N(EO)}PO_{N(PO)}EO_{N(EO)}$ in aqueous and organic solutions have permitted us to clarify the main parameters that control the rate of copolymer exchange and of micelle formation/breakdown. The Aniansson and Wall model appears to also apply to these systems with the appropriate changes for the fact that block copolymers and not surfactants are involved. Far too few studies report values of the exchange rate constants k^+ and k^-. Most reported studies used either too large perturbations of the system or methods that only give access to relaxation times that are not directly related to the rate constants. More work based on the same method as in Reference 35 should be performed on copolymers with varying molecular weight, composition, and architecture as this method affords the determination of values of copolymer exchange rate constants. This method is based on NRET. It involves the difficult synthesis of appropriate copolymers that are appropriately labeled and the use of the appropriate solvents. This is probably why it has not been more widely used yet.

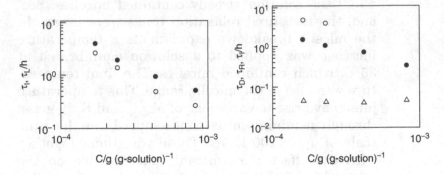

Figure 4.13 Variations of the relaxation time with the concentration of the poly(α-methylstyrene)-poly(vinylphenethyl alcohol) copolymer. Left: relaxation time for the micelle formation process at a temperature of 35°C, for a starting temperature of 60°C (o, the solution contains only unimers) and of 45°C (•, the solution contains unimers and micelles). Right: relaxation times for micelle recovery τ_R (•) and breakdown τ_D (△) measured after a temperature jump from 35°C to 45°C; (o) relaxation time for micelle formation measured at 35°C following a temperature quench from 60 to 35°C. All relaxation times are expressed in hours. Reproduced from Reference 82 with permission of the American Chemical Society.

VI. DYNAMICS OF EXCHANGE OF SOLUBILIZATES IN MICELLAR SOLUTIONS OF BLOCK COPOLYMERS

These studies were performed in aqueous solutions and used block copolymers containing at least one hydrophobic block: poly(propylene oxide), poly(styrene), poly(methyl methacrylate), or poly(ε-caprolactone), for instance. As for surfactant micelles, compounds that are sparingly soluble in water (because of their pronounced hydrophobicity), can be solubilized in the hydrophobic core of block copolymer micelles. This raised the possibility of using block copolymer micelles as carriers for the delivery of hydrophobic drugs. This possibility is enhanced by the fact that the core of block copolymer micelles can be large and/or in the glassy state. Both properties are expected to result in an increased retention of solubilizates into micelles and, thus, in a much longer solubilizate residence time in the micelles. This is a distinct advantage of block copolymer micelles with respect to surfactant micelles or vesicles. The residence time of a solubilizate in such assemblies is generally short and the assembly itself either has a limited lifetime (micelle) or is easily destroyed *in vivo* (vesicles). Also, the very low cmc of block copolymers ensures that they are not destroyed and that their content is released by simple dilution, as would be the case for surfactant micelles. Last, block copolymers can be made to be biocompatible and/or biodegradable.

Although this section is concerned with the dynamics of solubilizate exchange, a reminder is necessary concerning micellar solubilization. The solubilization of water-insoluble compounds by micelles, whether of surfactants or block copolymers, is generally characterized by the partition coefficient of the solubilizate between micelles and intermicellar solution. With surfactant micelles, the partition coefficient of a given solubilizate depends relatively little on the nature of the surfactant when it is referred to the micelle core volume because the nature of the micelle core is always the same, being generally made up of alkyl chains. Such is not the case with block copolymer micelles, where cores of very differing nature can be used.[66,87-91] As a result, the partition coefficient of a given solute can vary by two orders of magnitude and more, depending on the copolymer used (see values of the partition coefficient of pyrene in various copolymer systems listed in Reference 91). Recall that the partition coefficient can be related to the solubility parameter of the solubilizate and of the block making up the micelle core. To a first approximation, the highest value of the partition coefficient will be found when the values of the solubility parameter of the solubilizate and the core-forming block are nearly the same.[91] Other parameters that permit a tuning of the value of the partition coefficient are the nature and length of the hydrophilic block.

Many of the reported studies concern solubilizates of aromatic character such as pyrene and its derivatives as their entry into and exit from micelles can be easily monitored using UV absorption or fluorescence (sometimes involving NRET). The results obtained in these studies probably apply to more complex solubilizates such as drugs. Only typical results are presented next, and no attempt is made to give a comprehensive review of all reported studies.

A.   Rate of Release (Exit) of Solubilizates from Block Copolymer Micelles

Several studies showed that the release (exit) of solubilizates from block copolymer micelles occurs in two steps.[87–89,91–94] This is illustrated in Figure 4.14 for the release of pyrene from triblock poly(methacrylic acid)-poly(styrene)-poly(meth-acrylic acid) copolymers in aqueous solutions.[87] The copolymer was loaded with enough pyrene that pyrene excimer emission occurred. The release of pyrene upon dilution of the system resulted in a decrease of the excimer fraction and thus in a decrease of the intensity ratio Ie/Im (excimer over monomer). This behavior differs from that observed for the release of solubilizates from surfactant micelles, both qualitatively and quantitatively. Indeed, the exit of solubilizates from surfac-tant micelles usually takes place in one step, which is gener-ally fairly rapid. Thus the residence time of pyrene in micelles

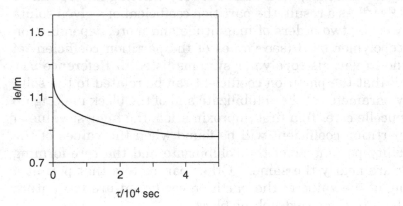

Figure 4.14   Time dependence of the ratio of the intensity of exci-mer to monomer fluorescence emission associated to the release of pyrene from pyrene-loaded micelles of poly(methacrylic acid)-poly(styrene)-poly(methacrylic acid). Reproduced from Reference 87 with permission of the American Chemical Society.

of a surfactant such as sodium dodecylsulfate is of the order of 0.2 ms.[95] Figure 4.14 shows that the largest part of the release occurs over hours. Thus, the residence time of pyrene in the block copolymer micelles used is in this range, which is some 7 to 8 orders of magnitude longer than in surfactant micelles. Even the fast release (time scale 10–20 min) is still much slower than the release from surfactant micelles. The authors explain their results by assuming that pyrene is solubilized in the poly(styrene) core and also in the inner part of the corona of poly(methacrylic acid). The pyrene solubilized in the corona is released fairly rapidly. Pyrene solubilized in the core diffuses out extremely slowly because the poly(styrene) core is in the glassy state at room temperature.

Other studies include the release of phenanthrene and pyrene from poly(styrene)-poly(methacrylic acid) diblock copolymer micelles[89,92] and of other copolymers such as poly(t-butyl acrylate)-poly(2-vinylpyridine)[89]; the release of benzopyrene and Cell-Tracker CM-DiI from poly(caprolactone)-poly(ethylene oxide) micelles[91]; the release of pyrene and estradiol from micelles of a $EO_{N(EO)}PO_{N(PO)}EO_{N(EO)}$ copolymer grafted with poly(acrylic acid).[93]

A theory was developed to account for the slow solubilizate release from the micelle core in terms of diffusion.[89,92,93] Figure 4.15 shows that the amount of solubilizate released during the slow process increases nearly linear with $(time)^{1/2}$ as expected from a diffusion theory.[89,92,93] Such plots yielded the diffusion coefficient D_S of the solubilizate in the core. The values of D_S were in the range of 10^{-18} to 10^{-16} cm²/s for micelles formed by a copolymer with a glassy poly(styrene) core.[89,92] The values of D_S were several orders of magnitude larger for other block copolymers such as those used in References 91, 93, and 94.

A very different behavior was reported for the release of bromoacetophenone from micelles of poly(styrene)-poly(ethylene oxide).[88] The exit rate constant was directly determined from measurements of luminescence intensity in the presence of increasing amount of quencher. The same method had been used to determine exit rate constants of solubilizates from surfactant micelles.[95] The value of k^- was very large, 9×10^3 s^{-1}, indicating a very fast release of the probe from the micelle,

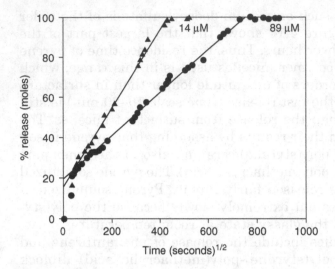

Figure 4.15 Time dependence of the fraction of benzopyrene released from benzopyrene-loaded poly(caprolactone)-poly(ethylene oxide) micelles. Reproduced from Reference 91 with permission of the American Chemical Society.

in the millisecond time scale.[88] Besides, the values of k⁻ showed only little dependence on the micelle size and copolymer architecture (diblock or triblock). This last result was interpreted as indicating that the probe was mainly located in the corona,[88] very near to the micelle surface. This explanation is supported by the fact that pyrene is known to interact with poly(ethylene oxide).[96] If such is the case, the peculiar behavior noted in Reference 88 would be specific to copolymers with a poly(ethylene oxide) block. Moreover, one cannot exclude that previous investigations failed to detect this very fast release of pyrene simply because the experimental procedures used had dead time of tens of seconds. More studies by means of the method used in Reference 88 should be performed to check on a possible very fast release such as that observed with bromoacetophenone. Note that the value of k⁻ for bromonaphthalene exit from sodium dodecylsulfate micelles is around 3×10^4 s⁻¹.[95] Bromonaphthalene is chemically close to bromoacetophenone and has a similar solubility in water.

Creutz et al. reported that no pyrene diffuses out poly((dimethylamino)alkyl methacrylate) block copolymer micelles over a period of up to 2 hours.[71,72] This result differs from that in References 87 and 89 where part of the pyrene was shown to be released fairly quickly. This peculiar behavior may reflect the existence of a strong interaction between pyrene and the core.

There have been many reports of kinetics of release of various compounds of biological importance from biocompatible block copolymer micelles, as for instance indomethacin release from $EO_{N(EO)}PO_{N(PO)}EO_{N(EO)}$-poly(caprolactone) micelles,[97] doxorubicin from poly(benzylaspartate)-poly(ethylene oxide) micelles,[98] or clonazepam from poly(benzylglutamate)-poly(ethylene oxide) micelles.[99] In the three studies, the release was extremely slow (time scale of many days) and incomplete.

B. Rate of Entry of Solubilizates into Block Copolymer Micelles

The rate of association (also referred to as capture or entry) of various solubilizates to block copolymer micelles has been investigated,[88–90,100–03] often by means of fluorescence techniques. All of these studies showed that the association is slow compared to what is found with surfactant micelles. However, the difference is not as large as in the case of the release (exit) of solubilizates. In most instances, the capture was found to occur in two steps, just like the release of solubilizates from block copolymer micelles. This reflects a fast penetration of the solubilizate in the corona, followed by its slow penetration in the micelle core.

The increase of fluorescence intensity with time associated to the capture of pyrene by poly(styrene)-poly(methacrylic acid) micelles[89,100] or the capture of perylene by poly(2-cinnamoylethyl methacrylate)-poly(acrylic acid) micelles[90] revealed a two-step process. For the first system, the analysis of the slow process (probe penetration in the micelle core) using the same diffusion equation as for the release process yielded a value of the diffusion coefficient close to that inferred from the release of pyrene by the same micelles.[89]

Poly(styrene)-poly(methacrylic acid) copolymers (PS-PMMA) tagged with a fluorescent probe or containing a solubilized fluorescent probe were used to investigate the rate of incorporation of quenchers of the probe[87] or of compounds that give NRET with the probe.[101] Carbon tetrachloride was found to quench the fluorescence of micelle-solubilized pyrene in 10–15 s, indicating its rapid incorporation into the micelles owing to its small size.[87] The effect of dimethylaniline was much slower (time scale: 10^4 s). The difference with the results for CCl_4 is in part (factor of about 10), due to the much lower solubility of this compound in water. Note that the fluorescence emission of the pyrene/dimethylaniline exciplex was biphasic.[87] The penetration of 5-(N-octadecanoyl)amino fluorescein into PS-PMMA micelles became much faster as the pH was increased, that is, as the corona expands (about 1 s at pH 9.3 and 400 s at pH 7.1).[101]

The rate constant of entry of bromoacetonaphthone into poly(styrene)-poly(ethylene oxide) micelles was reported to be in the order of 10^{12} $M^{-1}s^{-1}$, a value 100 to 1000 times larger than for a diffusion-controlled process.[88] The authors concluded from this result that when the probe exits from a micelle it nearly always reenters the same micelle.

Briefly summarizing this section, the rates of entry/exit of solubilizates into/from block copolymers micelles have been found to be many orders of magnitude slower than for surfactant micelles. Entry and exit occur in two steps: a first step that corresponds to the entry in or exit from the micelle corona and a much slower step corresponding to the diffusion of the solubilizate into/out of the micelle core. There may also be a third and very fast step that would correspond to an association of the solubilizate to the micelle surface or to the outer part of the corona. More work is required to confirm the existence of this very fast process. One cannot discard that in fact these discrete relaxation times should be replaced by a spectrum of relaxation times. Nevertheless, the reported results have well established the interest of using block copolymer micelles for solubilizing strongly hydrophobic (poorly water soluble) drugs and slowly releasing them *in vivo*.

VII. DYNAMICS IN AQUEOUS SOLUTION OF ASSOCIATIVE POLYMERS

Associative polymers are water-soluble polymers made up of a hydrophilic backbone and a few hydrophobic groups (hydrophobes). The hydrophobes can be located at the ends of the hydrophilic chain, forming a telechelic polymer that can be considered as a triblock copolymer with two short hydrophobic blocks. The hydrophobes can also be randomly attached to the chain, forming a comb polymer that can be considered to be a random copolymer.[104] The hydrophobes have a tendency to self-associate intra- and intermolecularly, giving rise to micelle-like objects. Intermolecular associations result in the formation of transient networks, with the micelles acting as junctions between polymer chains. These systems display interesting rheological properties such as viscoelasticity, shear-thickening and shear-thinning. Gel formation is often observed at sufficiently high concentration. Associative polymers are used for controlling viscosity at different shear rates (rheology) in paints and cosmetics, for instance.

Various dynamic processes occurring in solutions of associative polymers explain to a large extent their rheological behavior. Most studies have been performed using rheological methods (see Chapter 2, Section VI, and Chapter 9) and NMR (see Chapter 2, Section IV). Results for telechelic and for comblike associative polymers are examined successively.

A. Telechelic Polymers

Most studies refer to associative polymers made up of a poly(ethylene oxide) backbone terminated at both ends by alkyl chains with 8 to 20 carbon atoms. In many instances, a diisocyanate group separates the poly(ethylene oxide) chain from the hydrophobe.

Several rheological studies showed that solutions of telechelic polymers submitted to oscillatory shear exhibit a Maxwellian behavior (single relaxation time).[105-108] This behavior was explained by assuming that the main stress relaxation mechanism is the exit of one hydrophobe from the micelle-like junction in which it is embedded. As the polymer chain is freed at one end, it can rapidly relax the stress through Rouse modes. The relaxation time was found to vary exponentially with the hydrophobe carbon number (see Figure 4.16). The slope of the plot in Figure 4.16 yielded an increment of free energy per methylene group of 1.03 kT, a value that is very close to that for the free energy of transfer of a methylene group from a surfactant micelle to water. This result strongly supports the assignment of the observed relaxation. Mixtures of two polymers with different hydrophobe length were shown to relax with two relaxation times. This result indicated that the chains relax independently.[105] Note that the relaxation time for the exit of a hydrophobe from a junction is much longer than the residence time of the same alkyl chain in a surfactant micelle. This may reflect the fact that the microviscosity of the hydrophobic microdomains present

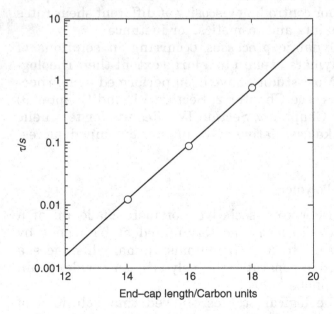

Figure 4.16 Variation of the relaxation time for hydrophobe exit in a 4 wt% solution of a telechelic poly(ethylene oxide) of molecular weight 35,000 with the hydrophobe carbon number at 298 K. Reproduced from Reference 105 with permission of the Society of Rheology.

in solutions of associating polymers or of polysoaps has been found to be much higher than that of surfactant micelles.[109] Also, the diffusive motion of a hydrophobe out of a junction is hindered (slowed down) because it is tethered to the hydrophilic polymer chain.

However, other rheological studies reported the existence of two relaxation processes.[110-112] Reference 112 presented an interpretation of the results that is very different from that in References 105–108. The slow process, which is that discussed in References 105–108, is now attributed to the network relaxation while the faster of the two processes, not seen in these references, is attributed to the exit of a hydrophobe from a junction. One of the difficulties with this interpretation is that the lifetime of a hydrophobe in a junction would increase with temperature.[112] The authors state that nonionic surfactants show such a behavior. Unfortunately, the references cited to back this point do not really refer to dynamic studies of micelles of nonionic surfactants. Such studies have been performed and show that the residence time/lifetime of nonionic surfactants in micelles decreases as the temperature is increased,[113,114] just as for ionic surfactants. Thus at the present time there appears to be no good evidence for the assignment of the fast relaxation observed in Reference 112 to the exit of a hydrophobe from a junction. In contrast, the available experimental results seem to indicate that it is the slow relaxation that is associated with this process.

B.   Comblike Polymers

Rheological studies of random copolymers of sodium acrylate (NaAA) or sodium 2-acrylamido-2-methylpropanesulfonate (NaAMPS) and methacrylate substituted with $HO(CH_2CH_2O)_nC_{12}H_{25}$ (DEnMA, with n = 2, 6, 25) have been reported.[115] This study showed the enormous importance of the chemical structure of the copolymer on the dynamics of the system. The relaxation time was found to decrease very much upon increasing m for the two types of copolymers. It was very small and independent of concentration for the NaAPMS-DEnMA copolymers but became concentration dependent and quite large, reaching 10^3 s, for the NaAA-DEnMA copolymers with n = 2 and 6. This difference of behavior was attributed to the possible existence of an interaction between methyl groups of NaAPMS and hydrophobes. Such interaction cannot occur with NaAA-based copolymers, favoring interchain interactions and slowing down motions.

Comblike polymers made up of a poly(sodium acrylate) backbone with various fluorocarbon or hydrocarbon pendant hydrophobes were studied by means of ^{13}C and ^{19}F NMR.[116,117] A preliminary study showed that the hydrophobes were slowly exchanging between the free and associated states and the hydrophobe lifetime in the associated state was reported to be longer than 1–10 ms, depending on the hydrophobe.[116] A more detailed study of similar polymers with $C(O)NHCH_2C_7F_{15}$ pendant groups showed that two processes spontaneously occurred in the system: a very fast one associated to the exit of hydrophobe from micellar junctions between polymer chains (time scale: 1 μs) and a much slower process corresponding to the association/dissociation of a whole polymer chain to/from an aggregate of polymer chains (time scale: > 1 s). This mechanism is similar to the AW model for the relaxation in micellar solutions of surfactants. The difference between the two time scales reflects the fact that the dissociation of a whole polymer chain from an aggregate requires the near simultaneous exit of all the hydrophobes on this chain from the aggregate. Obviously the value of the lifetime/residence time of a hydrophobe in a junction is much shorter than for a telechelic polymer (1 ms against 1 μs for a dodecyl chain or its equivalent). This large difference may be partly due to the different chemical structures of telechelic and comblike polymers.

In conclusion, the above results clearly call for more studies of solutions of associating polymers for reaching a better understanding of the dynamics of these systems that are so important in applications.

VIII. DYNAMICS OF MISCELLANEOUS PROCESSES OCCURRING IN BLOCK COPOLYMERS SYSTEMS

This section reviews studies that deal with the dynamics of various processes in solutions of block copolymers.

The time dependence of the pH in aqueous solutions of poly(styrene)–polymethacrylic acid upon addition of NaOH has been found to stretch over tens of days. This behavior has been interpreted as indicating that the inner part of the poly(methacrylic acid) corona is not easily accessible to OH⁻ ions because it is very compact.[60] This result supports those obtained for the dynamics of exchange of solubilizates in aqueous block copolymer solutions (see Section VII).

Yusa et al.[118] showed that the random copolymer of 11-aminododecanoic acid and sodium 2-(acrylamidopropane-sulfonate) forms unimolecular micelles in water at pH < 5. At pH > 5 the unimolecular micelle conformation is transformed into an open-chain conformation. Pyrene fluorescence measurements showed that the transformation is completed in less than 6 min (dead time of the measurements).

The poly(styrene)-poly(acrylic acid) copolymers with a short poly(acrylic acid) block (crew-cut copolymers) are capable of forming aggregates of different shapes, particularly spherical and rodlike micelles and vesicles.[67] The sphere-to-rod, rod-to-sphere, and rod-to-vesicle transformations can be induced in dioxane-water mixtures by solvent jumps (addition of water induces the phase sequence: sphere → rod → vesicle). The kinetics of these transformations was investigated for the $(styrene)_{310}(acrylic\ acid)_{52}$ copolymer. The sphere-to-rod transformation induced upon increasing water content was shown by means of transmission electron microscopy to first involve sticky collisions between spherical micelles that give rise to irregular pearl necklaces. The necklaces then reorganize into smooth rods (see Figure 4.17).[119] The rod-to-sphere transformation was induced by addition of dioxane. It proceeds via the breaking off of spherical micelles from the ends of the rods (see Figure 4.17).[119] The time scale of these processes is in 1000–2000 s. Rodlike

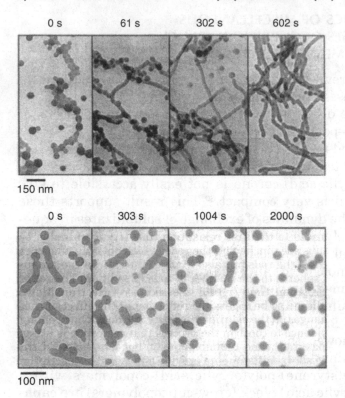

Figure 4.17 Micrographs showing the aggregates at different time points during the sphere-to-rod transformation (*top*) and the rod-to-sphere transformation (*bottom*) occurring in a 1.0 wt% solution of styrene$_{310}$acrylic acid$_{52}$ in dioxane–water mixtures. The sphere-to-rod transformation is induced by increasing the water content from 12 to 13.5%. The rod-to-sphere transformation is induced by decreasing the water content from 12 to 11.5%. Reproduced from Reference 119 with permission of the American Chemical Society.

micelles transformed into vesicles by first flattening into irregularly shaped or circular lamellae that closed into vesicles in a time scale of 1000–2000 s.[120]

In a later study, the kinetics of increase/decrease of vesicle size induced by an increase/decrease of the water content of the solvent mixture water/dioxane were investigated for the $(styrene)_{310}(acrylic\ acid)_{28}$ copolymer by stopped flow with turbidity detection.[121] The size changes were shown to occur via vesicle fusion or fission. The relaxation time characterizing the vesicle fusion increased with the water content of the system (time scale: 100–500 s).

Polymersomes are giant vesicles of varied shape, from spherical to tubular, formed by block copolymers in aqueous solution[122] and in organic solvent.[123] Poly(butadiene)-poly(ethylene oxide) tubular polymersomes in water have been shown to undergo beading upon a temperature quench.[124] Phase contrast microscopy showed that beading occurred from the ends to the center on the time scale of a few hours, following a temperature quench. Beading is an intermediate state between tubular and spherical polymersomes.[122]

The kinetics and mechanisms of disorder-to-order phase transitions have been investigated in concentrated solutions of block copolymers in selective solvents by small angle x-ray scattering (SAXS) or optical techniques as the kinetics was generally very slow. Two illustrative examples are cited without comment due to lack of space. First, the kinetics of the transition between hexagonally packed cylinders and the cubic gyroid phase of a poly(styrene)-poly(isoprene) block copolymer in di-*n*-butylphthalate, a selective solvent of poly(styrene), were investigated by SAXS.[125] Second, the kinetics of the transition between micelles and the body centered cubic phase of the copolymer poly(styrene)-poly(ethylene-*co*-butylene)-poly(styrene) in mineral oil were investigated by time-resolved SAXS.[126] In both studies, the time scale for the transitions was in the range of about 1 hour[126] to tens of hours.[125] The transition was found to occur in at least two stages and the results permitted the authors to gain some insight into its mechanism.

IX. GENERAL CONCLUSIONS

This chapter reviewed the dynamics of micellar solutions of amphiphilic block copolymers. The reported studies show that the triblock copolymers poly(ethylene oxide)-poly(propylene oxide)-poly(ethylene oxide) behave similarly to conventional surfactants. The copolymer exchange as well as the micelle formation/breakdown have been evidenced and investigated fairly completely. The entry of the copolymer into the micelles is slower than for a diffusion-controlled process and becomes increasingly slower as the length of the poly(propylene oxide) block is increased. The micelle formation/breakdown appears to occur via reactions of fragmentation/coagulation, as in the case of aqueous solutions of the nonionic poly(ethylene oxide) monoalkylether surfactants. The free energy of transfer of a propylene oxide unit from the micelle to the aqueous phase has been found to be in the range of values that can be inferred for this quantity from the variation of the cmc with the number of propylene oxide units. To a large extent, the Aniansson-Wall-Kahlweit model proposed for surfactant micellar solutions appears to hold for $EO_{N(EO)}PO_{N(PO)}EO_{N(EO)}$ copolymer micelles.

The situation is very different for other copolymers in solution in water and in hydroorganic or organic solvents. The copolymer exchange process is then very often frozen for copolymers where the insoluble block is long (thermodynamic freezing) or at a temperature below the glass transition temperature (kinetic freezing). The copolymer exchange can nevertheless be observed by selecting the proper solvent or solvent mixture. The range of solvent composition in which the dynamic equilibrium between unimers and micelles can be observed is generally relatively narrow, in accordance to dynamic Monte Carlo simulations. Whenever observable, the exchange process is much slower than for surfactants, by up to 10 orders of magnitude, owing most likely to the large size of the insoluble block making up the micelle core. Note that the relaxation signals associated with the copolymer exchange are in most instances not single exponential. This probably reflects the unavoidable polydispersity in composition and molecular weight of the copolymers. The lifetime of block copolymer micelles under near equilibrium conditions remains to be investigated. It is expected to be extremely long in view of the slowness of the copolymer exchange process.

In the case of aqueous solutions, the main reason for the strongly differing behaviors reported for the dynamics of micelles of $EO_{N(EO)}PO_{N(PO)}EO_{N(EO)}$ copolymers and of other copolymers appears to be the extremely weak hydrophobic character of the poly(propylene oxide) block.

Studies of the dynamics of exchange of compounds solubilized in the core of aqueous block copolymer micelles showed that the residence time of solubilizates in the core of block copolymer micelles is orders of magnitude longer than in surfactant micelles. This together with the long lifetime of block copolymer micelles are factors much in favor of the use of block copolymer micelles in drug delivery.

REFERENCES

1. Lindman, B., Alexandridis, P. In *Amphiphilic Block Copolymers*, Alexandridis, P., Lindman, B., Eds., Elsevier, Amsterdam, 2000, p. 1.

2. Sadron, C. *Pure Appl. Chem.* 1962, 4, 347.

3. Kalarakis, A., Havredaki, V., Yu, G.-A., Derici, L., Booth, L. *Macromolecules* 1998, 31, 944 and references therein.

4. Kikuchi, A., Nose, T. *Macromolecules* 1996, 29, 5321.

5. Chu, B. *Langmuir* 1995, 11, 414.

6. Gao, Z., Eisenberg, A. *Macromolecules* 1993, 26, 7353.

7. Matsuoka, H., Matsutani, M., Mouri, E., Matsumoto, K. *Macromolecules* 2003, 36, 5321.

8. Zana, R. In *Amphiphilic Block Copolymers*, Alexandridis, P., Lindman, B., Eds., Elsevier, Amsterdam, 2000, p. 221.

9. Alexandridis, P., Hatton, T.A. *Colloids Surf. A.* 1995, 96, 1.

10. Zhou, Z., Chu, B. *J. Colloid Interface Sci.* 1988, 126, 171.

11. Reddy, N.K., Fordham, P.J., Attwood, D., Booth, C. *J. Chem. Soc., Faraday Trans.* 1990, 86, 1569.

12. Aniansson, E.A.G., Wall, S.N. *J. Phys. Chem.* 1974, 78, 1024.

13. Aniansson, E.A.G., Wall, S.N. *J. Phys. Chem.* 1975, 79, 857.

14. Aniansson, E.A.G., Wall, S.N., Almgren, M., Hoffmann, H., Kielmann, I., Ulbricht, W., Zana, R., Lang, J., Tondre, C. *J. Phys. Chem.* 1976, 80, 905.

15. Kahlweit, M., Teubner, M. *Adv. Colloid Interface Sci.* 1980, 13, 1.

16. Kahlweit, M. *Pure Appl. Chem.* 1981, 53, 2069.

17. Kahlweit, M. *J. Colloid Interface Sci.* 1982, 90, 92.

18. Lessner, E., Teubner, M., Kahlweit, M. *J. Phys. Chem.* 1981, 85, 1529.

19. Halperin, A., Alexander, S. *Macromolecules* 1989, 22, 2403.

20. Dormidontova, E. *Macromolecules* 1999, 32, 7630.

21. Wang, M., Mattice, W.L., Napper, D.H. *Langmuir* 1993, 9, 66.

22. Haliloglu, T., Bahar, I., Erman, B., Mattice, W.N. *Macromolecules* 1996, 29, 4764.

23. Price, C. *Pure Appl. Chem.* 1983, 55, 1563.

24. Booth, C., Naylor, T., Price, C., Rajab, N., Stubbersfield, R. *J. Chem. Soc., Faraday Trans.* 1978, 74, 2352.

25. Malmsten, M., Lindman, B. *Macromolecules* 1992, 25, 5440.

26. Spacek, K., Kubin, M. *J. Appl. Polym. Sci.* 1985, 30, 143.

27. Spacek, K. *J. Appl. Polym. Sci.* 1986, 32, 4181.

28. Holmqvist, P., Nilsson, S., Tiberg, F. *Colloid Polym. Sci.* 1997, 275, 467.

29. Prochazka, K., Glöckner, G., Hoff, M., Tuzar, Z. *Makromol Chem.* 1984, 185, 1187.

30. Tuzar, Z., Kratochvil, P. In *Surface and Colloid Science*, Matijevic, E., Ed., Plenum Press, New York, 1993, 15, 1.

31. Michels, B., Waton, G., Zana, R. *Langmuir* 1997, 13, 3111.

32. Hecht, E., Hoffmann, H. *Colloids Surf. A* 1995, 96, 181.

33. Bednar, B., Edwards, K., Almgren, M., Tuzar, Z., Kratochvil, P., Tuzar, Z. *Makromol. Chem. Rapid Commun.* 1988, 9, 785.

34. Wang, M., Kausch, C.M., Chun, M., Quirck, R.P., Mattice, W.L. *Macromolecules* 1995, 28, 904.

35. Smith, C.K., Liu, G. *Macromolecules* 1996, 29, 2060.

36. Willner, L. Poppe, A., Allgaier, J., Monkenbusch, M., Richter, D. *Europhys. Lett.* 2001, 55, 667.

37. Honda, C., Hasegawa, Y., Hirununa, R., Nose, T. *Macromolecules* 1994, 27, 7660.

38. Rassing, J., McKenna, W., Bandyopadhyay, S., Eyring, E.M. *J. Mol. Liq.* 1984, 27, 165.

39. Waton, G., Michels, B., Zana, R. *J. Colloid Interface Sci.* 1999, 212, 593.

40. Thurn, T., Couderc-Azouani, S., Bloor, D.M., Holzwarth, J.F., Wyn-Jones, E. *Langmuir* 2003, 19, 4363.

41. Goldmints, I., Holzwarth, J., Smith, K., Hatton, T.A. *Langmuir* 1997, 13, 6130.

42. Kositza, M.J., Bohne, C., Alexandridis, P., Hatton, T.A., Holzwarth, J. *Langmuir* 1999, 15, 322.

43. Kositza, M.J., Bohne, C., Alexandridis, P., Hatton, T.A., Holzwarth, J. *Macromolecules* 1999, 32, 5539.

44. Kositza, M.J., Rees, G.D., Holzwarth, A., Holzwarth, J. *Langmuir* 2000, 16, 9035.

45. Waton, G., Michels, B., Zana, R. *Macromolecules* 2001, 34, 907.

46. Goldmints, I. PhD thesis, MIT, Boston, 1999.

47. Kositza, M.J., Bohne, C., Hatton, T.A., Holzwarth, J. *Prog. Colloid Polym. Sci.* 1999, 112, 146.

48. Alexandridis, P., Holzwarth, J., Hatton, T.A. *Macromolecules* 1994, 27, 2414.

49. Fleischer, G. *J. Phys. Chem.* 1993, 97, 517.

50. Hvidt, S. *Colloids Surf. A* 1996, 112, 201.

51. Wanka, G., Hoffmann, H., Ulbricht, W. *Colloid Polym. Sci.* 1990, 268, 101.

52. Booth, C., Attwood, D. *Macromol. Rapid Commun.* 2000, 21, 501.

53. Barreleiro, P., Alexandridis, P. *J. Colloid Interface Sci.* 1998, 206, 357.

54. Tanodekaew, S., Pannu, R., Heatley, F., Attwood, D., Booth, C. *Makromol. Chem. Phys.* 1997, 198, 927.

55. Crothers, M., Attwood, D., Collett, J.H., Yang, Y., Booth, C., Taboada, P., Mosqueran V., Ricardo, N.M., Martini, L.G. *Langmuir* 2002, 19, 8685.

56. Yang, Z., Crothers, M., Ricardo, N.M., Chaibundit, C., Taboada, P., Mosquera V., Kelarakis, A., Havredaki, V., Martini, L.G., Walder, C., Collett, J.H., Attwood, D., Heatley, F., Booth, C. *Langmuir* 2003, 19, 943.

57. Zana, R., Muto, Y., Esumi, K., Meguro, K. *J. Colloid Interface Sci.* 1988, 123, 502.

58. Auvray, X., Petipas, C., Anthore, R., Rico, I., Lattes, A., Ahmad-Zadeh Samii, A., de Savignac, A. *Colloid Polym. Sci.* 1987, 265, 925.

59. Munk, P., Ramireddy, C., Tian, M., Webber, S.E., Prochazka, K., Tuzar, Z. *Makromol. Chem., Makromol. Symp.* 1992, 58, 195.

60. Stepanek, M., Prochazka, K., Brown, W. *Langmuir* 2000, 16, 2502.

61. Rager, T., Meyer, W.H., Wegner, G., Winnik, M.A. *Macromolecules* 1997, 30, 4911.

62. Won, Y.-Y., Davis, H.T., Bates F.S. *Macromolecules* 2003, 36, 953.

63. Zhang, L., Barlow, R.J., Eisenberg, A. *Macromolecules* 1996, 28, 6065.

64. Jada, A., Hurtez, G., Siffert, B., Riess, G. *Macromol. Chem. Phys.* 1996, 197, 3697.

65. Wang, Y., Balaji, R., Quirk, R.P., Mattice, W.L. *Polym. Bull.* 1992, 28, 333.

66. Prochazka, K., Martin, T.J., Munk, P., Webber, S.E. *Macromolecules* 1996, 29, 6518.

67. Zhang, L., Eisenberg, A. *Science* 1995, 268, 1728.

68. Tian, M., Qin, A., Ramireddy, C., Webber, S.E., Munk, P., Tuzar, Z., Prochazka, K. *Langmuir* 1993, 9, 1741.

69. Rager, T., Meyer, W.H., Wegner, G. *Macromol. Chem. Phys.* 1999, 200, 1672.

70. van Stam, J., Creutz, S., De Schryver, F.C., Jérome, R. *Macromolecules* 2000, 33, 6388.

71. Creutz, S., van Stam, J., Antoun, S., De Schryver, F.C., Jérome, R. *Macromolecules* 1997, 30, 4078.

72. Creutz, S., van Stam, J., De Schryver, F.C., Jérome, R. *Macromolecules* 1998, 31, 681.

73. Stepanek, M., Podhajecka, K., Tesarova, E., Prochazka, K., Tuzar, Z., Brown, W. *Langmuir* 2001, 17, 4240.

74. Zhang, W., Shi, L., An, Y., Gao, L., He, B. *J. Phys. Chem. B.* 2004, 108, 200.

75. Schillen, K., Yekta, A., Ni, S., Winnik, M.A. *Macromolecules* 1998, 31, 210.

76. Pacovska, N., Procahzka, K., Tuzar, Z., Munk, P. *Polymer* 1993, 34, 4585.

77. Zhang, L., Shan, H., Eisenberg, A. *Macromolecules* 1997, 30, 1001.

78. Prochazka, K., Bednar, B., Mukhtar, E., Svobioda, P., Trnena, J. Almgren, M. *J. Phys. Chem.* 1991, 95, 4563.

79. Liu, G. *Can. J. Chem.* 1995, 73, 1995.

80. Underhill, R.S., Ding, J., Birss, V.I., Liu, G. *Macromolecules* 1997, 30, 8298.

81. Bednar, B., Karasek, L., Pokorny, J. *Polymer* 1996, 37, 5261.

82. Honda, C., Abe, Y., Nose, T. *Macromolecules* 1996, 29, 6778.

83. Iyama, K., Nose, T. *Macromolecules* 1998, 31, 7356.

84. Esselink, F.J., Dormidontova, E., Hadziioannou, G. *Macromolecules* 1998, 31, 2925.

85. Esselink, F.J., Dormidontova, E., Hadziioannou, G. *Macromolecules* 1998, 31, 4873.

86. Calderara, F., Riess, G. *Macromol. Chem. Phys.* 1996, 197, 2115.

87. Cao, T., Munk, P., Ramireddy, C., Tuzar, Z., Webber, S.E. *Macromolecules* 1991, 24, 6300.

88. Hruska, Z., Piton, M., Yekta, A., Duhamel, J., Winnik, M.A., Riess, G., Croucher, M.D. *Macromolecules* 1993, 26, 1825.

89. Teng, Y., Morrison, M.E., Munk, P., Webber, S.E., Prochazka, K. *Macromolecules* 1998, 31, 3578.

90. Wang, G., Henselwood, F., Liu, G. *Langmuir* 1998, 14, 1554.

91. Soo, P. L., Luo, L., Maysinger, D., Eisenberg, A. *Langmuir* 2002, 18, 9996.

92. Arca, E., Tian, M., Webber, S.E., Munk, P. *Int. J. Polym. Anal. Characterization* 1995, 2, 31.

93. Bromberg, L.E., Magner, E. *Langmuir* 1999, 15, 6792.

94. Cleary, J., Bromberg, L.E., Magner, E. *Langmuir* 2003, 19, 9162.

95. Almgren, M., Grieser, F., Thomas, J.K. *J. Am. Chem. Soc.* 1979, 101, 279.

96. Zana, R., Lianos, P., Lang, J. *J. Phys. Chem.* 1985, 89, 41.

97. Kim, S.A., Ha, J.C., Lee, M. Y. *J. Controlled Release* 2000, 65, 345.

98. Kwoon, G., Naito, M., Yokoyama, M., Okano, T., Sakurai, Y., Kataoka, K. *J. Controlled Release* 1997, 48, 195.

99. Jeong, Y.L., Cheon, J.B., Kim, S.H., Na, J.W., Lee, Y.M., Sung, Y.K., Akaike, T., Cho, C.S. *J. Controlled Release* 1998, 51, 169.

100. Fox, S.L., Chan, J.C., Kiserow, D.J., Ramireddy, C., Munk, P., Webber, S.E. In *Hydrophilic Polymers. Performances with Environmental Acceptance*, American Chemical Society, Washington, DC, 1996, p. 141.

101. Stepanek, M., Krijtova, K., Prochazka, K., Teng, Y., Webber, S.E. *Colloids Surf. A.* 1999, 147, 79.

102. Kriz, J. *Langmuir* 2000, 16, 9770.

103. Li, Y., Nakashima, K. *Langmuir* 2003, 19, 548.

104. Winnik, M.A., Yekta, A. *Curr. Opin. Colloid Interface Sci.* 1997, 2, 424.

105. Annable, T., Buscall, R., Ettelaie, R., Whittlestone, D. *J. Rheol.* 1993, 37, 695.

106. Annable, T., Buscall, R., Ettelaie, R. *Colloids Surf.* 1996, 112, 97.

107.Pham, Q.T., Russel, W.B., Thibeault, J.C., Lau, W. *Macromolecules* 1999, 32, 5139.

108.Laflèche, F., Durand, D., Nicolai, T. *Macromolecules* 2003, 36, 1331.

109.Anthony, O., Zana, R. *Langmuir* 1996, 12, 3590, and cited references.

110.Xu, B., Yekta, A., Li, L., Masoumi, Z., Winnik, M.A. *Colloids Surf.* A 1996, 112, 239.

111.Le Meins, J.-F., Tassin, J.-F., Corpart, J.-M. *J. Rheol.* 1999, 43, 1423.

112.Ng, W.K., Tam, K.C., Jenkins, R.D. *J. Rheol.* 2000, 44, 137, and references cited.

113.Hoffmann, H., Kielman, H.S., Pavlovic, D., Platz, G., Ulbricht, W. *J. Colloid Interface Sci.* 1981, 80, 237.

114.Frindi, M., Michels, B., Zana, R. *J. Phys. Chem.* 1992, 96, 6095.

115.Noda, T., Hashidzume, A., Morishima, Y. *Langmuir* 2001, 17, 5984.

116.Petit-Agnely, F., Iliopoulos, I. *J. Phys. Chem.* B 1999, 103, 4803.

117.Furo, I., Iliopoulos, I., Stilbs, P. *J. Phys. Chem.* B 2000, 104, 485.

118.Yusa, S., Sakakibara, A., Yamamoto, T., Morishima, Y. *Macromolecules* 2002, 35, 5243.

119.Burke, S.E., Eisenberg, A. *Langmuir* 2001, 17, 6705.

120.Chen, L., Shen, H., Eisenberg, A. *J. Phys. Chem.* B 1999, 103, 9488.

121.Choucair, A.A., Kycia, A.H., Eisenberg, A. *Langmuir* 2003, 19, 1001.

122.Discher, B.M., Won, Y.Y., Ege, D.S., Lee, J. C.-M., Bates, F.S., Discher, D.E., Hammer, D.A. *Science* 1999, 284, 1143.

123.Wittmann, J.-C., Lotz, B., Candau, F., Kovacs, A.J. *J. Polym. Sci. Polym. Phys. Ed.* 1982, 20, 1341.

124.Reinecke, A.A., Döbereiner, H.-G. *Langmuir* 2003, 19, 605.

125.Wang, C.-Y., Lodge, T.P. *Macromolecules* 2002, 35, 6997.

126.Nie, H., Bansil, R., Ludwig, K., Steinhart, M., Konak, C., Bang, J. *Macromolecules* 2003, 36, 8097.

5

Dynamic Processes in Microemulsions

CHRISTIAN TONDRE

CONTENTS

I.   INTRODUCTION

Microemulsions, unlike classical emulsions, are usually transparent and thermodynamically stable systems requiring a significant amount of surfactant. The addition of a cosurfactant, although not always required, often appears to be necessary so that microemulsions in their simplest compositions are ternary or quaternary systems. Microemulsions are thus some sort of swollen micellar systems in which the core of the droplets accommodates a more or less important amount of dispersed oil (or water in the case of reverse micelles). Research concerning such systems, which were first mentioned by Hoar and Schulman,[1] literally exploded in the late

1970s and early 1980s because of their important potential for tertiary oil recovery,[2,3] in relation to their ability to produce very low interfacial tensions. Since that time, many efforts have been made to fully understand the complex nature of these systems and to explore their applications. Many types of formulations have been tested, which were accompanied by the characterization of the corresponding phase diagrams.

A number of review articles[3-9] and books[10-16] have been devoted to microemulsions. The reader should refer to these publications (and to references therein) for more information regarding aspects of these systems that are not treated in this chapter and that extend from chemical formulation to physical properties, going through theoretical aspects, applications in analytical science and separation, biotechnology, pharmacology, chemical reactivity, polymer synthesis, and so forth.

The literature concerning the structural aspects of microemulsion systems is much more abundant than that devoted to their dynamic behavior. Scattering methods have been extensively used for this purpose, and the information that has been gained is very important for understanding the dynamic processes, even though they will not be reviewed here.

Far from having a totally static molecular organization, microemulsion systems are strongly dynamic in the sense that the constituting components can show different kinds of diffusion or exchange processes, and the cooperative fluctuations of the molecules constituting the amphiphilic film may result in droplet coalescence/decoalescence, percolation phenomena, etc. These dynamic processes probably play a major part in the stability of microemulsions, and they have significant implications regarding the use of these systems for applied purposes.

The dynamic behavior of microemulsions controls the exchange of solubilizate between droplets, and it has a strong impact on the chemical reactivity in such microheterogeneous systems. The latter aspect is reviewed in Chapter 10 of this book. We will try here to go from the macroscopic observations to their significance at the molecular level and from the global motions of the droplets to the dynamics of the individual components and of the interfacial film. Some of the implications of the dynamic behavior of microemulsions in specific domains will be discussed briefly.

II. GENERAL POINTS CONCERNING MICROEMULSIONS STRUCTURE, PHASE BEHAVIOR, AND DYNAMICS

The essential features responsible for the transparency of microemulsions is the small size of the droplets, whose diameters are typically from 5 to 50 nm, i.e., only a small fraction of the wavelength of visible light. However, the word "microemulsion" includes different kinds of systems depending on whether we are dealing with water-rich, oil-rich, or intermediate compositions. The continuous passage from the first one to the second one occurs through the formation of bicontinuous structures.[17] This implies a change of curvature of the amphiphilic film, which can be induced as well by a composition change as by a temperature change.

For compositions outside the monophasic microemulsion domains of the phase diagrams, polyphasic systems are obtained for which the Winsor terminology is widely adopted (see Chapter 1, Section IV)[18,19] In this terminology a homogeneous single-phase microemulsion is called Winsor IV. When excess oil separates from an oil-in-water (O/W) microemulsion, the diphasic system obtained is called Winsor I, whereas Winsor II refers to the reverse situation, in which excess water separates from a water-in-oil (W/O) microemulsion. A Winsor III describes a triphasic system in which a bicontinuous microemulsion is in contact with both excess oil and excess water. With ionic surfactant systems, an increase of salinity progressively changes the partitioning of the surfactant, initially in the aqueous phase, in favor of the organic phase. This may result in a Winsor I → Winsor III → Winsor II transition. A temperature change will generally have a similar effect in the case of nonionic surfactants.

Winsor systems are extremely useful for studying transfer processes from one phase to another. Due to the existence of well-defined liquid-liquid interfaces, they have been used to examine how solubilizates can be transferred in and transported by microemulsion droplets.[20,21]

Many physicochemical techniques are sensitive to the dynamics, including some techniques that are more specifically adapted to the study of structural aspects or particle size measurements, like, for instance, quasi-elastic light scattering. The dynamic processes occurring in microemulsions are so varied that they can take place on very different time scales. Several techniques are therefore required to cover all that variety of motions. Table 5.1 lists the main existing processes, their characteristic times, and the type of techniques best adapted to investigate them. These techniques are essentially the NMR, the fast chemical relaxation techniques (ultrasonic absorption; temperature-jump (T-jump); pressure-jump (p-jump); electric-

TABLE 5.1 Characteristic Times of Dynamic Processes Occurring in Microemulsion Systems

Dynamic Process	Characteristic Time	Technique	Reference
Alkyl chain motions in the aggregates	ps to ns	NMR	24, 54
Tumbling of the whole droplet (rotational correlation time)	ns to μs	NMR	25–28
Exchange of short chain alcohol (cosurfactant) between aggregates and bulk	order of 10 ns	Ultrasonic absorption	47, 48
Exchange of surfactant between aggregates and bulk water (in o/w microemulsions in the presence of cosurfactants)	order of 100 ns	Ultrasonic absorption	47, 48
Droplet coalescence	μs to ms	Time-resolved fluorescence quenching and stopped-flow	105–109 99–101
Clustering of aggregates	order of ms	Time-resolved luminescence quenching	117–119
Dissolution of water (oil) in w/o (o/w) microemulsions	ms to min (depending on mixing efficiency)	Stopped-flow	132–134
Change of film curvature	from 100 ms to μs	T-jump	126, 145

field jump (E-field jump); stopped-flow techniques), and time-resolved luminescence quenching techniques (see Chapter 2). A few other techniques also proved to be useful in the approach of the dynamics of microemulsion systems: electron paramagnetic resonance spectroscopy (EPR) gave some insight into the motions of probe molecules, and electrical conductivity has been widely used to study droplet percolation phenomena (see Section VI). Rheological measurements can also provide information on the dynamic behavior, but this constitutes a specialized field not considered here. These aspects have been recently reviewed.[8]

III. DIFFUSION AND ROTATION OF MOLECULES AND AGGREGATES

A. Global Motions of the Individual Aggregates

Before looking at what happens inside the droplets at the molecular level, the first motion a hypothetical observer would see (assuming he is able to visually follow a single droplet) would be a rotation of the aggregate as a whole. This kind of movement, and thus the droplet tumbling rate, is in principle accessible through the use of NMR or EPR spectroscopy. It has been pointed out[22-24] that in micellar and microemulsion systems ^{13}C NMR relaxation may be affected by two types of motions: (1) fast local motions in the interior of the aggregates characterized by a correlation time τ_c^f, and (2) slow molecular motions with correlation time τ_c^s, including the overall tumbling of the aggregate and the molecular lateral diffusion around the surface of the aggregates. Appropriate NMR experiments permit in some instances a discrimination between the different types of motions. However, the case of microemulsions is more complicated than that of simple micelles because the microstructure is not always known.

Assuming spherical microemulsion droplets, an upper limit of the correlation time (corresponding to the droplet tumbling rate) can be calculated from[25,26]

$$\tau_c^{tumbling} = 4\pi\eta(R_H)^3 / 3kT \qquad (5.1)$$

where η is the viscosity and R_H the droplet hydrodynamic radius. The value τ_c = 1.4 µs was obtained in the case of a four-component system water/hexadecane/sodium hexadecyl sulfate/pentanol with R_H = 12.7 nm.[25] For the sake of comparison, a τ_c value of 4.1 ns has been calculated[26] for the rotation of a micelle of decylammonium chloride, DAC (R_H = 16.9 Å). A study of the four-component systems SDS/toluene/water/alcohol (with alcohol chain lengths from C_3 to C_{10}) yielded values of the slow correlation time ranging from 1.7 to 15.0 ns when going from propanol to decanol. This effect was attributed to a change of the structure of the microemulsions. For the shorter-chain alcohols, the motions are comparatively rapid, with a rate comparable to that in DAC micelles. When the chain length of the cosurfactant is increased, the situation becomes more complex, with motions occurring on different time scales and including shape fluctuations of the droplets. The time constants of these fluctuations have been studied for the AOT/water/decane microemulsion system using quasi-elastic light scattering and neutron spin-echo spectroscopy.[27] The time constant that governed the relaxation mechanism reflected the relative strengths of the interfacial forces and the viscous forces, and an effective microscopic interfacial tension for the microemulsion droplet could be deduced for the first time ($\gamma \approx 0.05$ mN/m).

Information concerning the tumbling rate of microemulsions in AOT/water/isooctane was obtained from EPR spectroscopy using 4-hydroxy-TEMPO, a water soluble nitroxide radical.[28] In that case the rotational correlation time normally indicates the tumbling rate of the probe molecule, but after reducing the size of the microemulsion droplet through clathrate hydrate formation (which removes water from the droplets), the spin probe motion becomes much faster than in a larger droplet at the same temperature. This was interpreted as indicating that the motions of the probe, assumed to be rigidly held inside the microemulsion droplet, expresses the proper movement of the droplet. Fixing the water/AOT molar ratio (w_o) at 20, rotational correlation times from 0.2 to 0.7 ns were found when varying the temperature from 295 to 263 K. Hydrate formation at 268 K induced a drop of τ_c from 0.65 to 0.27 ns.

B.   Diffusion of Single Components

The self-diffusion coefficients of the different components of a microemulsion (surfactant, cosurfactant, water, and oil) can be measured using pulsed field gradient Fourier transform NMR (PG-FT-NMR) or radioactive tracer methods.[29–32] The values obtained for these molecular self-diffusion coefficients depend on obstruction factors and on confinement; for this reason, they are closely related to the microemulsion structure. Molecular components belonging to the continuous phase will experience faster diffusion than the components confined in a closed aggregate. Thus by comparing the self-diffusion coefficients of water and oil, it is possible to identify the probable structural organization of the system, and to distinguish between O/W, W/O, or bicontinuous systems. For the components confined in the aggregates and for the surfactants constituting their amphilic layers, the dominating diffusion process is expected to be the proper diffusion of the aggregate itself (with a diffusion coefficient much smaller than those relative to the components constituting the continuous phase). This is illustrated in Figure 5.1, which clearly shows the inversion of the diffusion coefficients of water and toluene, respectively, when moving the overall composition from the water apex to the oil apex in the phase diagram.[29] The self-diffusion coefficients are of the order of, say, 10^{-9} m^2 s^{-1} for the diffusion of free single solvent molecules, and of 10^{-10} to 10^{-11} m^2 s^{-1} for the aggregates constituting the dispersed phase.

The part played by the so-called "obstruction factor" should not be ignored in the interpretation of diffusion measurements. Indeed, it has been shown[33] that in micellar systems, using D_2O as the solvent, the observed coefficient of heavy water, D_{obs}, is related to the diffusion coefficient of the micelle, D_{mic}, to the fraction of bound water P_b and to the diffusion coefficient of pure water, D_w:

$$D_{obs} = f_{ob} (1 - P_b) D_w + P_b D_{mic} \qquad (5.2)$$

where the factor f_{ob} is the obstruction factor (the micelles give rise to an excluded volume hindering the diffusion of water). A major point is that the obstruction factor is sensitive to the

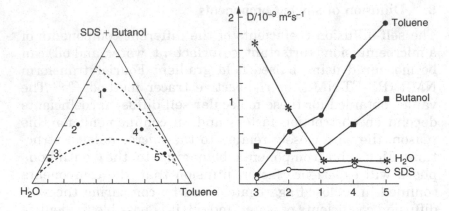

Figure 5.1 Composition diagram and measured diffusion coefficients for five different formulations in the system SDS (1/3)–n–butanol (2/3)/toluene/water. Reprinted from Reference 29, copyright 1981, with permission from Elsevier.

aggregate shape. By studying the change of the obstruction factor as a function of the obstructing volume, a discrimination between spheres, oblates, and prolates becomes possible. Typical obstruction factors range from approximately 1.0 to 0.7.

In conclusion, this section has shown how closely related the structural aspects and the dynamic aspects are, as mentioned earlier.

IV. DYNAMICS OF EXCHANGE OF SURFACTANTS AND COSURFACTANTS

A. Reminder of the Case of Simple Micelles

The continuous exchange of the surfactant molecules (as well as cosurfactant molecules in case of mixed micelles) constitutes a major dynamic process in micellar systems.[34] The situation in microemulsions, although more complex, directly derives from that in simple micelles. For this reason, we will briefly recall here the main conclusions that have been established concerning the micellar dynamics, reviewed in detail in Chapter 3.

Chemical relaxation studies[34-40] have shown that in a quasi-general manner relatively dilute micellar solutions are characterized by two relaxation processes: a fast process with a relaxation time τ_1 characterizing the exchange of surfactant molecules between the micellar aggregates and the bulk aqueous phase, and a slower process, with relaxation time τ_2, characterizing the formation/breakdown of the aggregates. For a surfactant having an alkyl chain length in C_{12}, τ_1 and τ_2 are typically of the order of 1–10 µs and 1–10 ms, respectively. These values increase with the alkyl chain length.

Aniansson and Wall[41] derived the first expressions of τ_1 and τ_2, which strictly apply to nonionic surfactants. Later theoretical treatments accounted for the presence of counterions in the case of ionic surfactants,[42] and for the contribution of reversible coagulation-fragmentation reactions of micelles to the slow relaxation process.[43,44] This last contribution, predominant at high surfactant concentration, can be represented by the reaction[44]

$$S_k + S_j \rightleftharpoons S_s \qquad (5.3)$$

where S_k and S_j are two submicellar aggregates and S_s a micelle of aggregation number s.

The effect of alcohol on the dynamic properties of micellar systems has been considered as a first approach toward the understanding of microemulsion systems. In mixed alcohol + surfactant micelles, the theory[45] predicts the existence of three relaxation processes, which have been experimentally observed using chemical relaxation techniques:[46] a slow process associated with the formation/breakdown of mixed micelles and two fast processes associated with the exchange of the surfactant and alcohol, respectively, between the mixed micelles and the bulk aqueous phase. With $M_{a,s}$ representing a mixed micelle with a alcohol (A) molecules and s surfactant (S) molecules, these two exchange reactions can be written in the form

$$M_{a-1,s} + A \rightleftharpoons M_{a,s} \quad \text{and} \quad M_{a,s-1} + S \rightleftharpoons M_{a,s} \quad (5.4)$$

Information concerning the rate constants associated with these equilibria is available in the literature. For instance, the three characteristic relaxation times of the tetradecyltrimethylammonium bromide (TTAB)/pentanol system have been measured using T-jump, p-jump (shock-tube), and ultrasonic absorption. The results showed that (1) the rate of association of alcohol molecules to TTAB micelles is diffusion-controlled and the residence time of pentanol in the mixed micelles is of about 50 ns; (2) the addition of alcohol reduces the rate constant for the dissociation of a surfactant from a mixed micelle; and (3) the lability of the micellar structure is increased in the presence of alcohol.[46]

B. Exchange Dynamics in Microemulsions

Microemulsions often contain relatively high surfactant concentrations, which allow the dissolution of appreciable amounts of alcohol and oil. The exchange dynamics in these complex systems, which involve four components in a range of concentrations outside the usual diluted domain considered so far, have been essentially studied in the groups of Zana and Lang[47–49] and of Hoffmann.[50] Their results have shown that the addition of alcohols (cosurfactants) and oils to pure micellar systems can change considerably the rate of the exchange processes, although the mechanisms involved remain of the same nature. Their studies also concerned the influence of the additives on the structural characteristics of the micelles (the word being taken in its wide sense and including microemulsion droplets) and the correlations between dynamic behavior and structural evolution (micelle shape and size, surfactant aggregation number, polydispersity, location of the solubilized species in the micelle, for instance).[49–51]

In a first approach, Lang et al. have studied the ultrasonic relaxation of microemulsions whose compositions were chosen so as to follow specific pathways in the ternary or pseudoternary phase diagrams.[47,48] Many different systems were studied, varying the nature of the components. The ultrasonic relaxation spectra were measured in the range 1 to 156 MHz, showing the existence of one or two relaxation processes:

$$\frac{\alpha}{f^2} = \frac{A_1}{1+(f/f_{R1})^2} + \frac{A_2}{1+(f/f_{R2})^2} + B \qquad (5.5)$$

In Equation 5.5 α/f^2 is the ultrasonic absorption (α = absorption coefficient in cm^{-1}; f = frequency in Hz), B is a constant, A_1 and A_2 are the relaxation amplitudes, and f_{R1} and f_{R2} the relaxation frequencies. The latter are related to the relaxation times τ by

$$\tau_{1, 2} = (2\pi f_{R1, 2})^{-1} \tag{5.6}$$

The two relaxation processes observed in O/W microemulsions were attributed to the exchange of the surfactant ion and of the alcohol between the interfacial film and the water-rich phase. It was found that in the water/sodium dodecylsulfate (SDS)/1-butanol system, the exchange of the alcohol (short-chain alcohol in that case) is very fast, with an exchange rate of about 10^8 s^{-1}. Upon addition of toluene to the preceding ternary system, the relaxation frequencies decreased and the low frequency relaxation process, associated with the surfactant exchange, vanished when toluene became the continuous phase. The disappearance of the surfactant exchange process when going from O/W to W/O microemulsions can be explained by the fact that the surfactant is insoluble in the oil phase. This observation has suggested the possibility of monitoring the W/O to O/W microemulsion transition by means of ultrasonic absorption, i.e., by simply following the change of the dynamic behavior. On the other hand, for solubility reasons, the exchange of the alcohol can occur between the interfacial film and the oil phase, but also between this film and the dispersed aqueous phase. In fact, it is believed that the two exchange equilibria involving the alcohol occur at comparable rates and are detected as a single relaxation process. Lang et al.[48] have attempted to explain the decrease of both f_{R1} and f_{R2} upon addition of cyclohexane to the ternary system water/SDS/1-pentanol, in the O/W domain, on the basis of Aniansson's theory, using the expression derived for the surfactant exchange process in simple micelles:

$$2\pi f_R = \frac{k^-}{N} \frac{C_M}{C_{free}} + \frac{k^-}{\sigma^2} \tag{5.7}$$

where N is the number of exchanging species per aggregate, σ is the standard deviation of N, k^- is the rate constant for the dissociation of one exchanging species from the aggregate, and C_{free} and C_M are the concentrations of free and aggregated species, respectively. Upon addition of oil, N and thus σ increase and C_{free} decreases. Remembering that $C_{free} \approx k^-/k^+$ (k^+ is the diffusion-controlled association rate constant of an exchanging species to an aggregate), k^- is also decreased and thus f_R is expected to progressively decrease, as experimentally observed for both the surfactant and alcohol.

In a second approach, chemical relaxation techniques were used to follow the kinetic changes associated to the process of micelle formation-breakdown (relaxation time τ_2), when the composition of the system was changed continuously from micelles to microemulsions.[49,50] Starting from concentrated micellar solutions of ionic surfactants (tetradecyl and hexadecylpyridinium chloride (TPyC, HPyC); hexadecyltrimethylammonium chloride (HTAC); and dodecyl and tetradecyltrimethylammonium salicylate (DTASal and TTASal), different alcohols and/or oils were added, finally leading to the formation of O/W microemulsions. Figures 5.2a and b and Figure 5.3 show the variation of $1/\tau_2$ upon additions of alcohols or oils of different chain lengths to 0.3 M HPyC solutions and to 0.3 M HPyC + 0.2 M 1-pentanol solutions.[49] Very similar observations have been reported for mixtures of sodium salicylate and alkyltrimethylammonium bromide.[50]

As shown in Figure 5.2 and Figure 5.3, the changes of τ_2 are very sensitive to the nature, chain length, and concentration of both the added alcohol and oil. Addition of 1-pentanol to HPyC can increase $1/\tau_2$ by a factor as large as 10^4. The increase of $1/\tau_2$ upon alcohol addition appears to be quite general, at least in the low concentration range. For alcohols longer than 1-pentanol, a more complex behavior is observed, with $1/\tau_2$ going through a maximum. Remembering that the average micelle lifetime T_M increases with τ_2,[52] the results of Figures 5.2a and b indicate a decrease of T_M when mixed surfactant/alcohol micelles are formed. The specific behavior of long-chain alcohols has been explained by the fact that at low concentrations they are mainly solubilized in the palisade layer with their polar heads in contact with the aqueous phase, whereas at high concentrations they also solubilize in the micellar core. The variation of $1/\tau_2$ then becomes similar

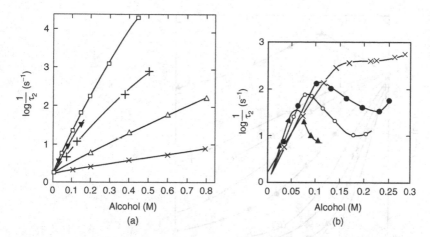

Figure 5.2 Effect of the concentration of added alcohol on $1/\tau_2$ for a 0.3 M HPyC solution at $T_f = 25°C$. (a): ($\times$) ethanol; ($\triangle$) 1-propanol; (+) 1-butanol; ($\blacktriangledown$) benzylalcohol; ($\square$) 1-pentanol; (b): ($\times$) 1-hexanol; ($\bullet$) 1-heptanol; ($\circ$) 1-octanol; ($\blacktriangle$) 1-decanol. Reprinted with permission from Reference 49, copyright 1986, American Chemical Society.

to that observed in Figure 5.3 for oil additions. Thus the solubilization in the micelle palisade increases the dynamics of the system, whereas solubilization in the micellar interior stabilizes the droplets by decreasing their dynamics.

Lang and Zana[49] have attempted to interpret qualitatively their data on the basis of the theoretical treatment of micellar solutions proposed by Kahlweit et al.,[44,53] which accounts for the contribution of reversible coagulation/fragmentation reactions to the process of micelle breakdown-formation. A careful examination of the different parameters led the authors to conclude that the average dissociation rate constant must largely contribute to the observed changes of $1/\tau_2$, which are also related to the variations of concentration of the species around the minimum of the size distribution curve of the aggregates (see Chapter 3, Section III. C). This means that the micelle breakdown-formation is affected by variations of polydispersity of the system, which were evidenced by quasi-elastic light-scattering experiments. The solubilization of alcohols and their penetration into the micelles also modify the aggregation number, the ionization degree, or the shape of the particle, all parameters that affect the value of τ_2 in a complex manner.

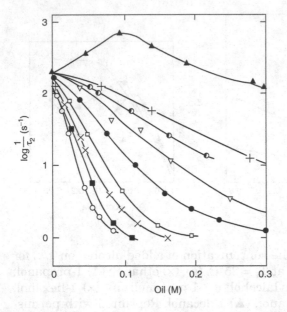

Figure 5.3 Effect of the concentration of added oil to the mixed micellar solution 0.3 M HPyC + 0.2 M 1-pentanol at 25°C: (▽) n-hexane; (●) n-heptane; (□)n-octane; (×) n-nonane; (■) n-decane; (○)n-dodecane; (+) cyclohexane; (▲) toluene; (◐) n-butylbenzene. Reprinted with permission from Reference 49, copyright 1986, American Chemical Society.

Hoffmann et al.[50] have shown that the changes of τ_2 upon addition of alcohol and/or oil are strongly correlated with the structural modifications of the aggregates, using for this purpose dynamic light scattering, electric birefringence, and viscosity measurements. Their results were especially demonstrative since, in this case, the initial solution (alkyltrimethylammonium bromide/sodium salicylate) contained rodlike micelles. The different techniques used consistently suggested the transformation of the rodlike micelles into globular aggregates upon addition of short-chain alcohols, which is in line with the strong acceleration observed for the reequilibration process. Besides, the addition of alicyclic hydrocarbons (toluene and cyclohexane) and long-chain alcohols was shown to have little effect on the size of the rods, whereas the addition of alkanes like decane, in the presence of pentanol, resulted in rodlike micelles being transformed back to globular aggregates. All these observations give substantial support to understand the complex variations of $1/\tau_2$. The authors have also drawn interesting information from the comparison of the value of τ_2 obtained from chemical relaxation techniques, with the value of the correlation time τ_c (related to the size and shape of the particles, through the diffusion coefficient) obtained from dynamic light scattering.

V.   AMPHIPHILE CHAIN DYNAMICS; WATER AND COUNTERIONS MOBILITY

A.   Motion of Hydrophobic or Hydrophilic Chains of Surfactant Molecules

The local motion of the alkyl chains of surfactant molecules is extremely fast, as indicated by longitudinal ^{13}C NMR relaxation time (T_1) measurements in microemulsions. It has been demonstrated that these internal motions are only slightly influenced by aggregation and by changes between different aggregate geometries.[24] The fact that cis-trans reorganizations in micellar aggregates are extremely rapid is in agreement with the liquid-like nature of the interior of micelles. The time scale of these fast motions is in the order of picoseconds.[24,54] In the L_2 domain of the sodium octanoate/octanoic acid/water system at different water contents, it has been shown that the mobility of the alkyl chains increase in going from the surface to the interior of the aggregates, the region around the head group of the surfactant being the least mobile.[24] The fast correlation time is relatively insensitive to the amount of water present in the system. Similar conclusions have been obtained for other three- or four-component systems of the W/O type.

The same approach has been used to gain some insight into the motions of the hydrophilic chains of nonionic surfactants. For instance, we have looked[55] at the ^{13}C NMR relaxation times of the carbon atoms constituting the hydrophilic moiety of tetraethyleneglycoldodecylether $(C_{12}EO_4)$ in decane/water/1-hexanol/$C_{12}EO_4$ W/O microemulsions. The values of T_1 for the carbon atoms along the hydrophilic chain were measured, leading to values between 29 to 68 ps for the effective rotational correlation time of the C-H vector considered, depending on carbon number and water content. The results were analyzed along the lines of the model described earlier (see Section III). It was concluded that the chain mobility goes through a minimum when increasing the water content. At low water content, the hydrophilic tails of the surfactant molecules can move freely. When more water is added, they lose some freedom due to hydrogen-bonding. When one exceeds the amount of water necessary for the simple hydration of the ethylene oxide (EO) units (about 2.5 water molecules per EO group), the hydrophilic tails recover a certain mobility in the resulting water pool.

It is more difficult to obtain information concerning local motions in cosurfactant molecules (usually short-chain alcohols) due to their very fast exchange rate between the different pseudo-phases.

B. Mobility of Confined Water

Many techniques have been used to characterize the state of water and the dynamics of water molecules in W/O microemulsions.[56,57] It is usual in that case to consider the molar concentration ratio w_o = [water]/[surfactant]. All the techniques used demonstrate that for low values of w_o, the water molecules are essentially bound to the counterions and to the surfactant polar heads. It is only when w_o exceeds a certain value (around 6–10 for AOT-based systems for instance) that free water can be detected. Properties such as the local microviscosity,[58] or the local dielectric constant,[59] are strongly affected by the state of water.

The mobility of water molecules confined in W/O microemulsions can be significantly reduced in two different situations: (1) when w_o is small enough so that only bound water exists, and (2) when water molecules are situated near the internal wall of the droplets, independently of the value of w_o. However in this last case fast exchange between bound and free water is known to take place,[57] with characteristic times of the same orders of magnitude as those measured for simple ionic solutions. The residence time of a water molecule around an ion strongly depends on its nature. For alkali metal ions, this time is about 1 ns.[60]

Again NMR spectroscopy is a choice technique for evaluating the water mobility. T_1 relaxation measurements have been used to evaluate the rotational correlation time τ_c of water molecules in situation of confinement. The values of τ_c, as a function of the water content, in $AOT/H_2O/C_7D_{16}$ reverse micelles, demonstrate that the water molecules are highly immobilized in small water pools due to strong ion-dipole interactions: τ_c is about 70 times longer than in bulk water (3 ps) when w_o is kept very low.[57] Upon increasing w_o, τ_c decreases and finally approaches the value measured in pure water, when the amount of bound water becomes negligible relatively to the quantity of free water.

C. Mobility of Counterions

It is well known that in charged colloidal systems (micelles, microemulsions, polyelectrolytes, etc.), a fraction of counterions are dissociated whereas another fraction is more or less strongly associated with the colloidal particles. Here again exchange processes are believed to occur between free and bound counterions or between different binding sites on the surface of the particles.[57] In microemulsions, not many quantitative data are available concerning the rate of such processes, which should depend on the strength of the electrostatic interactions. For monovalent ions like Na^+ (in the case of the anionic surfactant AOT) or I^- (in the case of the cationic surfactant cetylpyridinium iodide), some indications of the characteristic times involved have been reported. The temperature dependence of the resonance line width of ^{23}Na indicated that the exchange of Na^+ counterions is fast on the NMR time scale (i.e., well below 0.1 ms).[57] On the other hand, the residence time of I^-, evaluated from electric field jump experiments, was found to be in the order of 0.1 µs.[61]

VI. PERCOLATION PHENOMENA AND KINETICS OF DROPLET COALESCENCE

A. The Percolation Concept and Film Rigidity

Conductivity percolation is a phenomenon that has been reported in hundreds of studies of microemulsions, especially of the W/O type. It refers to a sudden increase of the conductivity of the system by orders of magnitude once a specific threshold is passed,[62-67] which means that the ion transport properties are drastically modified.[68] Such changes are usually induced by varying the temperature of the system or its water content. The salinity of the aqueous phase and/or the presence of additives also play a major role in favoring or, conversely, in making more difficult the percolation of conductivity (see below).

Percolation of a population of hard spheres occurs when their volume fraction reaches about 15%.[62] Statistically, at least one continuous path (like a string of beads) will exist above this threshold. This is based on a purely geometrical effect, which ignores the possibility of having either attractive or repulsive interactions between the particles, as is obviously the case in microemulsion systems. The dynamic properties of the amphiphilic film separating oil from water is thus an important parameter to take into account, and depending on whether it is rigid or fluid it will facilitate or make more difficult the transfer of charges between droplets and give rise to very different behaviors.[69,70] This means that the percolation threshold may considerably vary from that predicted from the simple hard sphere model, the situation being referred to as "dynamic percolation," in contrast to static percolation.[71]

The variation of conductivity K can be represented by two separate asymptotic power laws corresponding to systems above or below the percolation threshold[64,69,72]:

$$K = k \, (\Phi_c - \Phi)^{-s} \text{ when } \Phi < \Phi_c \text{ and}$$
$$K = k' \, (\Phi - \Phi_c)^{\mu} \text{ when } \Phi > \Phi_c \tag{5.8}$$

where k and k' are constants related to the conductance of the dispersed phase; Φ is the volume fraction of the dispersed phase, and Φ_c is the percolation threshold.

Similar equations apply for temperature percolation at constant composition:

$$K = p \, (T_c - T)^{-s} \text{ when } T < T_c \text{ and}$$
$$K = p' \, (T - T_c)^{\mu} \text{ when } T > T_c \tag{5.9}$$

where p and p' are constants, T the temperature of the system, and T_c the critical temperature.

Two different theoretical approaches have been proposed for the mechanisms involved in percolation[72]: (1) a static percolation approach, in which the sharp increase of conductivity is attributed to the appearance of long water channels, with exponents $\mu \approx 1.9$ and $S \approx 0.6$, and (2) a dynamic percolation approach, in which attractive interactions between droplets permit a hopping of charges across the surfactant layers; the critical exponents in that case are $\mu \approx 1.9$ but $S \approx 1.2$ (in fact, the experimental values of S range between 1 and 1.6).

The effect of globule dynamics on the conductivity in microemulsions was considered in the "stirred percolation" model.[69] According to this model, the conductive droplets that move with a diffusion coefficient D are assumed to interact through a hard-sphere type potential. A cluster existing at a time t will disappear after the time $t + \tau_R$ and rearrange in different clusters due to Brownian motion. Three different situations have been distinguished, depending on the value of the reordering time τ_R (which is in the order of R^2/D, where R is the droplet radius) and of the typical times of charge migration in and between the clusters. This concept has been shown to be consistent with the measured conductivity of microemulsions for $\Phi < \Phi_c$.

Monte Carlo simulations[71] led to the conclusion that the interaction strength, which can be modified either by temperature or globule size, shifts the value of Φ_c, whereas the dynamical effects are responsible for the change of the exponent. The reality of the two possible mechanisms (charge hopping and bicontinuity) when $\Phi > \Phi_c$ is considered in Section VI.E.

B.   Effect of Temperature

The effects of volume fraction of the dispersed phase and of temperature on the conductivity behavior are intimately related. An increase of temperature can induce a conductivity percolation in a situation where Φ is not high enough for percolation to occur.[63,64] This is clearly demonstrated in a study of AOT/decane/aqueous 0.5% NaCl system.[63] The authors pointed out the symmetrical roles of oil and water relatively to percolation processes: using viscosity and electric birefringence measurements, they were able to show that the threshold value of the oil volume fraction in water-continuous systems is very similar to the corresponding value for water in oil-continuous systems.

It is generally believed that the increase of temperature by accelerating the droplet Brownian motion facilitates the transport of charges, which results in an increase of the conductance. This is true for ionic surfactants. However, opposite behaviors have been reported, especially with nonionic surfactants, where a temperature increase is associated with a decrease of conductivity.[73,74] The word "antipercolation" has been used to refer to the increase of conductivity associated with a decreasing water content.[75]

C. Effect of Salinity

An increase of salinity can produce an O/W to W/O microemulsion transition through the formation of bicontinuous structures (see Section II). It is therefore no surprise that a system can go through a percolation threshold by changing the salinity. The effects associated with addition of electrolytes are manifold: change of the Debye screening length and thus of the interactions between droplets, change of surfactant partition between the water/oil phases, and possible change of the bending rigidity of the droplet film and/or of its curvature. Increasing the amount of electrolytes added to water/AOT/alkane systems of the W/O type results in a decrease of the conductivity by several orders of magnitude.[64] The analysis of the variations of conductivity with salinity in terms of percolation scaling laws yielded exponents different from those when the concentration of globules was varied.

We have studied[76] the extent of aqueous electrolyte solubilization in AOT/decane solutions when the electrolyte concentration was increased from 0 to 0.3 M. The solubility goes through a sharp maximum, which appeared to be very well correlated with a dramatic decrease of the conductivity measured near the phase limit (Figure 5.4). The origin of a maximum in water solubilization has been well investigated.[77,78]

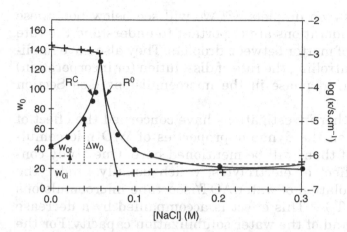

Figure 5.4 Variations of the maximum water solubilization capacity per AOT molecule (●) (left scale; w_o = [H$_2$O]/[AOT]) and conductivity measured near the phase limit (+) (right scale) with the NaCl concentration. Initial AOT/decane = 25/75 (wt/wt). T = 25°C. See Section VI.C for the meaning of R^c and R^o and Section VII.C for w_{oi}, w_{of}, Δw_o. Reprinted with permission from Reference 134, copyright 1998, American Chemical Society.

It has been demonstrated that the solubility limit is governed by two distinct phenomena: (1) at low-salt concentration the interfacial film is fluid, highly deformable, and the phase limit is governed by a critical droplet radius R^c above which demixing occurs; the high conductivity indicates strong interactions between these droplets and a percolation behavior can indeed be observed when a titration with an aqueous electrolyte is performed; and (2) when the salt concentration is above the solubilization maximum, the phase limit is governed by the radius of spontaneous curvature R^o; the droplets behave like hard spheres and when they reach this radius an excess water phase appears (Winsor II system); the system in that case is no longer conducting and no percolation can be observed upon aqueous electrolyte titration. The capacity of water solubilization is thus limited by either R^c or R^o depending on their respective values, with important consequences regarding droplet interactions, as indicated by conductivity measurements. R^c and R^o are dependent on parameters such as the rigidity constant of the amphiphilic film or the interaction potential between droplets.[17,79] We will see below how these types of considerations are important to understand the rate of exchange of matter between droplets. They also have implications in controlling the rate of dissolution (or incorporation) of the dispersed phase in the microemulsions (see Section VII.C).

Many other investigations have concerned the effect of electrolytes on the dynamic properties of W/O microemulsions. Two of them will be mentioned here. One study concerns the effect of electrolytes, which usually hinder the electric percolation of water/AOT/isooctane microemulsions (increase of T_c).[80] This effect is accompanied by a decrease of viscosity and of the water solubilization capacity. For the alkali metal chloride series, the relative efficacies of cations in increasing T_c were in the order $Li^+ < Na^+ \leq K^+ < Rb^+ < Cs^+$, i.e., the order predicted by the radii of the hydrated cations. By contrast, no clear trend was detected when comparing the effects of various salts sharing a common cation. This means that the larger the association of cations to the interface, the greater the shift of percolation threshold. According to the authors, the association of the cations to the surfactant polar heads results in an increase of the natural negative curvature of the surfactant film; it increases the interfacial rigidity, decreases attractive interactions between droplets, and thus hinders the exchange of matter between droplets. Assuming the passage of cations through transient channels formed during droplet collisions, and considering that these channels constitute regions of positive surfactant film curvature, the effect of added cations will make more difficult the formation of these channels. A second study involved conductance measurements in the water/AOT/heptane system in the presence of different added electrolytes: $NaCl$, Na_2SO_4, Na_3PO_4, $CaCl_2$, $BaCl_2$, $(C_2H_5)_4NBr$, potassium phthalate.[81] In all cases the authors observed a shift of the percolation threshold toward higher droplet volume fractions compared to the system with no additive. They tried to describe the conductivity curves using a classical scaling law (see Section VI.A) and found exponents always close to 1, in the presence and absence of salt.

D. Effect of Additives

The effect of additives other than simple electrolytes on the percolation behavior of W/O microemulsions has been much investigated recently. Significant efforts have been made to understand how they can influence the stability, film dynamics, droplet coalescence ability, etc. Depending on the type of additive considered and on its effects on the film dynamics, its presence could either facilitate or hinder the conductance percolation by decreasing or, in contrast, by increasing the threshold values, respectively. Two classes of compounds are of special interest in this regard[70,82]: those that increase membrane permeability and those that increase membrane rigidity. Additives of the first category, like propanol or gramicidin, brought about a significant lowering of the threshold temperature of water/AOT/isooctane systems, whereas cholesterol, which belongs to the second category, shifted the percolation threshold to a higher temperature.[70] Increasing the molar ratio of cholesterol to AOT up to 0.3 resulted in a complete disappearance of conductivity percolation in the studied temperature range. Trace amounts (0.005 molar ratio with respect to AOT) of gramicidin, a well-known channel-forming compound, were enough to induce a dramatic decrease of the percolation threshold. These results were interpreted as clearly indicating that the percolative conduction does not take place through the formation of a continuous open structure but rather through reversible and highly dynamic droplet aggregation processes.

The effect of urea has been much investigated because urea can modify the properties of compartmentalized water.[81,83] Added to water/AOT/heptane system at low concentration (0.01 M), it hardly changed the conductivity variation as a function of the droplet volume fraction.[81] However, when urea was added to water/AOT/hexane reverse micelles at concentrations up to 10 M, the percolation threshold was shifted to very low water volume fractions Φ_w: 0.05 and 0.02 for urea concentrations 3 M and 5 M, respectively.[83] Urea is assumed to preferentially bind at the micelle/water interface, thus inducing a counterion dissociation from surfactant head groups. This effect, which is analogous to the effect of short-chain alcohols, is associated with an increase of membrane fluidity and permeability (long-chain alcohols having an opposite effect as discussed in Section IV.B).

The solubilization of hydrosoluble polymers, like polyethyleneglycol (PEG), in the water pools of reverse micelles is known to produce a decrease of the interdroplet attractive interactions, at least when the polymer chains are smaller than the water droplets.[84-88] Addition of PEG (molecular weight 2000 or 10,000 at maximum) to water/AOT/n-decane systems shifts the threshold of conductivity percolation to higher water volume fractions, as expected, because of the polymer/surfactant stabilizing interactions. Similar observations were reported for water/AOT/isooctane reverse micelles in the presence of triblock copolymers[87] or of PEG of varied molecular weights.[86]

The temperature-induced percolation of W/O microemulsions (water/AOT/isooctane and water/AOT/decane) in the presence of different "hydrotropes" has been investigated by Hait et al.[89,90] who have attempted to propose schematic models to help achieve understanding of how the processes of droplet fusion, mass transfer, and droplet separation can be affected by the presence of these additives, depending on their specific nature. When the added compound helps bridge the droplets, their fusion becomes easier and the presence of additive assists the conductivity percolation process. Other additives are hindering contact between droplets, which in that case resist fusion to some extent. At a fixed $[H_2O]/[AOT]$ molar ratio $w_o = 25$ and fixed additive concentration (10 mM), the values of the temperature at the percolation threshold T_c follow the order sodium cholate < hydroquinone < pyrogallol < resorcinol < catechol < sodium salicylate. Hydrotropes with one hydroxyl group retard percolation whereas those having more than one hydroxyl group facilitate it.

E. Droplet Percolation and Bicontinuity

As already mentioned earlier, the mechanism at the basis of the conductivity percolation has been much debated and the high conductivity measured at and above the percolation threshold has been explained in two different manners.[64,72]

For some authors, it is the result of the clustering of water droplets (which can maintain their closed shell structure) and of the hopping of the charge carriers across the surfactant layers; for others, the formation of open structures (water channels) leads to a bicontinuous organization of water and oil. In all likelihood there is no unique answer, but one should rather expect different possible explanations depending on the characteristics and specific properties of the system under consideration.

In support of this conclusion, we will refer to two studies that aimed at distinguishing between the existence of closed or open structures. Maitra et al.[91] have compared the variations of the electrical conductivity and of the self–diffusion coefficients of water with temperature in AOT/water/isooctane microemulsions. They observed that the relative self-diffusion coefficient of water, unlike electrical conductivity, does not show a sudden increase at the percolation threshold temperature. This result clearly shows that although the connectivity of the droplets is developed enough to permit the transport of charges, the translational motions of water is very much restricted and remains similar to that in the pre-percolation region. The authors concluded that hopping of charge carriers is the prevailing mechanism in this system. Meier[92] has investigated the behavior of the system 1 mM NaCl/pentaethyleneglycol monotetradecylether ($C_{14}EO_5$)/n-octane, when the oil volume fraction, Φ_o, was varied. The clustering of droplets is directly reflected by a remarkable increase of viscosity, which shows a sigmoidal variation with an inflection point at $\Phi_o \approx$ 0.85. In spite of this percolation, the conductivity remains low in this region. It starts increasing at $\Phi_o < 0.5$. This value coincides with the phase transformation of the nanodroplet microemulsion into a lamellar L_α phase. Similar trends were observed by varying the temperature in place of Φ_o. The author concluded that the percolation transition of a W/O microemulsion is not necessarily reflected by an increased conductivity. These two examples show that one should be cautious when interpreting conductivity data and that comparison with results obtained from other techniques is often required to correctly analyze the experiments. The next subsection will show how the control of the exchange of matter between droplets constitutes an appropriate means to further approach the problem of droplet coalescence.

F. Kinetics of Droplet Coalescence and Exchange of Matter between Droplets

A large number of studies have demonstrated that exchange of matter can take place when droplets collide independently of the occurrence of percolation.[93-111] The latter is obviously expected to facilitate the exchanges that will be easier in a cluster, but percolation is not a requirement. Efforts have concentrated on the understanding of the mechanism of these processes using probe molecules and fast kinetic techniques. The exchange of matter confined within dispersed water droplets involves several steps: collision of droplets, transient droplet coalescence, and separation. Depending on the droplet membrane dynamics, the whole process will be, of course, more or less efficient. The knowledge of the rate-limiting step is an important information, which should tell us if the merging of the droplets (with formation of transient water channels in the case of W/O microemulsions) can instantaneously follow their collision or if, in contrast, there is a resistance to coalescence.

Menger et al.[93] in 1973 reported results suggesting that W/O microemulsion droplets are able to communicate through coalescence followed by rapid separation. Starting with two separate AOT/octane/water W/O microemulsions, one containing imidazole and the other containing p-nitrophenyl-p'-guanidinobenzoate hydrochloride, hydrolysis was observed to take place at a significant rate upon mixing. This meant that a rapid exchange of the droplet content had occurred in one way or another. Similar observations were reported by Eicke et al.,[94,95] giving further support to this interpretation. Abe et al.[96] have used a submicron particle analyzer to demonstrate that exchange of water takes place between large and small size W/O microemulsion droplets of an AOT/water/cyclohexane system, even without any stirring. The initial bimodal droplet size distribution curves were progressively replaced by unimodal curves. This was taken by the authors as a proof of water exchange through droplet coalescence.

In order to evaluate the rate of droplet coalescence, a common approach used by several groups[99–111] has consisted of the use of model reactions, as schematically represented in Figure 5.5. A and B are two water-soluble reagents, and their rate of reaction to give the product C must be known in homogeneous aqueous solutions. When the two reagents are initially separated in different water pools, two different situations may be produced after droplet collision and merging, depending on whether the whole process is reaction-controlled (the reaction inside the droplet takes place at its proper rate, with possibly some specific effects due to the confined environment) or merging-controlled (the reaction takes place instantaneously as soon as the droplets have coalesced and the fusion of the droplet membranes constitutes the rate-limiting step). Note that the collision, instead of involving the total droplets, may also involve probe-containing subaggregates if fragmentation of the droplets occurs (see Section IV.B), but this constitutes only a small variation of the scheme depicted here, which remains globally valid, although the kinetic equations will not be the same. More details can be found in References 97 and 98.

If a very rapid reaction is selected, with a possibility to easily follow the formation of the reaction products, it is expected that the measured rate will be that of the droplet merging process. Fletcher et al.[99–101] have used for this purpose three types of reactions: proton transfer, metal-ligand complexation, and electron transfer, all with second-order rate constants larger than at least 2×10^7 $M^{-1}s^{-1}$ and approaching the diffusion-controlled limit ($> 10^{10}$ $M^{-1}s^{-1}$) for the proton transfer. These reactions were studied in AOT/water/alkane systems with the stopped-flow method. Mathematical treatments of the kinetic data for all three reactions[99] led to very similar values of the exchange rate constants k_{ex}. For the AOT/water/heptane system, the k_{ex} values ranged between 1.0×10^6 and 14×10^6 $M^{-1}s^{-1}$ depending on the values of w_o and temperature. k_{ex} decreased when increasing w_o or when decreasing temperature or alkane chain length. The exchange rates of solubilizates between water droplets are thus two to four orders of magnitude slower than the droplet encounter

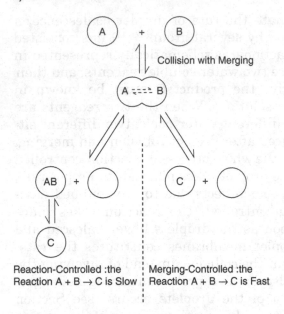

Reaction-Controlled :the | Merging-Controlled :the
Reaction A + B → C is Slow | Reaction A + B → C is Fast

Figure 5.5 Schematic representation of the limiting situations for the kinetics of reaction between A and B reagents in microemulsion droplets. Adapted with permission from Reference 97, copyright CRC Press, Boca Raton, FL, and from Reference 102 by permission of the Royal Society of Chemistry.

rate as predicted from simple diffusion theory, which implies that there is a free energy barrier to the exchange process. However, no energy barrier was found to slow down the exchange of material between water pools when using a much slower reaction, the Ni^{2+}/Murexide complexation.[102] Table 5.2 lists values of k_{ex} reported for various systems. Atik and Thomas[103] had previously reported similar observations using pulsed laser photolysis techniques and fluorescence life-time measurements of pyrenetetrasulfonic acid upon addition of quenchers such as Cu^{2+} or the nitroxyl radical $(KSO_3)_2NO$. Analysis of the decay kinetics in the presence of different additives enabled the determination of the exchange rate of solubilizates.

TABLE 5.2   Examples of Values of Rate Constants k_{ex} Associated with the Exchange of Material upon Collisions Between Droplets and Influence of Important Parameters

Parameter	System	w_o	T (°C)	Reaction	$10^{-6} \times k_{ex}$ ($M^{-1}s^{-1}$)	Reference
Droplet size[a]	AOT/heptane/H_2O	10	15	Electron transfer	4.9 ± 0.3	
		20	15	between $IrCl_6^{2-}$ and	3.5 ± 0.3	99
		30	15	$Fe(CN)_6^{4-}$	2.7 ± 0.1	
Temperature	AOT/heptane/H_2O	20	5	Electron transfer	1.0 ± 0.1	
		20	10	between $IrCl_6^{2-}$ and	2.0 ± 0.2	
		20	15	$Fe(CN)_6^{4-}$	3.5 ± 0.3	99
		20	20		7.5 ± 1.5	
		20	25		14 ± 2	
Oil chain length	AOT/alkane/H_2O (0.1 M AOT in alkane)			Electron transfer between $IrCl_6^{2-}$ and $Fe(CN)_6^{4-}$		
	Alkane = hexane	20	5		0.66 ± 0.04	99
	octane	20	5		1.7 ± 0.1	
	decane	20	5		4.1 ± 0.7	
	dodecane	20	5		7.0 ± 0.5	

Surfactant chain length	AlkylBDMAC/H$_2$O/ClBz[b]				
	Alkyl = dodecyl	20	20	Quenching of fluorescence of Ru tris(bipyridyl) ion by methyl viologen	5200[c]	107
	tetradecyl	20	20		1000[d]	
	hexadecyl	20	20		200	
	octadecyl	20	20		100	
Additives	AOT/heptane/H$_2$O/additive					
	No additive	11	25 ?[e]	Quenching of the triplet emission of pyrene tetrasulfonic acid by (KSO$_3$)$_2$NO	13	103
	0.3 M hexanol	11	25 ?		7.5	
	0.3 M benzylalcohol	11	25 ?		330	
	40% benzene	11	25 ?		2.5	

[a] Droplet size in the AOT/heptane/H$_2$O system[99] : r (nm) = 0.175 w$_0$ + 1.5

[b] AlkylBDMAC = Alkylbenzyldimethylammonium chloride. ClBz = chlorobenzene. Total surfactant concentration = 0.27 M

[c] Above conductivity percolation threshold (CPT)

[d] Close to CPT

[e] The temperature is not explicitly indicated in the original paper

Source: Reproduced with permissions from the Royal Society of Chemistry and the American Chemical Society

Bommarius et al.[104] have studied solubilizate exchange in the system dodecyltrimethylammonium chloride (DTAC)/hexanol/n-heptane/water using electron transfer indicator reactions and a Continuous Flow Method with Integrating Observation (CFMIO). The exchange rate constants were obtained through a mathematical treatment of the kinetic data, which considers a population balance model with two different scenarios, depending on whether the distribution of probe molecules is assumed to be reaction-dependent or decoalescence-dependent. The values of k_{ex} for the two extreme scenarios differ only by a factor of about 2 and ranged between 10^6 and 10^7 $M^{-1}s^{-1}$. Comparison with the rate constant for molecular diffusion (3.2×10^9 $M^{-1}s^{-1}$) indicates that the fraction of successful collisions (i.e., with exchange) between droplets is approximately 0.1 to 1%. The opening of the surfactant layer upon coalescence was thus recognized as the rate-limiting step.

Time-resolved fluorescence quenching (TRFQ) has also been used by Lang, Zana, and coworkers to clarify the correlations between electrical conductivity percolation and rate of exchange of materials between droplets.[105–109] For W/O microemulsions, the ruthenium tris-(bipyridyl) ion was generally used as fluorescent probe and methylviologen ion (or $Fe(CN)_6^{3-}$) as quencher. The probe and quencher should, of course, not be soluble in the organic continuous phase. The analysis of the fluorescent decay curves obtained in the presence or absence of quencher gives access to the mean surfactant aggregation number per droplet (N), the pseudo-first order rate constant for intradroplet fluorescence quenching (k_Q), and the rate constant for exchange of material between droplets k_{ex}, in which we are interested in this section. The principle and limitations of the TRFQ method for the study of microemulsions have been discussed in detail[98] (see also Chapter 2 of this book). The effects of different parameters on the exchange rate constant — nature of surfactant (anionic or cationic), droplet size, type of oil, alcohol chain length, temperature — have been systematically studied.[105–109] The authors arrived at the same important conclusion that the electrical conductivity percolation occurs only when the value of the rate constant k_{ex} becomes larger than $(1-2) \times 10^9$ M^{-1}s^{-1}. This threshold value was found independent of the parameter, which is varied to induce percolation. These observations support the idea that, above percolation threshold, the conductivity is mainly due to the motion of counterions through water channels. A typical result[105] is shown in Figure 5.6, where the variations of the rate constant k_{ex} and of the electrical conductivity as a function of w_o ([H$_2$O]/[surfactant] ratio) have been represented for three microemulsions constituted of water/chlorobenzene/alkylbenzyldimethylammonium chloride, with alkyl = dodecyl (C$_{12}$), tetradecyl (C$_{14}$) or hexadecyl (C$_{16}$). The conductivity behavior, which is dramatically modified when changing the surfactant chain length, appears to be well correlated with the measured values of k_{ex}. The effect of addition of water-soluble polymers (PEG, poly(acrylamide), poly(vinyl alcohol)) to AOT/decane/water/alcohol W/O microemulsions was also examined as a way to modify the interdroplet interactions.[84] At a given temperature and water volume fraction, k_{ex} decreases upon PEG addition or when increasing the polymer molecular weight. These variations, which indicate a decrease of the attractive interactions, are again well correlated with the variations of the electrical conductivity.

Very large values of k_{ex} were also found for other systems: Geladé and De Schryver[110] have measured a value in the order of 10^9 M^{-1}s^{-1} in inverse micelles of dodecylammonium propionate/water/cyclohexane; Lianos et al.[111] reported k_{ex} values in the same range for a series of cationic microemulsions containing cetyltrimethylammonium bromide (CTAB). They also found that k_{ex} increases with the droplet size and with the oil chain length, and they suggested the existence of a correlation between k_{ex}, droplet interactions and critical-like character of microemulsion. When the value of k_{ex} becomes close to the droplet diffusion rate, even if still slightly lower, as was the case here, this can be taken as an indication that the opening of the interfacial film is not the rate-limiting step of the exchange process.

To summarize the preceding features we can conclude, as did Fletcher and Robinson,[101] that (1) for reactions on the nanosecond time scale droplets can be considered isolated and

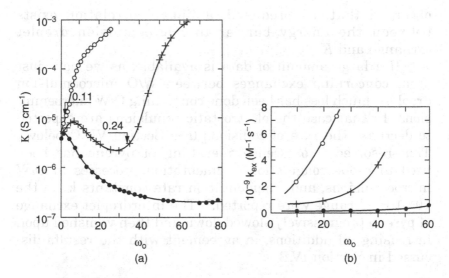

Figure 5.6 Variations of electrical conductivity K (a) and rate constant k_{ex} (b) with w_o for water/chlorobenzene/alkylbenzenedimethyl ammonium chloride microemulsions: surfactant alkyl chain C_{12} (○); C_{14} (+); C_{16} (●). Numbers on the curves indicate volume fraction of dispersed phase for onset of electrical percolation. Surfactant concentration 0.27 M; T = 20°C. Reprinted with permission from Reference 105, copyright 1989, American Chemical Society.

no exchange occurs; (2) for reactions on the microsecond to millisecond time scale, the droplet fusion can be rate limiting; and (3) for reactions taking place on longer times the communication between droplets is no longer rate limiting and the dispersed phase can be apparently regarded as if it was continuous. This discussion would not be complete without mentioning the role of the monolayer bending elasticity in controlling interdroplet exchange rates. Fletcher and Horsup[112] noted that the energy cost of forming a pore of radius r_p is of the order of $Kr_p/2\delta$ where K is the monolayer bending elasticity constant and δ the monolayer thickness. The resistance to coalescence should thus correlate with K for different surfactant monolayers. They investigated the properties of three different W/O microemulsions stabilized by polyethyleneglycol monoalkylethers nonionic surfactants and observed that, as predicted, a linear correlation exists between the energy barrier to microemulsion droplet exchange and K.

If a large amount of data is available, as we have just seen, concerning exchanges between W/O microemulsion droplets, much less has been done concerning O/W microemulsions. In that case the electrostatic repulsions are expected to decrease the rate of collisions (see Section VII.C below). Time-resolved fluorescence quenching of pyrene has been used to probe coagulation-fragmentation processes in O/W microemulsions, and fragmentation rate constants k^- in the 10^5–10^6 s^{-1} range were reported.[97] The interdroplet exchange of pyrene progressively slowed down and then vanished upon increasing oil additions, in agreement with the results discussed in Section IV.B.

G. Droplet Clustering and Rate of Exchange between Clusters

Above the percolation threshold infinite clusters are formed, with a significant droplet merging probability, as discussed in the preceding sections. Some research has focused on the early stages of the droplet clustering process (well before reaching the percolation threshold), where the droplets are expected to preserve their integrity. Guéring et al.[113] studied the benzylhexadecyldimethylammonium chloride/benzene/water system and proposed a first quantitative analysis of this clustering process by using electric birefringence. This technique proved to be a good tool for such investigations because only the anisotropic clusters are contributing to the signal. The birefringence decay could be analyzed with a sum of two exponentials with time constants in the microsecond range. The faster decay was expected to be associated with the rotational motion of droplet dimers (which was confirmed by theoretical predictions), and the slower decay was assigned to the rotational motion of the larger clusters.

Time-resolved luminescence quenching has been used to investigate the exchange rates of solubilizates in the cluster regime of AOT/water/oil systems.[114,115] Intermicellar exchange rate constants k_{cq} were found in the range 0.8×10^6 to 3.2×10^6 s^{-1}, depending on the quencher concentration and nature of the oil (dodecane, isooctane, or octane). The migration of a probe and a quencher between micelles in a cluster was approximated by a first-order quenching constant, k_{cq}. According to the authors, k_{cq} is related to k_{ex}, the exchange rate of a quencher or excited probe between any two micelles in a cluster by

$$k_{cq} = k_{ex}/(N_c - 1) \tag{5.10}$$

where N_c is the weight-average number of micelles in a cluster, and $(N_c - 1)^{-1}$ is the probability that the probe and quencher (initially in different micelles) will be in the same micelle after exchange. This results in a decrease of k_{cq} with cluster size, which was effectively observed. The authors emphasized the fact that the choice of probe is crucial and has to be adapted to the time scale of interest (different types of exchange processes will be observed with a short living probe or a long-living one). They also suggested that the exchange between different clusters may be much slower than the intracluster exchange, in the millisecond time scale compared to the microsecond time scale. Johannsson et al.[114] pointed out that the second-order rate constant used by Lang et al.[116] (see the preceding section) to characterize the exchanges between micelles *within a cluster* should rather be replaced by a first-order rate constant. Lang et al. partially agreed with this remark but argued that, even in the case of collisions between discrete droplets, the rate-limiting step remains the opening of the two surfactant layers separating two water pools. This step is characterized by a first-order rate constant k_{op}:

$$k_{op} = k_{ex} [M] \tag{5.11}$$

where $[M]$ is the droplet concentration. When considering opening of surfactant layers instead of collisions, the critical value of about 10^9 M^{-1}s^{-1} (see above) remains valid. It simply means that the critical value of the first-order rate constant is in the order of $(4 - 10) \times 10^5$ s^{-1}, taking into account the values of $[M]$.

Mays et al.[117-119] have studied droplet clustering and intercluster exchange rates in ionic and nonionic W/O microemulsions using time-resolved luminescence quenching techniques with a terbium complex $(Tb(pda)_3^{-3}$ where pda = pyridine-2, 6-dicarboxylic acid) as lumophore and bromophenol blue as quencher. This system is especially well suited for measurements in the millisecond time range. Their work summarizes the different quenching and exchange processes involving the participation of clusters and gives an indication of the time ranges characterizing them. The second-order rate constant k_{cc} for the quenching reaction between different clusters and between clusters and remaining single droplets has been obtained for different microemulsion systems from an analysis of the decay curves based on an extension of the theory proposed by Johannsson et al.,[114] introducing an additional term that accounts for the slow exchange between clusters.[117] The values of k_{cc} increase by almost one order of magnitude with the percolation transition, approaching 10^9 M^{-1} s^{-1}, in agreement with the results of Jada et al.[105] Some typical results have been collected in Table 5.3, which include nonionic microemulsions obtained with Igepal CO-520 (a nonionic amphiphile similar to $C_{12}EO_5$ with, however, a distribution of polar tail lengths).

TABLE 5.3 Values of the Rate of Intercluster Exchanges k_{cc} (Reprinted with Permission from the Work of Mays et al.,[117] copyright 1995, American Chemical Society)

System	T°C	State	k_{cc} $(M^{-1}s^{-1})$
AOT-water-n-decane	20	Nonpercolated	1.85×10^8
$w_o = 21$, $C_T = 0.1$ M	34	Percolated	7×10^8
AOT-water-n-heptane	30	Nonpercolated	4.62×10^7
$w_o = 55$, $C_T = 0.089$ M	45.5	Percolated	9.47×10^8
Igepal-water-n-hexane	28	Nonpercolated	5.1×10^7
$w_o = 24.5$, $C_T = 0.173$ M	21.5	Percolated	2.3×10^8
Igepal-water-n-decaline	23	Nonpercolated	3.0×10^7
$w_o = 24.5$, $C_T = 0.205$ M	21	Percolated	1.2×10^8

C_T = total surfactant concentration

VII. DYNAMIC PROCESSES INDUCED BY AN EXTERNAL PERTURBATION

A. Interfacial Dynamics and Droplet Deformation

In several recent papers[120-122] the dynamics of the amphiphilic film of W/O microemulsion droplets has been studied through the response of the system to perturbations caused by a sudden increase of temperature. The authors have used the iodine laser temperature-jump (ILTJ) technique,[123] where the relaxation of the system was monitored by the change of scattered light intensity with time (see Chapter 2, Section III). After the fast heating of the water pool, during the ILTJ pulse, the volume of the droplet increases due to thermal expansion, and this results in a deformation of the amphiphilic film from its equilibrium shape. The measured relaxation time is thus attributed to the disturbance and reorganization of the oil-surfactant-water interface and is inversely proportional to the bending/rigidity modulus K of the surfactant layer.[124] Alexandridis et al.[120] have deduced an expression for the characteristic time τ for the droplet shape relaxation in the form

$$\tau \propto \frac{6\pi\eta R^3}{K} \tag{5.12}$$

where η is the viscosity and R is the droplet characteristic length.

AOT/water/isooctane systems were investigated for w_0 = 5 to 50 in the temperature range 10-50°C. A single relaxation time of 2–10 µs was measured depending on experimental conditions, yielding a value of K of about 0.4 kT. This value decreased by a factor of 2 upon addition of 0.05 M NaCl (corresponding to longer relaxation times compared to those measured in the absence of added salt), which means that the interface would be less rigid. This result is in agreement with some previous theoretical predictions,[120] but it is in contradiction with the work of Hou et al.,[125] who reported that an increase of salinity, by decreasing the interfacial area per polar head of surfactant molecules, makes the interface more rigid and less penetrable. The strength of the attractive interactions was thus decreasing. However, it should be noted that the salt concentration used in Hou et al. was larger than 0.05 M. The effect of salinity on the rigidity (and thus on the bending modulus) of the surfactant film remains an open question (see also Section VI.C and VII.C).

The ILTJ technique was also used to show that the scattering signal is sensitive to the percolation threshold of the AOT W/O droplets.[121] In fact, due to the high sensitivity of the technique to the formation of aggregates, the percolation thresholds obtained in this manner are smaller than from conductivity experiments: the sign of the deviation of the optical density changes from negative to positive when going from below to above the percolation process. This is related to the fact that the relaxation below percolation threshold is due to an intramicellar perturbation, whereas above the percolation threshold it is related to the kinetics of aggregation of the reverse micelles, thus becoming an intermicellar process. The authors also pointed out that the increase of droplet concentration induces a small decrease of the bending elastic modulus.

Nazario et al.[122] have extended the preceding studies in order to see how the AOT film dynamics is affected by the presence of cosurfactants such as alkanols (hexanol or decanol) or poly(oxyethylene)monoalkylether ($C_{10}EO_4$ or $C_{10}EO_8$). The alcohols, which are supposed to be localized within the tail region of the AOT surfactant film, favored a higher curvature of this film toward the water pool; the C_mEO_n surfactants have their polar heads within the water pools and tend to curve the interface in the reverse direction. The former were shown to increase the interfacial rigidity, whereas the latter decreased it. A new relaxation process, which was previously undetected, was observed in the millisecond range, in addition to the fast relaxation already reported before (μs range). This new process was observed only above a certain water content and temperature. Its dependence on the reverse micelle concentration suggested that it is related with reverse micelle–reverse micelle interactions such as clustering and/or coalescence. Assigning this slow relaxation to the formation of droplet dimers, its relaxation time τ_2^{ILTJ} would be given by

$$1/\tau_2^{ILTJ} = 4\,k_2[M] + k_{-2} \qquad (5.13)$$

where k_2 and k_{-2} are the association (coalescence) and dissociation (decoalescence) rate constants, respectively, and [M] is the micelle concentration. The linear variation of $1/\tau_2^{ILTJ}$ with [M] was indeed verified at 17.5°C (deviations to linearity showed up at higher temperatures). Rate constants $k_2 = 3.1 \times 10^6\,M^{-1}\,s^{-1}$ and $k_{-2} = 2000\,s^{-1}$ were obtained for $w_o = 55$. The first value compares satisfactorily with the coalescence rates reported by Fletcher et al.[99] (see Table 5.2) and the second value yields a dimer lifetime on the order of 0.5 ms. This was the first direct (i.e., in the absence of any chemical reaction) measurement of the dissociation constant for the dimer formed by coalescence of two reverse micelles.

B. Temperature-Induced Change of Film Curvature

In the preceding subsection we have seen that the temperature-jump technique may be very powerful for investigating the dynamic processes taking place in the monophasic W/O microemulsion domain of ternary or quaternary systems. We would like to show here that this kind of technique can also give information on the kinetics of phase transformations that can be induced by a simple temperature change. Such experiments have been performed in our group,[126] using the Joule-heating temperature-jump technique, which, in contrast to the ILTJ technique, requires the presence of an added electrolyte (see Chapter 2, Section III. Note that temperature-jump and pressure-jump techniques had been used before to investigate the kinetics of phase transitions in binary amphiphile systems[127] or the kinetics of phase separation in fluid mixtures[128]). The reorganization of the system after a temperature-jump was followed by turbidity and/or birefringence measurements. It is advantageous that the equilibrium temperature of the system be not too far away from room temperature. For this reason the fluorinated system water/$C_8F_{17}CH=CH_2$/$C_6F_{13}CH_2(EO)_5$ was investigated. The temperature-composition phase diagram for the fixed ratio of surfactant/fluorinated oil = 0.247 is shown in Figure 5.7a. The phase limits indicated were shifted down by about 3°C by the addition of NaCl at concentration 0.1 M. The one-phase region of this phase diagram is constituted of two interconnected channels, and it shows strong similarity with the reported diagram for the water/tetradecane/$C_{12}(EO)_5$ system.[129] In the oil-rich region W/O microemulsions are present in the system, whereas in the water-rich region (not fully represented in Figure 5.7a) O/W microemulsions are formed. Transition from W/O to O/W droplets is expected in the intermediate region, where bicontinuous structures should exist. A temperature-jump of a few degrees (approximately from 1.5 to 5°C) is enough to bring the system in a new equilibrium state, as indicated by the arrows in Figure 5.7a. Three different types of experiments have been carried out.[126]

In type I experiments (path 1 → 2 in Figure 5.7a, which brings the system from an isotropic region lower channel to

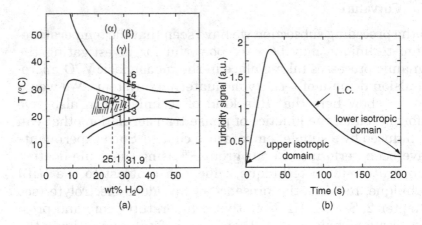

Figure 5.7 (a) Temperature composition phase diagram obtained for the water/$C_{18}F_{17}CH=CH_2$/$C_6F_{13}CH_2(EO)_5$ system when water is added to a solution of 24.7% surfactant in fluorocarbon. T-jump experiments, indicated by the arrows, were performed for the compositions α-γ. The full line delineates the isotropic domain.
(b) Change of turbidity versus time during the thermal return along path 4 → 3. Reprinted with permission from Reference 126, copyright 1988, American Chemical Society.

the liquid crystalline phase), the change of birefringence with time, following the temperature-jump, was strongly dependent on the value of the final temperature (i.e., the location of point 2 in the LC domain of Figure 5.7a), which could be chosen by adjusting the temperature-jump amplitude. The appearance of maximum birefringence could be fitted with an exponential function with a relaxation time of 25–30 ms. When turbidity was used to monitor the kinetics, instead of birefringence, conditions being otherwise similar, much faster changes were observed, with characteristic times on the order of 2.5 ms. On the whole, the formation of the liquid crystalline phase thus appears to be quite fast in that case, although different processes were obviously monitored when using turbidity or birefringence signals, the latter in fact including a turbidity component. A similar observation was made by Mayer and Woermann,[130] who have studied by T-jump 2,6-dimethylpyridine/water binary mixtures of near-critical composition. The signal of transmitted light intensity (turbidity) yielded a much shorter relaxation time than the signal of scattered-light intensity. This was explained by the fact that the turbidity caused by critical opalescence is mainly determined by the Fourier components of composition fluctuations of short wavelength, which relax faster. The important point in the two studies[126,130] is that the signal of interest (birefringence or scattering intensity, respectively) will be influenced by the change of turbidity only at short times after the temperature-jump. It should also be noted that much slower rates than those observed here have been reported for the transition between fluid isotropic amphiphile solutions and liquid crystalline phases in binary systems[128] or in ternary systems including two amphiphiles[131] (see Chapter 7).

In type II experiments (path 3 → 4 in Figure 5.7a, which brings the system from the lower isotropic channel to the higher one), a transition is induced between two isotropic solutions separated by a liquid crystalline phase.[126] If the temperature was raised at a sufficiently slow rate that the equilibrium was established all the time, we could see, using adequate methods, the formation of a liquid crystal. When a fast temperature-jump is applied so that the final temperature corresponds to point 4 shown in the phase diagram, we may have three different situations: (1) the transition requires first the buildup of the liquid crystal, which should be seen as an intermediary state; (2) the time allotted before the temperature goes back to its initial value is too short for the system to reorganize; or (3) the transition is extremely rapid and the reorganization is instantaneous compared to the time scale of the experiment. The results in Figure 5.7b support the third possibility, that is, a very fast transition (faster than the heating time of the solution, which in this case was of 15 μs). Figure 5.7b shows the change of turbidity recorded during the slow thermal return along path 4 → 3. It follows exactly the curve that has been independently obtained when varying the temperature in equilibrium conditions.[126] The fact that, on the way back, we go through the liquid crystalline phase is evidently proving that the system truly attained the structural organization characterizing point 4. By referring to the work of Olsson et al.,[129] who have used NMR self-diffusion coefficient measurements to characterize the structure of microemulsions whose temperature-composition phase diagram strongly resembled that in Figure 5.7a, the very fast structural rearrangement can be supposed to correspond to a simple inversion of the preferred curvature of the surfactant layers from a concavity toward the fluorinated oil to a concavity toward water.

Finally, in type III experiments (path 5 → 6 in Figure 5.7a, which takes the system from the upper isotropic channel to the biphasic region), our aim was to get information on the kinetics of appearance of an excess water. The kinetic curve shows an induction period because the nuclei of the new phase must attain a certain size or concentration before they can be detected by turbidity. By fitting the portion of the curves following the induction period with an exponential function, relaxation times in the range 0.25 to 0.5 s were obtained.[126] This last process is thus considerably slower than the preceding ones.

C. Rearrangements Following a Fast Change of Composition: Rate of Dissolution of Excess Dispersed Phase Below the Saturation Limit

The study of the rate of dissolution of oil in O/W microemulsions or of water in W/O microemulsions,[132–135] besides its importance in purely applied domains, may be relevant to obtain information on the interfacial dynamics of the amphiphilic film, on the mechanism of solubilization of the dispersed phase, or on the reorganization of the system upon a composition change. Stopped-flow techniques or other mixing methods with turbidity detection were used to follow the kinetics of dissolution.[132–134] Unfortunately, the absolute values of the measured rates are largely dependent on the mixing conditions (especially on the state of dispersion of the injected phase), and because of this, it is reasonable to compare only data obtained from a same technique.

The first report in this field has concerned the rates of dissolution of water and dodecane in water/sodium dodecylsulfate (SDS)/1-pentanol/dodecane microemulsions.[132] In the stopped-flow apparatus used for the kinetic measurements, one of the standard syringes was replaced by a microsyringe permitting the fast injection of very small volumes, so that the mixing volume ratios could range from 1/20 to 1/77. In a typical experiment, the mixture was highly turbid just after mixing and the turbidity progressively vanished as solubilization went along. The decay curves were fitted with a single exponential function from which k_{obs}, the first-order dissolution rate constant, was obtained. Figure 5.8a shows the lines in the pseudo-ternary phase diagram, along which experiments have been performed. The points representing the final equilibrium compositions are located by the intersection of two straight lines defined by the percent weight ratios R = oil/(oil + water), R' = water/(water + amphiphiles) and R" = amphiphile/(amphiphile + oil). The rate of dissolution was too fast to be measured when water was injected in W/O microemulsions, or when the domain of bicontinuous microemulsions (R' < 71) was approached. Figure 5.8b shows the variation of k_{obs} as a function of R, for different values of R'. The values of k_{obs} range from about 0.1 s^{-1} to > 10^2 s^{-1} (5 × 10^2 s^{-1} being the reciprocal of the estimated mixing time). They decrease when approaching the phase limit at constant R' and they increase with decreasing R', R" being fixed. Increasing the carbon number in a series of n-alkane resulted in a decrease of k_{obs}. The results have been tentatively accounted for in terms of a dissolution process based on diffusion-controlled collisions between oil (or water) droplets and

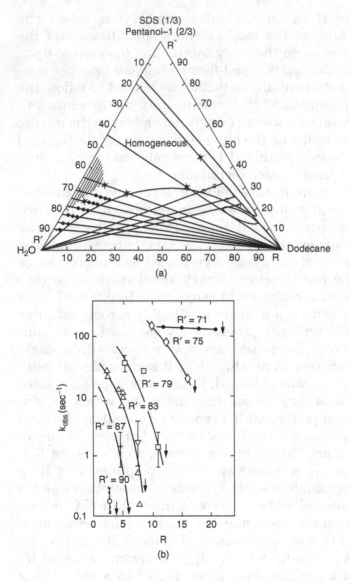

Figure 5.8 (a) Pseudo-ternary phase diagram of the system SDS (1/3)-pentanol (2/3)/dodecane/water at 20°C: (●) systems with measurable rate of dissolution; (○, ×) systems where the rate was very fast. (b) Dissolution rate k_{obs} for dodecane injection in micellar solution or microemulsions in the water-rich side of the monophasic region versus R at constant R'. See text for meaning of R and R'. Vertical arrows indicate where phase separation occurred. From Reference 132.

microemulsion droplets, which led to predicted k_{obs} values in fairly good agreement with the experimental observations.[132]

The dissolution of water in W/O microemulsions can be viewed as the collision of two neutral entities (water droplets and microemulsion droplets) at a rate controlled by diffusion, followed by rapid coalescence. Upon successive collision and coalescence steps, a redistribution of SDS and pentanol between microemulsion and water droplets is expected, leading to complete dissolution. Such a process is consistent with the droplet merging phenomena described in Section VI.F. An upper limit of the rate of dissolution could be estimated from

$$k_{obs} = k_d^\circ[M] \qquad (5.14)$$

where k_d° is the second order diffusion-controlled rate constant and [M] the concentration of microemulsion droplets. k_d° can be calculated from the Schmoluchowski equation

$$k_d^\circ = 4 \times 10^{-3}\ \pi a(D_M + D_W)N_A \qquad (5.15)$$

(a is the "reaction" distance, D_M and D_W the diffusion coefficients of the microemulsion and water droplets, N_A Avogadro's number) leading to k_{obs} values on the order of 10^5 s^{-1}. This is two orders of magnitude larger than the reciprocal mixing time, which appeared consistent with the fact that the dissolution process of water droplets could not be resolved (we will see below that the case of AOT-based systems is totally different[134]). Extending the model to the dissolution of oil in O/W microemulsions, one has to take into account the fact that the colliding particles are no longer neutral: not only the microemulsion droplets are now strongly negatively charged but in addition the free surfactant ions are expected to rapidly adsorb on the injected oil droplets, which will thus acquire negative charges. Due to the electrostatic repulsions, the diffusion-controlled rate of collision becomes

$$k_d = k_d^\circ\ X/(exp X - 1) \qquad (5.16)$$

$$\text{with } X = \frac{Z_M Z_o e^2}{\varepsilon k T a} \qquad (5.17)$$

(e is the electron elementary charge, Z_M and Z_o the charge numbers of the microemulsion and oil droplets, ε the dielectric constant). The proposed model led to k_{obs} values ranging from 0.2 to 10^3 s^{-1}, i.e., in agreement with the experimental observations.[132]

Battistel and Luisi[133] have studied the rate of formation of water pools in the isooctane/water/AOT system after injection of water, or water containing different solubilizates, in the sample cuvette of a spectrophotometer. The turbidity decay curve was found to be bimodal. For pure water solubilization, the measured rate constants k_1 and k_2 were respectively of the order of 1.4-3 s^{-1} and 1-0.2 s^{-1}, in the range of w_0 studied. The effects of adding salts, buffer, and proteins showed that only the fast process (corresponding to k_1), which accounts for more than 80% of the turbidity change, is markedly affected by the concentration of these compounds. Salts and buffers may accelerate or decrease the solubilization rate, whereas small amounts of proteins can significantly increase this rate, which was found proportional to the number of preexisting micelles. The first step was attributed to the initial uptake of water leading to random sized nonequilibrium microemulsion droplets; the second step was assumed to correspond to the readjustment of the micellar parameters through intermicellar collision and coalescence leading to the final equilibrium, associated with narrowing of the droplet size distribution. The much larger time constants observed in these experiments compared to those reported for the previous system are obviously related to the experimental conditions, where, instead of fast mixing in a few milliseconds, a gentle stirring avoiding vortex formation was utilized.

More recently extensive work has been carried out in our group in order to establish some correlations between the rate of water uptake by W/O microemulsions and the dynamic properties (fluidity or rigidity) of the amphiphilic film.[134] Different additives known to modify the film dynamics (see Sections VI.C and VI.D) were introduced in different proportions in the AOT/n-decane/water system: alcohols with different chain lengths (from 1-butanol to 1-decanol), salt (up to 0.3 M), and hydrophilic polymers (PEG 2000 and 10,000). A versatile stopped-flow apparatus (Biologic SFM-3) was used, which provided an excellent dispersion of the solutions by their passage through two successive mixers, and which allowed the adjustments of the mixing volume ratios (1/20 was selected) and of the rate of injection. The kinetics was analyzed in many different situations. The turbidity decay curves were best fitted with biexponential functions, in agreement with the observations of Battistel and Luisi mentioned above. The fast process, which controls the initial uptake of water, gave k_1 values varying from about 1 s^{-1} to 10^3 s^{-1}, depending on the nature and quantity of additives. Recall that these rate constants depend on the mixing conditions, especially the flow rate, which was here a variable parameter.

Conductivity measurements were systematically used to characterize the existence (or absence) of droplet percolation in the region of the phase diagram where a jump of water concentration was induced.[134] Figure 5.9 gives an example of the kinetic curves obtained for water solubilization in alcohol-containing microemulsions, starting with w_o = 25 and finishing with w_{of} = 32.6. The observed kinetics is extremely sensitive to the length of the alcohol alkyl chain, which is known

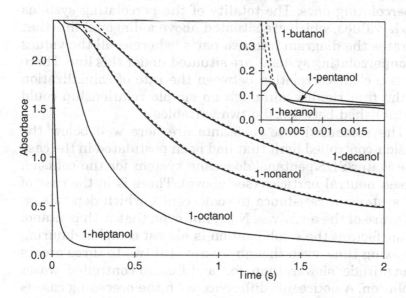

Figure 5.9 Change of absorbance (turbidity) with time for water solubilization in alcohol-containing microemulsions. Initial decane/AOT = 25/75 (w/w); 1-alkanol/AOT molar ratio = 0.5. Best theoretical fits with biexponential functions. Reprinted with permission from Reference 134, copyright 1998, American Chemical Society.

to be responsible for considerable changes in droplet interactions. Water incorporation is much faster when the film is fluid (short chain alcohols) than when it is more rigid (longer chain alcohols), as it could be expected. The observed rate constants decreased when approaching the phase limit as reported before (see Figure 5.8b), which means that the distance of the final composition with respect to the phase limit is also an operating parameter, which cannot be neglected. We have defined Δw_o as representing the gap between w_{oi} (value of w_o at the point of injection) and w_{olimit} (value of w_o corresponding to the phase limit) (see Figure 5.4). The Δw_o parameter takes into account the fact that the limits of the monophasic domain can drastically change when different additives are considered, which otherwise would prevent us from making direct comparisons between results obtained with the same initial w_o. Figure 5.10 shows the values of k_1, which were measured in a variety of conditions, as a function of Δw_o. By reference to their conductivity behaviors we have distinguished in this figure the percolating systems from the nonpercolating ones. The totality of the percolating systems gives k_1 values, which are situated above a diagonal line that separates the diagram into two parts, whereas all the values for nonpercolating systems are situated under this line. There is thus a clear correlation between the rate of solubilization and the film rigidity, although no simple relationship could be established between the two variables.

The measured rate constants are here well below the diffusion-controlled limit that had been postulated in the case of the water/SDS/pentanol/dodecane system for the collision between neutral particles (see above). There is in the case of AOT systems a resistance to coalescence, which depends on the nature of the additives. Note, however, that with pentanol as cosurfactant the solubilization is almost completed during the mixing time, even though we are still two to three orders of magnitude slower than for a diffusion-controlled water dissolution. A noticeable difference with the preceding case is that the AOT surfactant, contrary to SDS, has a measurable solubility in the organic phase where the monomer concentration may attain the millimolar range; whether this is enough for the colliding entities to become electrically charged

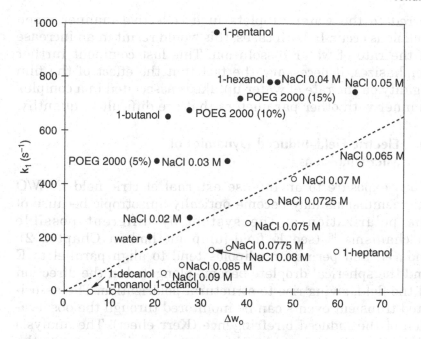

Figure 5.10 Measured k_1 values versus Δw_o (see text), with different additives. The diagonal dashed line separates the percolating systems (upper part) and the nonpercolating systems (lower part). Reprinted with permission from Reference 134, copyright 1998, American Chemical Society.

due to surfactant adsorption (which would reduce the diffusion-controlled rate constant as discussed earlier) is an open question. In any case, the mechanism by which the system goes back to transparency after water injection is necessarily not a simple one because it implies a reorganization of the droplets with a diminution of their number. The k_1 values obtained in the presence of PEG 2000 were surprisingly high, if one considers the fact that the polymer is retarding the onset of the conductivity percolation. Possible explanations have been proposed for this observation[134]: either the confined polymer may be responsible for an increase of the osmotic pressure, which would favor water penetration, or due to the association of AOT molecules with the polymer chain, large portions of the microemulsion surfactant film may be transferred to the water droplets in a collective manner when collisions occur. In both cases, this would result in an increase of the rate of water dissolution. This last comment further emphasizes, if necessary, the fact that the effect of the film rigidity on the rate of water uptake is associated in a complex manner with other parameters that are difficult to quantify.

D. Electric Field–Induced Dynamics of Microemulsions

Upon exposure to an intense external electric field E, W/O microemulsions may become optically anisotropic because of the polarization of the system by different possible mechanisms[136] (see E field-jump method in Chapter 2). Induced and permanent dipoles tend to align parallel to E and the spherical droplets become elongated in the direction of the field, giving rise to structural polarization. The associated transient events can be monitored through the observation of the induced birefringence (Kerr effect). The analysis of the time-dependent birefringence signal gives access to the dynamics of different structural changes, provided that any contribution due to turbidity components has been removed by appropriate devices. Symmetrical behaviors are usually observed upon application of the electric field and after the field is turned off, at least in the absence of phase separation or percolation phenomena (in the latter cases the return to the initial equilibrium species cannot simply follow the reverse route[136b]).

Tekle and Schelly[137,138] have studied the electric birefringence of AOT/isooctane/water W/O microemulsions, which revealed two distinct relaxation processes on timescales of the order of 10 and 100 μs, respectively. The fast relaxation was attributed to the polarization/alignment of the individual reverse microemulsion droplets; the slow relaxation, of smaller amplitude, was assigned to the linearization/reorientation of the micellar clusters. The rates of both processes became slower when w_o, or the AOT concentration, or the temperature was increased. Transient phase separation could occur beyond some threshold values of the preceding parameters, when applying electric fields of sufficient magnitude and duration.

Ilgenfritz and coworkers[139-143] have used the E field-jump technique to investigate the dynamics of field-induced percolation in both ionic and nonionic W/O microemulsions. They showed that high electric fields can shift the structure of microemulsions from a nonconducting to a highly conducting state within times ranging from microseconds to milliseconds.[139-141] In the case of AOT-stabilized microemulsions, for which prior reports were due to Eicke and Naudts,[144] the transition curve is shifted to lower temperatures whereas for nonionic systems based on Igepal CO-520 the reverse is true.[140] Apart from this difference related with the temperature dependence of the electrical conductivity, a strong similarity has been noted for both the dynamic behavior and the field strength dependence in the two types of systems. The influence of additives on the electro-optic properties associated with the dynamics of clustering of the water droplets has also been investigated.[88,140] The addition of long PEG (degree of polymerization > 200) to large AOT microemulsion droplets had no effect, but the addition of gelatin resulted in a significant slowing down of the field-off birefringence decay. It has been proposed[88] that whereas the droplet structure would remain unchanged in PEG containing microemulsions, in case of gelatin, the rebuilding of the triple helix that characterizes such systems, would force the microemulsion to adopt a completely different structure.

Schwarz et al.[142] have investigated the dynamics of the "sponge" L_3 phase in the ternary $H_2O/C_{10}EO_4/n$-decanol system. The time constants measured from T-jump (with scattering light detection) and from p-jump (with conductivity detection) were found identical within the experimental accuracy and varied from seconds to milliseconds depending on the volume fraction Φ of total bilayer. They obeyed a scaling law of the form

$$\tau_{T,p}^{-1} \propto \Phi^{9.2} \tag{5.18}$$

A similar exponent was found by Le et al.[145] for the $C_{12}EO_5/n$-decane/brine system. The time constants from electric birefringence were found in the range from milliseconds to microseconds and showed a weaker dependence on Φ:

$$\tau_E^{-1} \quad \propto \quad \Phi^{2.8} \tag{5.19}$$

The dynamic processes revealed by these two kinds of techniques are obviously not the same. Varying the temperature or the pressure is expected to change the spontaneous mean curvature of the monolayers, which in the conditions of the experiments reported here, should become more strongly curved toward water. This process may lead to the formation of new passages inside the bicontinuous structure. The fast process that is seen in E field-jump experiments has a different origin. The authors interpreted the observed relaxation dynamics as the viscoelastic response of the L_3 structural units to the effect of the electric field on the surfactant molecule dipoles. The measured time constant would thus be related with the distortion of the L_3 structure. A proportionality between τ_E^{-1} and the bending rigidity constant K has been proposed,[142] suggesting that time-resolved birefringence measurements could be a convenient method to determine K values of surfactant monolayers and lipid membranes. Note the similarity of this approach with that described in section VII.A, where we discussed the results obtained from the ILTJ technique.

VIII. IMPLICATIONS OF THE DYNAMIC BEHAVIOR OF MICROEMULSIONS IN SOME SPECIFIC DOMAINS

The dynamics of microemulsion droplets has important consequences in some chemical, technological, and biomedical applications of these systems. Much work has been done concerning reaction kinetics in microdisperse systems, taking advantage of the compartmentalization of reagents or of the possibility of carrying out chemical reactions involving lipophilic reagents in an aqueous environment. Chapter 10 in this book reviews such studies, whose theoretical aspects have been examined by Barzykin and Tachiya.[146] The dynamics of the microemulsion system itself cannot be ignored when looking for specific effects associated with reagent compartmentalization. The effects of intermicellar exchange rate have, for instance, been reported in the controlled polymerization of monomers to give latexes[147] or inorganic polymers,[148] and in the preparation of metallic colloids.[149–151]

In the case of the polymerization of acrylamide in AOT reverse micelles, where the initiation rate is small compared to the polymerization rate, only a reduced number of micelles are nucleated by the initiator and the nonnucleated particles act as reservoirs of monomer.[147] Collisions between the droplets are responsible for the growth of the latex particles, but a narrow size distribution is still obtained. The same kind of observation was made in a study of the formation of silica particles by the ammonia-catalyzed hydrolysis of tetraethylorthosilicate (TEOS) in nonionic W/O microemulsions.[148] In this case the growth of silica particles is rate-limited by the extremely slow TEOS hydrolysis. Due to this very slow rate and to fast interdroplet matter exchange, the silica particles are nucleated in the early reaction period and they grow up to their final size through the continuous collection of reacting species. Correlations could be established between size distribution of silica particles and the size, connectivity, and stability of microemulsion droplets. However, examples exist where the rate of chemical reactions in microheterogeneous medium were found insensitive to droplet percolation.[152]

The dynamics of microemulsions is also important when using microemulsions to perform selective extraction or transport of different types of chemical compounds (biomolecules, metal ions, pollutants, etc.). Experimental works in this direction have been carried out in our group using liquid membrane techniques.[20,21,153,154] The microemulsion phase in Winsor I or Winsor II systems was taken as the liquid membrane, the carriers being nanodroplets of oil (Winsor I) or of water (Winsor II); the excess phase, either oil or water, was used in place of the source (S) and receiving (R) phases. The liquid/liquid interfaces between the microemulsion and the excess phase are expected to be coated with a monolayer of surfactant molecules. This flat monolayer is of the same nature as the curved interfaces of the nanodroplets. Depending on the nature of the surfactant involved and the film rigidity, processes of coalescence involving the interfacial film (see the scheme below) are more or less likely. This effect, also

 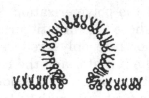

Scheme 5.1

studied by Nitsch et al.,[155–157] is undoubtedly very important in extraction and transport processes, since it assumes that a direct (and supposedly non selective) communication can be established between the source phase and the membrane phase as well as between the membrane and receiving phases.

A demonstrative example was reported for the transport of a series of Ni^{2+} salts between the S and R compartments in a U-shaped tube using two different microemulsion systems.[154] When the Ni^{2+} ions are transported by reverse micelles of $C_{12}EO_4$/1-hexanol/water in n-decane, the transfer rate in ppm.min^{-1} is strongly dependent on the nature of the anion associated with Ni^{2+} (see Figure 5.11a). The magnitude of this rate follows the well-known Hofmeister ion series with $ClO_4^- > NO_3^- > Cl^- \approx Ac^- > SO_4^{2-}$. This has been interpreted as giving clear evidence that the interfacial transfer of Ni^{2+} involves a first step in which the ion-pair formed by Ni^{2+} and its associated anion is transferred in the decane continuous phase, and a second step in which this ion-pair is solubilized in the reverse micelle. When the Ni^{2+} ions are transported by reverse micelles of AOT/water (0.25 M KBr) again in n-decane, the transfer rates were found almost totally insensitive to the nature of the anion (Figure 5.11b). After examination of different possible explanations, this led to the conclusion that in that case there is a direct transfer of the Ni^{2+} salts in the AOT water pools, through droplet coalescence with the inter-

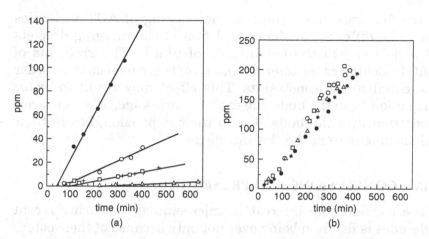

Figure 5.11 Concentration (in ppm) of Ni^{2+} transported versus time. (a) $C_{12}EO_4$/1-hexanol/n-decane/water system: ($\triangle$) SO_4^{2-}; (+) Cl^-; ($\square$) Ac^-; ($\circ$) NO_3^-; ($\bullet$) ClO_4^-; (b) AOT/n-decane/water (0.25 M KBr) system: ($\triangle$) SO_4^{2-}; ($\circ$) Cl^-; ($*$) Ac^-; ($\square$) NO_3^-; ($\bullet$) ClO_4^-. Reprinted with permission from Reference 154, copyright 1990, American Chemical Society.

facial film, followed by droplet separation in a reverse process. In other studies the picrate ion was used for probing the transport processes for the two types of systems just discussed. Theoretical models permitted us to predict the variations of the picrate flux as a function of different parameters. These models have provided additional possibilities of differentiation between the two postulated mechanisms.[153]

Extensive studies have been performed in view of a better understanding of the detailed mechanisms of biphasic transfers between an aqueous phase and AOT reverse micelles, using Winsor II systems. They have shown that the mechanism by which ions and water are transferred into AOT droplets is quite complex because of the electrostatic interactions between ions and amphiphilic films. Exchanges between the initial AOT counterions and the ions to be transferred will affect the compactness of the amphiphilic film of the nanodroplets opened at the interface, in a different manner depending on the respective valencies of these ions. The concept of packing gradient has been introduced to describe the fact that in some instances, the packing of AOT molecules may be different in the curved film of the merging droplets ("buds") and in the flat interfacial film.[157] The gradients of AOT packing cause lateral motions of these molecules in order to equalize the constraints. This effect may result in water rejection from the buds due to their shrinkage, or water incorporation into the buds, due to their expansion, before their detachment from the flat interface.

IX. CONCLUSIONS AND PROSPECTS

The considerable interest in microemulsions during recent decades is far from being over not only because of their potential applications in many different fields, but also because they constitute fascinating objects for fundamental research, especially due to their complex morphologies and dynamics. We have attempted in this chapter to give a comprehensive picture of the various dynamic phenomena that have been reported to exist in microemulsion systems. The attribution of the measured time constants, obtained from a variety of techniques, to specific mechanisms and processes, is not always an easy task. Some of these processes have been clearly identified, but others remain to be completely solved. For instance, this is the case of the processes related to the change of curvature of surfactant monolayers (including the creation of continuous passages) for which discrepancies exist about the characteristic times involved. More systematic experiments are needed to fully understand such processes.

Besides, one can already observe a trend toward investigating even more complex systems combining microemulsions and polymers.[88,158-161] It can be assumed that the interest in such systems, having original rheological properties, will increase in view of developing new materials.

REFERENCES

1. Hoar, T.P., Schulman, J.H. *Nature* 152, 102–103, 1943.

2. Salager, J.L., Morgan, J.C., Schechter, R.S., Wade, W.H., Vasquez, E. *Soc. Petr. Eng. J.* 19, 107–115, 1979.

3. Bavière, M. *Les Microemulsions, Rev. Inst. Fr. Pet.* 29, 41–72, 1974.

4. Holt, S.L. *J. Disp. Sci. Technol.* 1, 423–464, 1980.

5. Friberg, S.E. *J. Disp. Sci. Technol.* 6, 317–337, 1985.

6. Paul, B.K., Moulik, S.P. *J. Disp. Sci. Technol.* 18, 301–367, 1997.

7. Langevin, D. *Annu. Rev. Phys. Chem.* 43, 341–369, 1992.

8. Moulik, S.P., Paul, B.K. *Adv. Colloid Interface Sci.* 78, 99–195, 1998.

9. Zana, R. *Heter. Chem. Rev.* 1, 145–157, 1994.

10. Friberg, S.E., Bothorel, P., Eds., *Microemulsions: Structure and Dynamics*, CRC Press, Boca Raton, FL, 1987.

11. Rosano, H.L., Clausse, M., Eds., *Microemulsion Systems, Surfactant Science Series* Vol. 24, Marcel Dekker, New York, 1987.

12. Robb, I.D., Ed., *Microemulsions*, Plenum Press, New York, 1982.

13. Prince, L.M., Ed., *Microemulsions: Theory and Practice*, Academic Press, New York, 1977.

14. Mittal, K.L., Ed., *Micellization, Solubilization, and Microemulsions*, Vols. 1 and 2, Plenum Press, New York, 1977.

15. Chen, S.–H, Rajagopalan, R., Eds., *Micellar Solutions and Microemulsions: Structure, Dynamics and Statistical Thermodynamics*, Springer-Verlag, 1990.

16. Shah, D.O., Ed., *Micelles, Microemulsions and Monolayers Science and Technology*, Marcel Dekker, New York,1998.

17. De Gennes, P.G., Taupin, C. *J. Phys. Chem.* 86, 2294–2304, 1982.

18. Winsor, P.A. *Solvent Properties of Amphiphilic Compounds*, Butterworths, London, 1954.

19. Winsor, P.A. *Trans. Faraday Soc.* 44, 376–382, 1948.

20. Xenakis, A., Tondre, C. *J. Phys. Chem.* 87, 4737–4743, 1983.

21. Tondre, C., Xenakis, A. *Faraday Discuss. Chem. Soc.* 77, 115–126, 1984.

22. Staples, E.J., Tiddy, G.J.T. *J. Chem. Soc. Faraday Trans.* 1, 74, 2530–2541, 1978.

23. Wennerström, H., Lindman, B., Söderman, O., Drakenberg, T., Rosenholm, J.B. *J. Am. Chem. Soc.* 101, 6860–6864, 1979.

24. Ahlnäs, T., Söderman, D., Hjelm, C., Lindman, B. *J. Phys. Chem.* 87, 822–828, 1983.

25. Tricot, Y., Kiwi, J., Niederberger, W., Grätzel, M. *J. Phys. Chem.* 85, 862–870, 1981.

26. Söderman, O., Canet, D., Carnali, J., Henriksson, U., Nery, H., Walderhaug, H., and Wärnheim, T., in Rosano, H.L., Clausse, M., Eds., *Microemulsion Systems, Surfactant Science Series* Vol. 24, Marcel Dekker, New York, 145–161, 1987.

27. Huang, J.S. in *Surfactants in Solution*, K.L. Mittal, Ed., Vol. 10, pp. 45–59, Plenum Press, New York, 1989.

28. Kommareddi, N.S., McPherson, G.L., John, V.T. *Colloids Surf.* A, 92, 293–300, 1994.

29. Lindman, B., Stilbs, P., Moseley, M.E. *J. Colloid Interface Sci.* 83, 569–582, 1981.

30. Guéring, P. and Lindman, B. *Langmuir* 1, 464–468, 1985.

31. Fabre, H., Kamenka, N., Lindman, B. *J. Phys Chem.* 85, 3493–3501, 1981.

32. Lindman, B., Stilbs, P., in *Microemulsions: Structure and Dynamics*, Friberg, S.E., Bothorel, P., Eds., CRC Press, Boca Raton, FL, 119–152, 1987.

33. Néry, H., Söderman, O., Canet, D., Walderhaug, H., Lindman, B. *J. Phys. Chem.* 90, 5802–5808, 1986.

34. Aniansson, E.A.G., Wall, S., Almgren, M., Hoffman, H., Kielman, I., Ulbricht, W., Zana, R., Lang, J., Tondre, C. *J. Phys. Chem.* 80, 905–922, 1976.

35. Graber, E., Lang, J., Zana, R. *Kolloid, Z.Z. Polymere* 238, 470–478, 1970.

36. Graber, E., Zana, R., *Kolloid, Z.Z. Polymere* 238, 479–485, 1970.

37. Lang, J., Tondre, C., Zana, R., Bauer, R., Hoffmann, H., Ulbricht, W. *J. Phys. Chem.* 79, 276–283, 1975.

38. Kahlweit, M., Teubner, M. *Adv. Colloid Interface Sci.* 13, 1–64, 1980.

39. Tondre, C., Zana, R., *J. Colloid Interface Sci.* 66, 544–558, 1978.

40. Hoffmann, H., *Ber. Bunsenges. Phys. Chem.* 82, 988–1001, 1978.

41. Aniansson, E.A.G., Wall, S.N. *J. Phys. Chem.* 78, 1024–1030, 1974.

42. Hall, D.G. *J. Chem. Soc. Faraday Trans.* 2, 77, 1973–2006, 1981.

43. Lessner, E., Teubner, M., Kahlweit, M., *J. Phys. Chem.* 85, 1529–1536, 1981.

44. Lessner, E., Teubner, M., Kahlweit, M., *J. Phys. Chem.* 85, 3167–3175, 1981.

45. Aniansson, E.A.G. in *Techniques and Applications of Fast Reactions in Solution*, p. 249, Reidel, Dordrecht, Holland, 1979.

46. Yiv, S., Zana, R., Ulbricht, W., Hoffmann, H. *J. Colloid Interface Sci.* 80, 224–236, 1981.

47. Lang, J., Djavanbakht, A., Zana, R. *J. Phys. Chem.* 84, 1541–1547, 1980.

48. Lang, J., Djavanbakaht, A., Zana, R., pp. 233–254 in Robb, I.D., Ed., *Microemulsions*, Plenum Press, New York, 1982.

49. Lang, J., Zana, R. *J. Phys. Chem.* 90, 5258–5265, 1986.

50. Hoffmann, H., Platz, G., Ulbricht, W. *Ber. Bunsenges. Phys. Chem.* 90, 877–887, 1986.

51. Zana, R., Yiv, S., Strazielle, C., Lianos, P. *J. Colloid Interface Sci.* 80, 208–223, 1981.

52. Aniansson, E.A.G. *Progr. Colloid Polym. Sci.* 70, 2–5, 1985.

53. Kahlweit, M., (a) *Pure Appl. Chem.* 53, 2069–2081, 1981; (b) *J. Colloid Interface Sci.* 90, 92–99, 1982.

54. Ahlnäs, T., Söderman, O., Walderhaug, H., Lindman, B. in *Surfactants in Solution*, K.L. Mittal, B. Lindman, Eds., Plenum, New York, Vol. 1, pp. 107–127, 1982.

55. Tondre, C., Xenakis, A., Robert, A., Serratrice, G. in *Surfactants in Solution*, K.L. Mittal, P. Bothorel, Eds., Plenum, New York, Vol. 6, pp. 1345–1355, 1986.

56. Bhattacharyya, K., Bagchi, B. *J. Phys. Chem.* A 104, 10603–10613, 2000.

57. Wong, M., Thomas, J.K., Nowak, T. *J. Am. Chem. Soc.* 99, 4730–4736, 1977.

58. Fendler, J.H. *Membrane Mimetic Chemistry*, John Wiley and Sons, New York, 1982, p. 59.

59. Leodidis, E.B., Hatton, T.A. *Langmuir* 5, 741–753, 1989.

60. Eigen, M. *Pure Appl. Chem.* 6, 97–115, 1963.

61. Gruenhagen, H. *J. Colloid Interface Sci.* 53, 282–295, 1975.

62. Lagües, M., Ober, R., Taupin, C. *J. Phys.-Lett.* 39, L487-L491, 1978.

63. Borkovec, M., Eicke, H.-F., Hammerich, H., Gupta, B.D. *J. Phys. Chem.* 92, 206–211, 1988.

64. Moha-Ouchane, M., Peyrelasse, J., Boned, C. *Phys. Rev.* A 35, 3027–3032, 1987.

65. Lagües, M., Sauterey, C. *J. Phys. Chem.* 84, 3503–3508, 1980.

66. Cazabat, A.M., Chatenay, D., Langevin, D., Meunier, J. *Faraday Discuss. Chem. Soc.* 76, 291–303, 1983.

67. Chatenay, D., Urbach, W., Cazabat, A.M., Langevin, D. *Phys. Rev. Lett.* 54, 2253–2256, 1985.

68. Eicke, H.-F., Thomas, H. *Langmuir* 15, 400–404, 1999.

69. Lagües, M. *J. Phys.-Lett.* 40, L331-L333, 1979.

70. Mathew, C., Patanjali, P.K., Nabi, A., Maitra, A. *Colloids Surf.* 30, 253–263, 1988

71. Safran, S.A., Grest, G.S., Bug, A.L., Webman, I. pp. 235–243 in Rosano, H.L. and Clausse, M., Eds., *Microemulsion Systems, Surfactant Science Series* Vol. 24, Marcel Dekker, New York, 1987.

72. Meier, W., Eicke, H.-F. *Curr. Opinion Colloid Interface Sci.* 1, 279–286, 1996.

73. John, A.C., Rakshit, A.K. *Langmuir* 10, 2084–2087, 1994.

74. Eicke, H.-F., Meier, W. *Biophys. Chem.* 58, 29–37, 1996.

75. Testard, F., Zemb, T. *Langmuir* 16, 332–339, 2000.

76. Derouiche, A., Tondre, C. *J. Disp. Sci. Technol.* 12, 517–530, 1991.

77. Hou, M.J., Shah, D.O. *Langmuir* 3, 1086–1096, 1987.

78. Leung, R., Shah, D.O. *J. Colloid Interface Sci.* 120, 320–329 and 330–344, 1987.

79. Di Megglio, J.-M., Dvolaitzky, M., Taupin, C. *J. Phys. Chem.* 89, 871–874, 1985.

80. Garcia-Rio, L., Leis, J.R., Mejuto, J.C., Peña, M.E. *Langmuir* 10, 1676–1683, 1994.

81. Ray, S., Bisal, S.R., Moulik, S.P. *J. Chem. Soc. Faraday Trans.* 89, 3277–3282, 1993.

82. Thakur, L.K., De, T.K., Maitra, A. *J. Surface Sci. Technol.* 13, 39–47, 1997.

83. Amaral, C.L., Brino, O., Chaimovich, H., Politi, M.J. *Langmuir* 8, 2417–2421, 1992.

84. Suarez, M.-J., Levy, H., Lang, J. *J. Phys. Chem.* 97, 9808–9816, 1993.

85. Suarez, M.-J., Lang, J. *J. Phys. Chem.* 99, 4626–4631, 1995.

86. Meier, W. *Langmuir* 12, 1188–1192, 1996.

87. Eicke, H.-F., Gauthier, M., Hammerich, H. *J. Phys. II France* 3, 255–258, 1993.

88. Schübel, D., Ilgenfritz, G. *Langmuir* 13, 4246–4250, 1997.

89. Hait, S.K., Sanyal, A., Moulik, S.P. *J. Phys. Chem.* B 106, 12, 642–12,650, 2002.

90. Hait, S.K., Moulik, S.P., Rodgers, M.P., Burke, S.E., Palepu, R. *J. Phys. Chem.* B 105, 7145–7154, 2001.

91. Maitra, A., Mathew, C., Varshney, M. *J. Phys. Chem.* 94, 5290–5292, 1990.

92. Meier, W. *Colloids Surf.* A 94, 111–114, 1995.

93. Menger, F.M., Donohue, J.A., Williams, R.F. *J. Am. Chem. Soc.* 95, 286–288, 1973.

94. Eicke, H.-F., Shepherd, J.C., Steinemann, A. *J. Colloid Interface Sci.* 56, 168–176, 1976.

95. Eicke, H.-F., Zinsli, P.E. *J. Colloid Interface Sci.* 65, 131–140, 1978.

96. Abe, M., Nakamae, M., Ogino, K., Wade, W.H. *Chem. Lett.* 1613–1616, 1987.

97. Zana, R., Lang, J., . in *Microemulsions: Structure and Dynamics*, Friberg, S.E. and Bothorel, P. Eds., CRC Press, Boca Raton, FL, 153—172, 1987.

98. Lang, J. in *The Structure, Dynamics and Equilibrium Properties of Colloidal Systems*, Bloor, D.M., Wyn-Jones, E., Eds., Kluwer Acad. Publ., Dordrecht, 1990, pp. 1–38.

99. Fletcher, P.D.I., Howe, A.M., Robinson, B.H. *J. Chem. Soc. Faraday Trans.* 1 83, 985–1006, 1987.

100. Fletcher, P.D.I., Howe, A.M., Perrins, N.M., Robinson, B.H., Toprakcioglu, C., Dore, J.C. in *Surfactants in Solution*, Mittal, K.L., Lindman, B., Eds., Plenum, New York, Vol. 3, pp. 1745–1758, 1984.

101. Fletcher, P.D.I., Robinson, B.H. *Ber. Bunsenges. Phys. Chem.* 85, 863–867, 1981.

102. Robinson, B.H., Steytler, D.C., Tack, R.D. *J. Chem. Soc. Faraday Trans.* 1, 75, 481–496, 1979.

103. Atik, S.S., Thomas, J.K. *J. Am. Chem. Soc.* 103, 3543–3550, 1981.

104. Bommarius, A.S., Holzwarth, J.F., Wang, D.I.C., Hatton, T.A. (a) in *The Structure, Dynamics and Equilibrium Properties of Colloidal Systems*, Bloor, D.M., Wyn-Jones, E., Eds., Kluwer Acad. Publ., Dordrecht, 1990, pp. 181–199; (b) *J. Phys. Chem.* 94, 7232–7239, 1990.

105. Jada, A., Lang, J., Zana, R. *J. Phys.Chem.* 93, 10–12, 1989.

106. Lang, J., Lalem, N., Zana, R. *J. Phys. Chem.* 96, 4667–4671, 1992.

107. Jada, A., Lang, J., Zana, R., Makhloufi, R., Hirsch, E., Candau, S.J. *J. Phys. Chem.* 94, 387–395, 1990.

108. Lang, J., Lalem, N., Zana, R. *J. Phys. Chem.* 95, 9533–9541, 1991.

109. Lang, J., Lalem, N., Zana, R. *Colloids Surf.* 68, 199–206, 1992.

110. Geladé, E., De Schryver, F.C. *J. Am. Chem. Soc.* 106, 5871–5875, 1984.

111. Lianos, P., Zana, R., Lang, J., Cazabat, A.M. in *Surfactants in Solution*, K.L. Mittal, P. Bothorel, Eds., Plenum, New York, Vol. 6, pp. 1365–1372, 1986.

112. Fletcher, P.D.I., Horsup, D.I. *J. Chem. Soc. Faraday Trans.* 88, 855–864, 1992.

113. Guéring, P., Cazabat, A.M., Paillette, M. *Europhys. Lett.* 2, 953–960, 1986.

114. Johannsson, R., Almgren, M., Alsins, J. *J. Phys. Chem.* 95, 3819–3823, 1991.

115. Johannsson, R. and Almgren, M. *Langmuir* 9, 2879–2882, 1993.

116. Lang, J., Jada, A., Malliaris, A. *J. Phys. Chem.* 92, 1946–1953, 1988.

117. Mays, H., Pochert, J., Ilgenfritz, G. *Langmuir* 11, 4347–4354, 1995.

118. Mays, H., Ilgenfritz, G. *J. Chem. Soc. Faraday Trans.* 92, 3145–3150, 1996.

119. Mays, H. *J. Phys. Chem.* B, 101, 10271–10280, 1997.

120. Alexandridis, P., Holzwarth, J.F., Hatton, T.A. *Langmuir* 9, 2045–2052, 1993.

121. Petit, C., Holzwarth, J.F., Pileni, M.P. *Langmuir* 11, 2405–2409, 1995.

122. Nazario, L.M., Crespo, J.P., Holzwarth, J.F., Hatton, T.A., *Langmuir* 16, 5892–5899, 2000.

123. Holzwarth, J.F., Schmidt, A., Wolff, H., and Volk, R. *J. Phys. Chem.* 81, 2300–2301, 1977.

124. Helfrich, W., *Z. Naturforsch.* 28, 693–703, 1973.

125. Hou, M.J., Kim, M., Shah, D.O. *J. Colloid Interface Sci.* 123, 398–412, 1988.

126. Burger-Guerrisi, C., Tondre, C., Canet, D. *J. Phys. Chem.* 92, 4974–4979, 1988.

127. Knight, P., Wyn-Jones, E., Tiddy, G.J.T. *J. Phys. Chem.* 89, 3447–3449, 1985.

128. Schneider, G.M., Dittmann, M., Metz, U., Wenzel, J. *Pure Appl. Chem.* 59, 79–90, 1987.

129. Olsson, H., Shinoda, K., Lindman, B. *J. Phys. Chem.* 90, 4083–4088, 1986.

130. Mayer, W. Woermann, D. *J. Phys. Chem.* 92, 2036–2039, 1988.

131. Tiddy, G.J.T., Wheeler, P.A. *J. Phys.* 36, C1 167–172, 1975.

132. Tondre, C., Zana, R. *J. Disp. Sci. Technol.* 1, 179–195, 1980.

133. Battistel, E., Luisi, P.L. *J. Colloid Interface Sci.* 128, 7–14, 1989.

134. Caillet, C., Hébrant, M., Tondre, C. *Langmuir* 14, 4378–4385, 1998.

135. Evilevitch, A., Olsson, U., Jönsson, B., Wennerström, H. *Langmuir* 16, 8755–8762, 2000.

136. Schelly, Z.A. (a) *Curr. Opinion Colloid Interface Sci.* 2, 37–41, 1997; (b) *J. Mol. Liq.* 72, 3–13, 1997.

137. Tekle, E., Ueda, M., Schelly, Z.A. *J. Phys. Chem.* 93, 5966–5969, 1989.

138. Tekle, E., Schelly, Z.A. *J. Phys. Chem.* 98, 7657–7664, 1994.

139. Runge, F., Röhl, W., Ilgenfritz, G. *Ber. Bunsenges. Phys. Chem.* 95, 485–490, 1991.

140. Schlicht, L., Spilgies, J.-H., Runge, F., Lipgens, S., Boye, S., Schübel, D., Ilgenfritz, G. *Biophys. Chem.* 58, 39–52, 1996.

141. Bedford, O.D., Ilgenfritz, G. *Colloid Polym. Sci.* 278, 692–696, 2000.

142. Schwartz, B., Mönch, G., Ilgenfritz, G., Strey, R. *Langmuir* 16, 8643–8652, 2000.

143. Mönch, G., Ilgenfritz, G. *Colloid Polym. Sci.* 278, 687–691, 2000.

144. Eicke, H.-F., Naudts, J. *Chem. Phys. Lett.* 142, 106–109, 1987.

145. Le, T.D., Olsson, U., Wennerström, H., Uhrmeister, P., Rathke, B., Strey, R. *J. Phys. Chem.* B 106, 9410–9417, 2002.

146. Barzykin, A.V., Tachiya, M. *Heter. Chem. Rev.* 3, 105–167, 1996.

147. Candau, F., Leong, Y.S., Pouyet, G., and Candau, S.J. (a) in *Physics of Amphiphiles: Micelles, Vesicles and Microemulsions*, Degiorgio, V., Corti, M., Eds., North-Holland, Amsterdam, pp. 830–841, 1985; (b) *J. Colloid Interface Sci.* 101, 167–183, 1984.

148. Chang, C.-L., Fogler, H.S. *Langmuir* 13, 3295–3307, 1997.

149. Pileni, M.P. *J. Phys. Chem.* 97, 6961–6973, 1993.

150. Bagwe, R.P. and Khilar, K.C. *Langmuir* 16, 905–910, 2000.

151. Quintillan, S., Tojo, C., Blanco, M.C., and Lopez-Quintela, M.A. *Langmuir* 17, 7251–7254, 2001.

152. Boumezioud, M., Kim, H.S., and Tondre, C., *Colloids Surf.* 41, 255–265, 1989.

153. Derouiche, A. and Tondre, C., (a) *J. Chem. Soc. Faraday Trans.* 1, 85, 3301–3308, 1989; (b) *Colloids Surf.* 48, 243–258, 1990.

154. Tondre, C. and Derouiche, A. *J. Phys. Chem.* 94, 1624–1626, 1990.

155. Plucinski, P. and Nitsch, W. (a) *J. Colloid Interface Sci.* 154, 104–112, 1992; (b) *Langmuir* 10, 371–376, 1994.

156. Bausch, T.E., Plucinski, P.K., and Nitsch, W. *J. Colloid Interface Sci.* 150, 226–234, 1992.

157. Nitsch, W., Plucinski, P., and Ehrlenspiel, J. *J. Phys. Chem.* B 101, 4024–4029, 1997.

158. Meier, W. *Colloid Polym. Sci.* 275, 530–536, 1997.

159. Beitz, T., Koetz, J., and Friberg, S.E. *Prog. Colloid Polym. Sci.* 111, 100–106, 1998.

160. Michel, E., Filali, M., Aznar, R., Porte, G., and Appell, J. *Langmuir* 16, 8702–8711, 2000.

161. Porte, G., Filali, M., Michel, E., Appell, J., Mora, S., Sunnyer, E., and Molino, F. *Studies in Surface Science and Catalysis* 132, 55–60, 2001.

6

Dynamic Processes in Aqueous Vesicle Systems

BRIAN H. ROBINSON AND RAOUL ZANA

CONTENTS

I. INTRODUCTION

Vesicles are, in the simplest case, micron-sized spherical structures containing an inner aqueous compartment separated from the external water phase by a lipid or surfactant bilayer. A representative structure, which gives an idealized vesicle structure and some closely related surfactant configurations, is shown in Figure 6.1 (see also Chapter 1, Section V).

Pioneering work on vesicles was carried out by Bangham and Horne in the early 1960s.[1] Vesicles can vary in diameter from 30 nm up to a micron, but giant vesicles, with sizes up to 20 microns, have also been made and studied over recent years.[2,3] The surfactants or lipids used to generate vesicles normally contain two alkyl chains, and their shape is such that they cannot properly pack into a small micelle structure. The methods of preparation of vesicles have been well described and illustrated (see also Chapter 1, Section V).[4] For instance, phospholipid vesicles (*liposomes*) can form when phospholipids are hydrated in hot water to form extended sheets, followed by application of ultrasound. To obtain vesicles of a narrow size distribution, extrusion through the cylindrical pores of polycarbonate membrane filters of a particular size under pressure is carried out. This results in a destabilization of larger vesicles.[5,6]

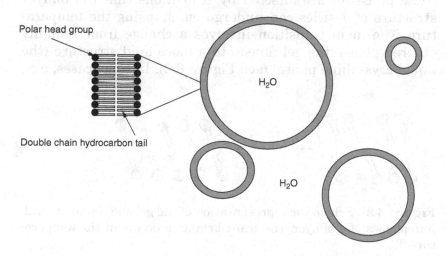

Figure 6.1 Schematic representation of a vesicle dispersion made up of a phospholipid or double-chain surfactant.

In addition to phospholipid vesicles, synthetic vesicles can be made, often by adjusting the ionic strength of the medium, as in the case of dialkylbenzenesulfonates or by simply dispersing in water quaternary ammonium surfactants with two equal or unequal alkyl chains, provided the total number of carbon atoms of the two chains is sufficient.[7,8] A review of the properties of vesicle-forming synthetic amphiphiles was published some years ago by Engberts and Hoekstra.[9] Combinations of phospholipids and fatty acids, and phospholipids with cholesterol, have also been studied. Recently, there has also been growing interest in phospholipids incorporating fluorocarbons. Semifluorinated vesicles and vesicles containing semifluorinated alkanes have greater stability and resistance to fusion than their hydrocarbon analogues.[10,11]

Of great current interest is the use of vesicles in drug-delivery systems, and in particular for targeting of drugs to particular environments, where the drug is released at a defined rate to a specific location. There have been some recent reports on this topic,[12,13] and a variety of techniques have been used to incorporate molecules into vesicles.[14]

Several dynamic processes can occur in vesicle systems. First, the alkyl chains of the lipids or surfactants making up the vesicle undergo conformational changes. The dynamics of these processes are affected by transitions that the bilayer structure of vesicles can undergo on changing the temperature. The main transition involves a change from a highly ordered phase (the gel phase) to a more fluid structure (the liquid-crystalline phase) (see Figure 6.2). In some cases, e.g.,

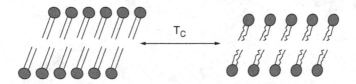

Figure 6.2 Schematic representation of the gel and liquid-crystalline phases of a bilayer. The transformation occurs at the temperature T_C.

with phosphatidylcholines, there is a pretransition preceding the main one, which is associated with changes in orientation of the hydrocarbon chains.[15] Second, the lipids or surfactants making up vesicles can diffuse laterally on the outer or inner layer of the vesicles. They can also diffuse transversally or *flip-flop* (a motion by which a lipid molecule crosses from one side of the bilayer to the other). These motions are illustrated in Figure 6.3. The transversal diffusion is several orders of magnitude slower than lateral diffusion in the fluid phase of the vesicles. Also, ions and other compounds can diffuse through from the inner compartment of the vesicle to the bulk phase and conversely (vesicle permeability). Third, vesicle fusion leads to instability of the vesicle system, and this may be induced by changes of temperature or pH, or addition of cations.[16,17,18] Papahadjopoulos[16] showed that mixing of the aqueous core contents occurs during vesicle fusion, indicated by the formation of the fluorescent chelation complex $Tb(DPA)_3^{3-}$ upon mixing a population of vesicles containing $Tb(citrate)_3^{6-}$ with another population containing dipicolinic acid (DPA). Other dynamic processes are those occurring when transforming vesicles into micelles or micelles into vesicles by appropriate means (changes of ionic strength, dilution, addition of a surfactant, etc.). Last, vesicles submitted to electrical fields can undergo various processes that have been recently investigated.

At this stage, a few reminders on micelle kinetics are appropriate (see Chapter 3 for details). Micelle kinetics has been studied since the 1960s. One of the first experiments was carried out by Jaycock and Ottewill,[19] who showed that

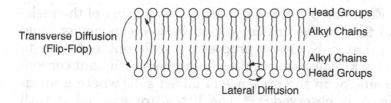

Figure 6.3 Schematic representation of the transverse and lateral diffusive motions of lipids or surfactants in a bilayer.

in aqueous solutions of sodium dodecylsulfate, on dilution from above to below the critical micelle concentration in a flow apparatus, the micelles decomposed in less than a few milliseconds, which was faster than the mixing time of the equipment. Further experiments, however, established that there were two kinetic processes of importance: a fast relaxation process (relaxation time τ_1, with a value often in the microsecond time scale) associated with the exchange of surfactant between free and micellar states[20,21] and a slow process (relaxation time τ_2, in the millisecond to second time scale) associated with a change in the concentration of micelles, usually through formation/breakdown of whole micelles.[22] Theoretical aspects of the micelle formation/breakdown process were reported earlier by Aniansson and Wall.[23] The fast process was mostly studied by ultrasonic relaxation and shock tube methods (see Chapter 2, Section II.D) and the expression for τ_1 is

$$1/\tau_1 = (k^-/\sigma^2)[1 + (\sigma^2/N)a] \tag{6.1}$$

where $a = (C - cmc)/cmc$.

In Equation 6.1, N is the average micelle aggregation number, k^- is the exit rate constant of a monomer from the micelle, and σ is the half-width of the micelle size distribution. Equation 6.1 predicts that the reciprocal of the fast relaxation time should vary linearly with surfactant concentration C above the critical micelle concentration (cmc).

The slow relaxation process was investigated by the T-jump, p-jump, and stopped-flow methods (see Chapter 2, Section IID). The expression for its relaxation time is given by[23]

$$1/\tau_2 \cong N^2/\{cmc \times R[1 + (\sigma^2/N)a]\} \tag{6.2}$$

In Equation 6.2, R is a quantitative measure of the resistance to flow in the critical region between monomers and micelles. Later, Kahlweit proposed a different equation to explain his kinetic data obtained at higher surfactant concentrations and/or in the presence of added salt, where a maximum in τ_2 is observed.[24-26] The literature associated with micelle exchange/breakdown has been recently reviewed.[27] The most useful derived data are the residence time of a surfactant molecule in a micelle, $T_R = N/k^-$, and the micelle lifetime $T_M \sim N\tau_2$ (see Chapter 3, Section III for details).

To date there appear to be no equations available to describe the corresponding situation in vesicle systems. One reason for this is that the equilibrium state of vesicles is still not clearly established. Many systems are unstable and the usual transition is to return (by fusion of vesicles) to a lamellar state from which the vesicle system is often formed in the first place, which can be a very slow process. However, this is certainly not always the case. The problem of vesicle stability has been discussed in general terms by Luisi[28] who suggests the following as some of the criteria for establishing whether or not a self-assembly system is at chemical equilibrium:

1. If two different-sized vesicle populations are mixed, does the system stay as a bimodal distribution or does the system relax to give an average (and more stable) size distribution?
2. Is the same final state reached regardless of the initial mixing conditions chosen to make the vesicle system?
3. What happens when the vesicle system is diluted with water? Is there a change in size or breakdown of the vesicles? Is there an equilibrium between vesicles and monomers or are other states involved, e.g., micelles?

When these criteria are applied to micellar systems, it is clear that micellar systems do indeed represent a true equilibrium state. The situation, however, is more complex with vesicles. One reason is that, in the case of phospholipids, changes could take a long time, e.g., weeks/months, as the monomer concentration external to the vesicle bilayer can be very low (of the order of nanomolar in the case of phospholipids) and this implies that any adjustment to a new equilibrium state will be very slow and hard to monitor. As a result, few systems would appear to conform to all the criteria indicated above. It seems that vesicle systems that have a larger monomer concentration or critical vesicle concentration (cvc) are often unstable in the direction of forming lamellar phases, a process that may take place over a period of hours, or even days. An example here would be the type of vesicle system formed by surfactants with two medium-length alkyl chains and with sulfonate head groups. Luisi also takes issue with the use of the term "metastable" in studies of transformations involving vesicles, which he believes is often misused. He concludes that several systems may exhibit a couple of the above criteria for equilibrium but that few, if any, fulfill all his criteria. In this, he is in line with the views expressed by Laughlin[29,30] and Lasic,[31] who have not been persuaded that vesicles represent the lowest energy state of an aggregated system, preferring the notion of vesicles as a dispersion of a lamellar liquid-crystalline phase. A different view was articulated by Yatcilla et al.,[32] who mixed cationic and anionic surfactants to form what they claim are thermodynamically stable vesicle systems, after an incubation period of several weeks. Recently stable dispersions that contained coexisting micelles and vesicles were also obtained by mixing cetyltrimethylammonium bromide (CTAB) with sodium perfluorohexanoate.[33] In reality, this question of thermodynamic stability of vesicles is a debate which, although raising the passions of investigators, is not always relevant to scientists primarily interested in dynamic properties, as the study of nonequilibrium metastable systems can still produce very useful information.

Based on the molecular structure and shape of the individual surfactants, the preferred aggregate structures formed in surfactant solutions was discussed many years ago in a much quoted paper by Israelachvili, Mitchell, and Ninham.[34] Their general thesis has been found to be useful, but the role of the surfactant tail has recently been reevaluated in a paper by Nagarajan.[35] For most systems, the concentration at which the system is studied is likely to be of importance. One might suppose that as the concentration is increased, the range of structures that can be classed as vesicular is likely to be broader, including spheres, worms, and interconnected vesicle structures with bridging sheets or rods, and this seems to be supported by results obtained by numerous workers (see e.g., Reference 36 and Chapter 1, p. 21). Therefore, to keep the system as close to an ideal vesicular system as possible, it is probably best to work at as low a concentration as possible.

Theoretical aspects relating to vesicle formation have been addressed in a series of papers by Lasic,[37] and his papers should be consulted for details since the main emphasis in this chapter is on experimental work. Indeed, until recently there have been very few kinetic studies of vesicle formation and/or breakdown, but the field is now being extensively investigated.

The chapter is organized as follows. Section II reviews studies that broadly refer to transport: lateral and transversal diffusion of molecules/lipids in vesicles, transport of molecules/lipids between vesicles (a process that is equivalent to the exchange of surfactants or of solubilizates between micelles), and transport of molecules through the vesicle bilayer. Section III reviews studies dealing with the dynamics of vesicle-to-micelle and micelle-to-vesicle transformations. Section IV reviews dynamic processes occurring in particular types of vesicles formed by fatty acids and mixtures of phospholipids and fatty acids. Section V reviews dynamic changes associated with vesicle phase transitions. The last section considers the effect of applied electric fields on the properties of vesicle bilayers (transient formation of pores).

This chapter is restricted to aqueous vesicles but reverse vesicles have also been reported to occur in aprotic organic solvents.[38]

II. KINETICS OF TRANSPORT IN AND BETWEEN AQUEOUS VESICLES

A very large number of studies have been published on this topic. This section makes no attempt to present a comprehensive review. Only selected examples are presented that illustrate the observed behavior.

A. Lateral and Transversal Diffusion

1. Flip-Flop

Pioneering work on surfactant/lipid motions in vesicles was carried out some 30 years ago by McConnell and coworkers at Stanford University using esr spin label probes. Kornberg and McConnell[39] measured the rate of transfer of a spin-labeled phosphatidylcholine (incorporating a nitroxide probe) between the inside and outside of a sonicated egg phosphatidylcholine (eggPC) vesicle system. Addition of sodium ascorbate destroys the paramagnetism of probes located on the external layer of the vesicle, whereas the paramagnetism of probes on the inner surface of the bilayer is not affected. The return to an equilibrium state of the vesicles was associated with a half-life for redistribution of the probe of 6.5 h at 30°C.

Fluorescence methods have been much used to study the transversal motion of a variety of molecules. Fluorescent probes and lipids with an attached fluorescent group have been used for this purpose. For instance Bai and Pagano[40] investigated the flip-flop motion of lipid analogues labeled with a fluorescent fatty acid in biologically relevant liposome systems. Data were reported for analogues of spinghomyelin (SM'), ceramide (CM'), phosphatidylcholine (PC') and diacylglycerol (DAG'). CM' and DAG' moved relatively quickly in flip-flop mode (halftime for probe redistribution of 22 min and 12 min, respectively), whereas SM' and PC' took several hours for this process. Likewise, the flip-flop time for a pyrene–substituted lecithin in vesicles of dipalmitoylphosphatidylcholine (DPPC) was around 8 h at 50°C.[41]

In comparison, several studies showed that the flip-flop times for surfactants solubilized in vesicle membranes are much shorter. Using a stopped-flow method with fluorescence detection, Almgren[42] reported a value of 100 s for the flip-flop time for cetylpyridinium chloride (CPCl) solubilized in vesicles of soybean lecithin. The vesicles contained some solubilized pyrene in the form of excimers. The excimer fluorescence was quenched by the CPCl. Figure 6.4 shows a typical time dependence of the fluorescence intensity recorded in the stopped-flow experiments. In another example, Cocera et al.[43] used the variation of the fluorescence intensity of the probe sodium 2-(p-toluidinyl)naphthalenesulfonate attached to the outer layer of large egg lecithin vesicles to determine a characteristic time of 10–90 min for the flip-flop of sodium dodecylsulfate (SDS) in the system. This surfactant incorporates very rapidly to the vesicle outer surface, thus quenching rapidly the probe fluorescence (less than 10 s). As SDS slowly diffuses from the

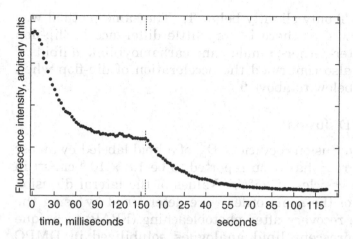

Figure 6.4   Stopped-flow/fluorescence relaxation signal from a 1.2 mM soybean lecithin-0.1 mM cetyltrimethylammonium bromide mixed with 1.2 mM lecithin-0.01 mM cetylpyridinium chloride. Both solutions also contained 0.15 M NaCl, 5 vol.% ethanol and 15 µM pyrene. The signal is biphasic (see break in time scale). The fast process corresponds to the transfer of pyrene between vesicles. The slow process is associated with the motion flip-flop of the cetylpyridinium chloride. Reproduced from Reference 42 with permission of the American Chemical Society.

outer to the inner vesicle layer, the fluorescence intensity increases.

Moss et al.[44,45] used a chemical method for studying flip-flop. A functionalized surfactant bearing a p-nitrophenyl benzoate group was solubilized in vesicles of various two-chain cationic surfactants. The p-nitrophenyl benzoate was hydrolyzed at pH 8 and the amount of hydrolyzed surfactant measured as a function of time. This yielded biphasic plots the slower part of which comes from the flip-flop of the functionalized surfactant located in the inner vesicle layer. All reported times for flip-flop were in the range of a few minutes, irrespective of the length of the alkyl chain of the functionalized surfactant or of the two-chain surfactant making up the vesicles. For a system investigated well below the surfactant gel-to-liquid crystal phase transition temperature, T_C, the flip-flop time was much longer, around 1 hr.[45] This time decreased to 2 min at T only slightly below T_C. The same method was used to show that there is very little difference in flip-flop times for ester-, ether-, amido-, and carbamoyl-linked lipids.[46] This study also confirmed the acceleration of flip-flop when going from below to above T_C.

2. Lateral Diffusion

The lateral diffusion coefficient D_{lat} of a lipid labeled by an esr (nitroxide) group has been reported to be 1.8×10^{-8} cm^{2}s^{-1} in egg lecithin vesicles at 25°C.[47] Values of the lateral diffusion coefficients of the same order were determined by using the fluorescence recovery after photobleaching (FRAP) technique for two fluorescent lipid analogues solubilized in DMPC (dimyristoylphosphatidylcholine) or DPPC vesicles[48] at $T > T_C$. An enormous decrease of the diffusion coefficient, by over two orders of magnitude over a temperature interval of about 3°C, was reported to occur as the temperature was decreased below T_C. Addition of cholesterol was found to decrease D_{lat} at $T > T_C$.[48] Similar behavior was observed when using one of these two probes in vesicles of DMPC and of a mixture of DMPC and a polymerizable butadiene lipid.[49] Figure 6.5 shows the very large decrease of D_{lat} when T becomes smaller than T_C and the effect of the polymerization of the butadiene lipid on D_{lat}.

A large increase of D_{lat} with temperature when going from below to above T_C is also observed when replacing lipids by hydrophobic molecules, particularly pyrene derivatives. For instance, Kano et al.[50] studied the quenching of pyrene and pyrenedecanoic acid fluorescence by aromatic amines in DPPC vesicles and reported diffusion coefficients of 1.6 and 1.3×10^{-8} cm^{2}s^{-1}, at $T < T_C$, respectively. At $T > T_C$ the reported values of D_{lat} were 4.7 and 3.1×10^{-7} cm^{2}s^{-1}, respectively. Likewise, Van den Zegel et al.[51] reported values $D_{lat} = 2 \times 10^{-6}$ at 70°C and 4×10^{-7} cm^{2}s^{-1} at 25°C for the lateral diffusion of 1-methylpyrene in small unilamellar vesicles of DMPC. Note that these values are comparable to those reported for lipid analogues. The addition of cholesterol was also found to decrease D_{lat} of pyrene in various model membranes.[52] Papahadjopoulos and Kimelberg[53] showed that lateral exchange of monomers in the vesicle is very fast and dependent on the

Robinson and Zana

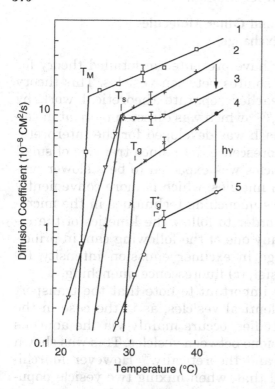

Figure 6.5 Temperature dependence of the lateral diffusion coefficient of the fluorescence probe diO18 (see Reference 49) in large vesicles of pure DMPC (curve 1), of a 1/1 mixture of DMPC and of a butadiene lipid before polymerization (curve 2), of the same mixture after polymerization of the butadiene lipid (curve 3), of the butadiene lipid before polymerization (curve 4) and of this lipid after polymerization (curve 5). T_M, T_S, T_g and T_g^* correspond to the transition temperatures of the lipids or lipid mixture. Polymerization is seen to reduce the lateral diffusion coefficient. Reproduced from Reference 49 with permission of American Physical Society.

length of the hydrocarbon chain, head group hydration, and temperature.

Note that the values of D_{lat} for pyrene and pyrene derivatives in erythrocyte membranes have been found to be in the range of 10^{-8} cm^{2}s^{-1},[54] that is, close to those measured in phospholipid vesicles.

B. Transport of Lipids and Other Molecules Between Vesicles (Exchange)

Tachiya and Almgren[55] have presented a detailed theory for migration dynamics of solutes between vesicles. This theory is closely related to earlier separate theoretical work by Tachiya[56] and Almgren,[57,58] which was directed more at micellar systems. The approach was developed for the interpretation of stopped-flow/fluorescence data since transfer of small molecules between vesicles was expected to be a slower process than that between micelles, which is more conveniently studied by luminescence quenching techniques in the microsecond time range. In order to follow the kinetics of the reequilibration process, any one of the following can, in principle, be used: (a) change in excimer emission intensity, (b) excitation energy transfer, (c) fluorescence quenching.

At the outset it is important to note that the transport of material between identical vesicles, as is the case in the vast majority of the studies, occurs mainly via the aqueous phase and not via collisions between vesicles. This was shown both experimentally[58] and theoretically.[59] However Marchi-Artzner et al.[60] showed that, when mixing two vesicle populations of low but opposite electrical charge, lipid exchange can occur via contact of the vesicles. In this case, lipid exchange results in the progressive neutralization of the electrical charge of the vesicles, without fusion of the vesicle inner compartments. Most of the studies reviewed below consider transport between identical vesicles.

An early study of molecular exchange between vesicles was reported by Sengupta and coworkers based on stopped flow/fluorescence.[61] They monitored the change in excimer fluorescence of pyrene and pyrenedecanoic acid in DPPC vesicles with added cholesterol. Exchange rates were comparable for the two molecules, and the exchange time between vesicles was reduced from 3 s to 0.5 s on increasing the temperature from 23°C to 68°C. Addition of cholesterol reduced the exchange rate, as did charging the vesicles by adding either europium chloride or dipalmitoylphosphatidic acid. The authors confirmed that probe transfer took place on a much faster time scale than either lipid exchange or vesicle fusion.

Almgren studied the migration of pyrene between unilamellar soybean lecithin vesicles (radius 32 nm) doped with 10% dicetylphosphate.[42,57,58] A typical result obtained with the stopped-flow/fluorescence method is shown in Figure 6.4.[42] Exit of pyrene from the vesicles was rate-limiting, and a value of 400 s^{-1} was found for the exit rate constant k^-. This compares with a value of 4000 s^{-1} for the exit rate constant of pyrene from SDS micelles. This large difference in exit rate constants was attributed to the large difference in size between micelles and vesicles.[57,58] Almgren used the same vesicle system to measure exit rate constants of perylene and pyrene, but using excitation energy transfer, since there is a dramatic sensitization of perylene fluorescence by pyrene. Values of 3 s^{-1} were measured for perylene, and 80 s^{-1} for pyrene. In a related study, static and time-resolved fluorescence quenching and stopped-flow/fluorescence were used to monitor dynamic processes involving the fluorophore pyrenebutyric acid in a didodecyldimethylammonium bromide vesicle system.[42] At low concentrations of iodide as quencher, only molecules on the vesicle outer surface were affected, but at higher iodide concentrations, the inside vesicle surface was also involved.

Almgren and Swarup[62] determined the relaxation time for the exchange (migration) of the pyrene-labeled carboxylic acids pyrenyl$(CH_2)_m CO_2 H$ with m = 3, 9 and 11, from soybean lecithin vesicles containing 10% dicetylphosphate. The results are represented in Figure 6.6. In the semi-logarithmic representation used, the plot is linear and its slope yields a free energy increment of 1 kT per methylene group. This value is rather close to that found for the free energy increment per CH_2 group from the same type of plot for the exit rate constant of surfactants from surfactant micelles (see Chapter 3, Section III.B, and Figure 3.8).

Table 6.1 lists some typical values of the exit rate constant of various molecules from vesicles.[57,58,61,63–68] The effect of alkyl chain length on the rate of transfer of N-alkylated alloxazines and isoalloxazines between DPPC vesicles is very similar to the results in Figure 6.6.[63] The very low values listed for lipid exchange rates are noteworthy. The listed results show that the value of k^- for a given exchanged mol-

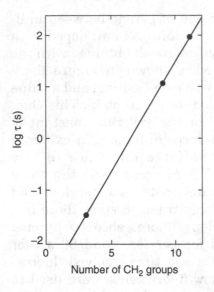

Figure 6.6 Variation of the relaxation time τ for the exchange (migration) of pyrene carboxylic acids pyrenyl$(CH_2)_m CO_2 H$ with the number of CH_2 groups m, in soybean lecithin + 10% dicetylphosphate vesicles at pH = 5.6. Adapted from Reference 62.

ecule depends relatively little on the nature of the lipid making up the vesicle but there is a much greater dependence on the alkyl chain length or hydrophobicity of the exchanged molecule. De Cuyper et al.[67] showed that the value of k^- for a given solute depends on the state of charge of the vesicle. However, again this effect is relatively small compared to that of the hydrophobicity of the exchanged molecule.

Recently, fluorescence measurements have been used to probe the motion of fluorescent-labeled molecules in more biologically relevant liposome systems.[40] Data were obtained for fluorescent-labeled analogues of spinghomyelin (SM'), ceramide (CM'), phosphatidylcholine (PC'), and diacylglycerol (DAG'). The half-lives for the exchange between vesicles was in the sequence SM' < PC' ~ CM' < DAG' with values ranging from 21 to 400 s for the first three molecules.

Bisby and Morgan[69,70] were interested in the release of solutes from liposomes under light illumination, and their

TABLE 6.1 Exit Rate Constant k^- for Various Molecules from Unilamellar Vesicles

Vesicle	Exchanged Molecule	k^- (s^{-1})	Experimental Conditions	Reference
Soybean Lecithin + 10% DCP[a]	Pyrene	77	18 °C; in 2 mM tris buffer and 8% ethanol	57
	Perylene	3		57
Egg lecithin	Pyrene	18	15 °C; in 2 mM tris buffer	57
	Pyrene	28	15 °C; in 0.15 M NaCl	
High-density lipoprotein	Pyrene	33	Buffer, pH = 7.4	64
DMPC	Pyrene	22	20 °C; 4% ethanol	57
	Perylene	1.2	20 °C; 4% ethanol	
DMPC	Pyrenenonanoic acid	5.4	28 °C; pH = 7.4	65
		0.08	28 °C; pH = 2.8	
DMPC	Pyrenelecithin	2×10^{-5}	36 °C	66
DMPC/DMPG	DMPC	1.5×10^{-4}	33 °C; 5 mM MES buffer, pH = 6.0; 10 mM KCl	67
DMPC	DMPC	1.4×10^{-4}	37°C	68
DPPC	Pyrenedecanoic acid	0.32	23°C; 1 mM NaCl	61
			68°C; 1 mM NaCl	
DPPC	Dibutylalloxazine	0.23	4°C; pH = 7.0	63
	Dioctylalloxazine	0.002	4°C; pH = 7.0	
	Didodecylalloxazine	very small	4°C; pH = 7.0	
	Dibutylisoalloxazine	0.1	4°C; pH = 7.0	

[a] DCP = dicetylphosphate

group has incorporated into low-temperature (the gel phase) phospholipids a photochromic lipid (with acyl chains and an azo group) as a membrane sensitizer. On UV illumination, the azobenzene species switches from the *trans* to the *cis* configuration, which destabilizes the immediate bilayer environment of the sensitizer, resulting in a loss of drug solubilized in the liposome interior to the external environment. In the papers, calcein release is monitored; upon release through the bilayer, there is a large increase in fluorescence intensity as calcein fluorescence is initially quenched inside the aqueous core of the liposome. At low sensitizer concentrations, the rate of solute release increases with temperature but then decreases above a threshold temperature. For example, for distearoylphosphocholine, the threshold temperature is 41°C, well below the phase transition temperature of 56°C.

Kleinfeld et al.[71,72] have been concerned with the problem of cellular fatty acid metabolism and the question of whether physiological transport of fatty acid across bilayers requires protein mediation or if it is a spontaneous process. Flip-flop of anthroyloxy fatty acids (AOFA) was found to be slower than 100 s. This result contrasted with data obtained by Kamp et al.,[73] who concluded that flip-flop of AOFA was a much more rapid process with a characteristic time of about 0.1 s. In a more recent paper, Kleinfeld et al.[74] looked again at the problem using a different method involving pyranine and carboxyfluorescein, but their conclusions were not changed.

C. Transport across Lipid Vesicle Bilayers

Transport across lipid vesicles is of the utmost importance as it mimics transport across biological cell membranes, which influences many biological processes essential in sustaining life. This problem has been much investigated, and only a few examples are given that illustrate the different mechanisms of transport of ions, fatty acids, and aminoacids. Note that the permeability of lipid bilayers to water and polar solutes has been reviewed.[75] In a fairly recent study Olson et al.[76] showed, perhaps surprisingly, that water crosses the bilayer of vesicles of the nonionic surfactant $C_{12}EO_4$ in less than 0.1 s (the resolution of the experiment).

Ion transport through lipid vesicles was reported in 1981 by Brock et al.[77] and was claimed to be the first application of a laser temperature-jump method for this type of work. The general background is described in earlier publications.[78,79] One approach was to employ a voltage pulse and monitor the current relaxation. Another method was to apply a temperature-jump during a series of short voltage pulses, with the aim of investigating the dynamics of adsorption/desorption of hydrophobic ions at the membrane interface. For the 2,4,6-trinitrophenolate ion, it was suggested that the permeation rate was limited by the barrier imposed by the membrane, whereas for the tetraphenylborate ion, the energy barriers for translocation and desorption from the membrane into water were of comparable magnitude.

Pyranine (8-hydroxy-1,3,6-pyrenesulfonate) has been used as a probe for proton concentration within vesicles. It is usually trapped inside the vesicle and proton concentration jumps then give rise to changes of fluorescence intensity.[80,81] Figure 6.7 shows the biphasic change of fluorescence intensity of vesicle-entrapped pyranine, observed following a rapid change of the pH external to the asolectin vesicles.[80] The fast process was attributed to a direct diffusion of protons through the vesicle bilayers. The slower of the two processes is attributed to a passive diffusion of protons across the bilayers. This is confirmed by the effect of addition of the nonionic surfactant Triton X100 prior to pH change (Figure 6.7c). The lysis of the vesicles then results in a very fast and monophasic change of fluorescence intensity upon the addition of hydrochloric acid. The slow component is strongly accelerated in the presence of valinomycin indicating a proton transport process via complexation of the valinomycin (Figure 6.7b). In the case of soybean phospholipid vesicles, the half-time for the intensity change was strongly decreased, from several minutes in the absence of ion carrier to 300 ms in the presence of valinomycin and to less than 1 ms in the presence of gramicidin.[81]

The rate of transfer of Na^+ and K^+ through bilayers of sodium 4-(1'-heptylnonyl)benzenesulfonate vesicles was increased upon the addition of the ionophore dicyclo-18-crown-6 (see Figure 6.8).[82] The rate increase was stronger for K^+ than Na^+. This completely artificial system thus shows

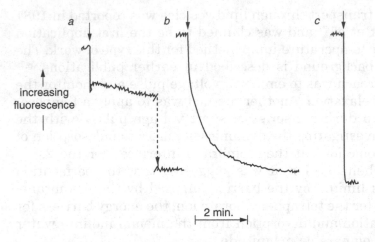

increasing
fluorescence

|← 2 min. →|

Figure 6.7 Change in the fluorescence intensity of pyranine entrapped in unilamellar vesicles of asolectin following a rapid decrease in external pH. The downward arrow indicates the addition of HCl which lowered the external pH to 6.2. (a) asolectin vesicles alone; the rapid drop of fluorescence intensity (direct diffusion of protons through the vesicle bilayers) is followed by a very slow one; (b) asolectin vesicles doped with valinomycin: the rapid change is not affected but the slow change is much accelerated; (c) asolectin vesicles + nonionic surfactant Triton X100: monophasic variation of intensity that reflects the lysis of the vesicles by the added Triton X100. Reproduced from Reference 80 with permission of the American Chemical Society.

selectivity in the same way as natural membrane systems. The observed rate increases were attributed to the formation of ionophore channels across the vesicle bilayers (see Figure 6.9). The same authors attributed the transport of alkali metal ions across vesicle bilayers of sodium 4-(1'-nonylundecyl)benzenesulfonate to the passage of the unhydrated ions through transient holes forming in the bilayers.[83] Transport as ion pairs or hydrated ions was excluded.

Ion transport can be strongly modified upon UV irradiation of vesicles that contain a photochemically active molecule. Thus Sato et al.[84] showed that the transport rate of K^+ ions through the bilayer of DMPC/dicetylphosphate (10/1) vesicles containing about 1% 4-octyl-4'-(5-carboxypentamethyle-

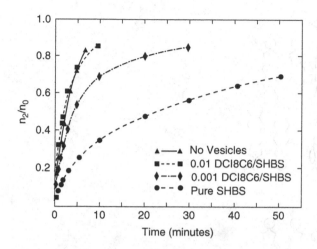

Figure 6.8 Effect of the ionophore dicyclohexano-18-crown-6 on the rate of K^+ ion leakage from vesicles of sodium 4-(1'-heptyl-nonyl)benzenesulfonate. At the highest concentration of ionophore K^+ diffuses through the dialysis bag containing the vesicle system as fast as if it is not entrapped in vesicles. Reproduced from Reference 82 with permission of Academic Press/Elsevier.

neoxy)azobenzene was increased by a factor of more than 5000 upon UV irradiation with respect to its value in the dark. Irradiation also increased the permeability of the negatively charged ion $Fe(CN)_6^{3-}$ through this negatively charged bilayer.

Additives, particularly surfactants, can have a very strong effect on the transport rate of compounds entrapped in the inner vesicle compartment, at concentrations well below that for which vesicle lysis occurs. The leakage rate of the probe 6-carboxyfluorescein (6CF) has been particularly used for that purpose.[85,86] Edwards and Almgren[85] measured the leakage rate of 6CF from sonicated egg lecithin vesicles upon addition of nonionic surfactants poly(ethyleneglycol) monododecylether $C_{12}EO_n$ (n = number of ethylene oxide units; see Figure 6.10). They attributed the increased permeation rate to the formation of transient channels that are coated with the added surfactant.

Ohno et al.[86] observed a rapid increase of leakage rate of 6CF at the main transition temperature of DMPC, DPPC,

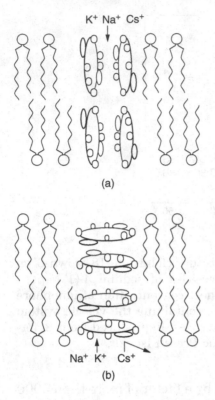

K^+ Na^+ Cs^+

(a)

Na^+ K^+ Cs^+

(b)

Figure 6.9 Possible conformations of the ionophore in the channel crossing the vesicle bilayer and through which the ions diffuse in and out. Reproduced from Reference 82 with permission of Academic Press/Elsevier.

and DSPC (S = stearoyl) vesicles, which they attributed to a transient structural disorder of lipid packing at the approach of the gel-to-liquid crystalline phase transition temperature of the lipid (see Section V).

Kleinfeld et al. worked at elucidating the rate-limiting step for transport of fatty acids (FA) across vesicle bilayers.[74,87] For this purpose they determined the rates at which FA flips-flops across, and binds to and dissociates from, the bilayers, using both large and giant unilamellar vesicles made up of eggPC and cholesterol. The rate constant for flip-flop was found to be less than the other two rates, and this process is therefore rate-limiting in the transport of FA.

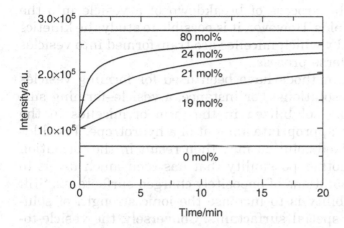

Figure 6.10 Leakage of 6CF from sonicated lecithin vesicles after addition of the indicated amounts of $C_{12}EO_8$. Reproduced from Reference 85 with permission of the American Chemical Society.

One last example illustrates the use of NMR for studying transport across vesicle bilayers. Viscio and Prestegard[88] were able to use proton NMR to follow the slow penetration of the neutral amine 5-hydroxytrytamine into the inner compartment of eggPC vesicles by following its reaction with adenosine triphosphate contained in this compartment. This technique has the advantage of not perturbing the system during the course of the reaction by taking samples.

III. KINETIC STUDIES OF THE VESICLE-TO-MICELLE AND MICELLE-TO-VESICLE TRANSFORMATIONS

As pointed out in Chapter 1, Section V of this book, vesicles are very long-lived structures. Indeed, the lipids or surfactants making up vesicles are much more hydrophobic than the surfactants making up micelles. Thus they have a very slow exchange rate as noted in the preceding section; see also Table 6.1. Besides, the number of lipids or surfactants making up a vesicle is 10^2 to 10^4 times larger than that for a usual surfactant micelle. Thus, under normal experimental conditions it will be difficult to observe the association of lipids into a vesicle or the process of breakdown of a vesicle into the constituting lipids. However, it is possible to study the kinetics of the process by which micelles are transformed into vesicles and of the reverse process.

Different methods have been used for forming vesicles from micellar solutions. For instance, a vesicle–forming surfactant may be solubilized in the form of micelles in the presence of an appropriate amount of a hydrotrope. The dilution of the mixed solution may then result in the formation of vesicles. Another possibility that has seen much use is to mix micellar solutions of oppositely charged surfactants. Still another possibility is to increase the ionic strength of solutions of some special surfactants. Conversely, the vesicle-to-micelle transformation can be induced by adding a micelle-forming surfactant, or a hydrotrope, or a hydrophobically modified water-soluble polymer to a vesicle system. This can also be achieved by decreasing the ionic strength. Note that the mechanism of the micelle-to-vesicle transformation has been investigated a great deal because it mimics the mechanism of reconstitution of biological membranes. Conversely, the vesicle-to-micelle transformation mimics the lysis of cell membranes. These transformations involve different intermediate structures as shown via transmission electron microscopy at cryogenic temperature (cryo-TEM)[89,90] and give vesicles of different sizes depending on the details of the method used to generate vesicles (see Reference 91, for instance). It is therefore expected that the kinetics of these transformations will also be much dependent on the method used to achieve the transformation in addition to a dependence on the nature of the compounds making up the system.

Early work was reported by Friberg and coworkers,[92,93] who studied the rate of transformation of disklike micelles of laureth 4 (Brij 30) containing the hydrotrope sodium xylenesulfonate into vesicles on a 1/1 dilution with water. Stopped-flow with light-scattering detection was used to follow the kinetics, and dynamic light scattering was used to characterize the initial and final states of the system. In another study, the sodium xylenesulfonate was replaced by dimethylisosorbide, a nonionic hydrotrope.[94] The relaxation signals were single exponential and relaxation times in the 0.1 to 10 s range were measured in these studies. Later work by the Friberg group was concerned with the lecithin/sodium xylenesulfonate system, and the kinetics of the micelle-to-vesicle transformation was again in the 1 to 10 s time range.[95] There was no detailed attempt to interpret the kinetics in terms of a mechanism for the conversion. However, the presence of a minimum in the variation of the reciprocal of the measured relaxation time with concentration in Reference 94 (see Figure 6.11) was noted by the authors to be reminiscent of the type of variation found for the formation/breakdown of surfactant micelles (see Chapter 3, Section III.C and Figure 3.7). In view of this behavior and the results for the variation of the measured relaxation time with the lipid concentration, the authors concluded that lecithin/sodium xylenesulfonate vesicles form by collision between aggregates.[95] It should be emphasized that in References 92–95 the perturbation brought to the initial micellar solution is very large and may affect its relaxation behavior.

Yatcilla et al.[32] investigated the conversion of a mixture of cationic and anionic surfactants to form vesicles. In this case, the reaction was very slow and a kinetic phase was associated with the evolution and growth of vesicles over a period of weeks (see Figure 6.12) to a final vesicular system, which, as already mentioned, was thought by the authors to be thermodynamically stable. A subsequent study explored the system CTAB mixed with sodium perfluorooctanoate.[96] Cylinders, disks, and spherical uni-lamellar vesicles were found to coexist at equilibrium by cryo-TEM. This observation confirms the importance of structural confirmation by cryo-TEM when this technique can be applied. From their analysis of the data, the mean curvature modulus, the Gaussian curvature modulus, and the spontaneous curvature could all be evaluated.

At around the same time that Yatcilla et al.[32] published their work, Robinson and coworkers[97] studied the formation and breakdown of vesicles formed from sodium 6-phenyltridecanesulfonate (S6PTDS). Light-scattering detection at 300 nm was used to monitor vesicle concentrations. In pure water, micelles are formed (cmc = 1.5 mM), but as the salt (NaCl) concentration is increased, vesicles are the preferred species.

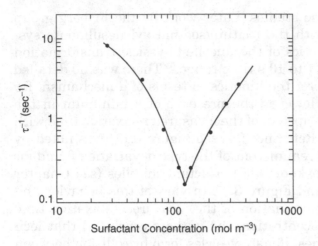

Figure 6.11 Variation of the reciprocal of the relaxation time measured upon 1/1 dilution of micellar solutions of laureth 4/dimethylisosorbide with water with the initial surfactant concentration. Reproduced from Reference 94 with permission of AOCS Press.

The critical vesicle concentration (cvc) decreases as the salt concentration is increased. Values of the cvc in the range 0.1 to 0.4 mM were measured for added NaCl concentrations in the range 0.025 to 0.10 M. The onset of vesicle formation can be monitored by measurement of a critical salt concentration (csc) for the micelle-to-vesicle transition. Vesicle breakdown was induced by rapid dilution of a vesicle system with an equal volume of water while maintaining the salt concentration constant. The rate of breakdown of vesicles at an initial surfactant concentration of 0.5 mM, and added ionic strength $I = 0.025$ M, showed that the half-life for breakdown varied from ~ 1s at 15°C to ~ 0.2 s at 35°C. The activation energy associated with the breakdown process was large, about +60 kJmol^{-1}. The reaction in the reverse direction, i.e., vesicle formation by monomer aggregation or micellar aggregation on adding salt, was slower with half-lives of the order of 1 min, and was not very dependent on the temperature. A mechanism was proposed that involved disk-like intermediate states (state D in Figure 6.13), followed by disjoining (or opening up) of the disks (state D' in Figure 6.13) to form stable vesicles.

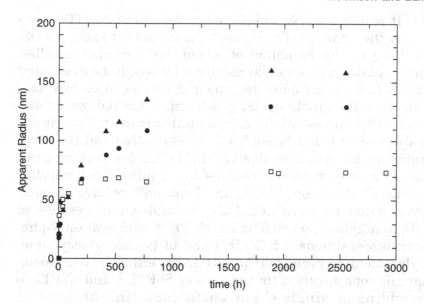

Figure 6.12 Time-dependence of the apparent aggregate radius for mixtures of CTAB and SOS. Micellar stock solutions with mixing ratio 20/80 CTAB/SOS are diluted to 2.5 wt% (□), 2.0 wt% (▲), and 1.5 wt% (•). The radius is measured by dynamic light scattering. After 3 months the apparent radius becomes constant at all 3 final concentrations. Reproduced from Reference 32 with permission of the American Chemical Society.

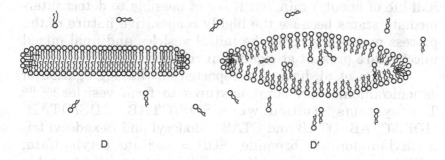

Figure 6.13 Disklike micelle D and disjoined disklike micelle D' postulated as intermediate states in the micelle-to-vesicle transformation. Reproduced from Reference 97 with permission of IOP Publ. Co.

It is also possible to induce breakdown of S6PTDS vesicles by the addition of a single-chain surfactant such as SDS, resulting in the formation of mixed S6PTDS/SDS micelles. The breakdown process was monitored through the associated change in light turbidity. In principle, there can be two components to the kinetics: a lag phase before breakdown occurs, followed by an essentially exponential transient. The kinetic analysis shows that breakdown is cooperative and the cooperative unit is based on about 8 SDS molecules working cooperatively to disrupt the vesicles by forming local micellar domains.[98] Robinson, Bucak, and Fontana[99,100] have recently investigated in more detail the breakdown of vesicles of sodium alkylbenzenesulfonates (S6PTDS and sodium 7-phenyltridecanesulfonate S7PTDS) and of hexadecyldecyldimethylammonium bromide ($C_{16}C_{10}DMABr$) either by decreasing the salt concentration (in the case of S6PTDS and S7PTDS) or adding a single-chain surfactant (in the case of $C_{16}C_{10}DMABr$). It was established that the greater the perturbation of the system beyond the micelle/vesicle transition region, the faster the rate of breakdown.[99,100] The approach was later extended to include studies with didodecyldimethylammonium bromide vesicles.[101] Very recently, the breakdown of S7PTDS vesicles on addition of SDS has been studied, detecting the structural changes by small angle neutron scattering detection in real time.[102] The breakdown process has a half-life of about 1 min, but it is not possible to detect intermediate states because the highly cooperative nature of the process means that only the initial vesicles and final mixed micelles are present at significant concentrations.[102]

Hatton et al. have also reported on the aggregation of cationic/anionic surfactant mixtures to form vesicles.[103–105] The systems studied were SOS/CTAB, SDS/DTAB, HDBS/CTAB (DTAB and CTAB = dodecyl and hexadecyl trimethylammonium bromide; SOS = sodium octylsulfate; HDBS = dodecylbenzenesulfonic acid). The transition to stable vesicles is rather slow for the SOS/CTAB and SDS/DTAB systems and complex with the observation of three relaxation processes with relaxation times of about 10, 100, and 2000 s in addition to a very fast process (< 4 ms, the dead time of the stopped-flow apparatus used).[103,105] In contrast, vesicle breakdown was very rapid (< 4 ms for the SOS/CTAB system and about 10 s for the SDS/DTAB system).[103] For the CTAB/HDBS system the micelle-to-vesicle transformation is more rapid and complete within a few minutes.[104] The progress of the transformation was monitored mainly by dynamic light scattering, but cryo-TEM was also used to characterize the evolving vesicle structures. Wormlike micelles, disks, and vesicles were all present in the first 2 h of reaction for CTAB/SOS system.[104] In a follow-up paper, Shioi and Hatton again use a combination of static and dynamic light scattering to follow the CTAB/SOS reaction in more detail.[105] Some 8 min after mixing, cryo-TEM shows that a mixture of closed vesicles and floppy bilayers are already present in solution. The first stage in the CTAB/SOS reaction is the formation of mixed micelles. It is supposed that SOS is incorporated into CTAB micelles in this initial step. It is entirely reasonable that this step will be very fast since the process is likely to involve an essentially diffusion-controlled insertion of SOS monomer into the CTAB micelles. The reaction scheme is therefore the following:[103]

CTAB micelles + SOS monomers → Mixed micelles	(in less than 1 ms)
Mixed micelles → Floppy bilayer aggregates	(in s to minutes)
Floppy bilayer aggregates → Small closed vesicles	(in minutes to hours)
Small closed vesicles → Equilibrium vesicles	(in weeks)

This series of reactions is represented schematically in Figure 6.14. The intermediate structures in this scheme are reminiscent but not completely identical to the intermediate structures visualized by electron microscopy or cryo-TEM. Shioi and Hatton[105] postulated that the small vesicles grow by fusing together. The progress to vesicle formation is complicated by the coexistence of bilayer aggregates and small closed vesicles in the reaction medium some minutes after mixing. It is possible that the system might behave in a simpler way if lower concentrations of the initial reactants would be used. At any rate, a reasonable theoretical model for the growth of the floppy bilayers (nonequilibrium disklike micelles) and of the small nonequilibrium vesicles was pro-

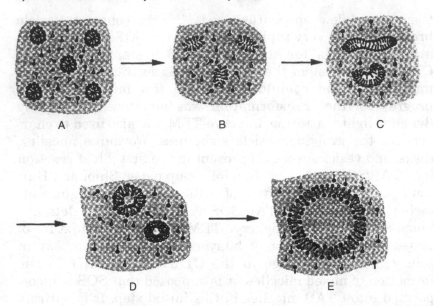

Figure 6.14 Possible intermediate structures occurring during the micelle-to-vesicle transformation for the SOS/CTAB system. (A) SOS monomers and CTAB micelles (In the experimental conditions used the number of SOS micelles was low and SOS micelles are not shown); (B) nonequilibrium mixed micelles; (C) floppy, irregularly shaped mixed micelles; (D) nonequilibrium vesicles; and (E) final vesicles. Reproduced from Reference 103 with permission of the American Chemical Society.

posed.[105] In this regard, the ideas of Lipowsky, concerning vesicle rigidity, were found to be helpful.[106] However, a quantitative interpretation of the changes in the initial stages of reaction remains somewhat elusive.

The system of SDS/DTAB has been also investigated by Wan et al.[107] The system involves the evolution of vesicles on mixing micelles of SDS and DTAB at a weight ratio of 1.6:1. Included in the reaction mixture was a low concentration of the solvatochromic probe 2,6-diphenyl-4-(2,4,6-triphenyl-1-pyridino) phenoxide ($E_T(30)$), which has been used previously to probe the interfacial properties of micelles and vesicles.[108,109] Wan et al.[107] suggest that a vesicle interfacial environment is established rather rapidly, on the time scale of seconds. Evolution of the initial vesicles to final stable structures was taking a period of days and the process was probed by cryo-TEM and video-enhanced microscopy.[107]

Blume and coworkers[110] have pioneered the use of isothermal titration calorimetry and differential scanning calorimetry in studies of micelle-to-vesicle transitions induced by increasing the temperature. Specifically the system DMPC/SDS has been studied, and some hysteresis effects were observed based on study of both the heating and cooling cycles.

Gradzielski, Grillo, and Narayanan[111] have studied the micelle-to-vesicle interconversion in two very different systems: (1) tetradecyltrimethylammonium hydroxide (TTAOH)/Texapon N_{70}-H ($C_{12}H_{25}(OC_2H_4)_{2.5}OSO_3H$), and (2) sodium oleate/1-octanol. The neutralization reaction in the first system was studied in real time by stopped-flow with small-angle x-ray scattering (SAXS) detection (ID-2 beam-line of the European Synchrotron Research Facility – ESRF, Grenoble, France). The second system was studied by stopped-flow with small-angle neutron scattering (SANS) detection (D22 instrument at the Institut Laue-Langevin–ILL, Grenoble). In the first experiment, 100 mM TTAOH was mixed with 100 mM Texapon N_{70}-H. For this equimolar ratio in the mixed system, stable vesicles are formed. Dynamic light scattering shows that the final vesicles are fairly monodisperse with a hydrodynamic radius of about 125 nm. In a first stage, parent micelles convert to bilayer structures within a few seconds, and the instrument resolution is such that this transformation can be followed, since a scattering intensity curve can in principle be measured in about 30 ms. The radius of the disk increases from 20 to 35 nm over the first 5 sec after mixing. The next stage of reaction for the bilayers, which are in the form of sheets, is to transform to vesicles. This growth process has a half-life of about 25 sec, and there is a decrease in polydispersity associated with this transition. The final vesicle size is comparable with the result from dynamic light scattering of 125 nm. In the second experiment, it is proposed that the intermediate structure is rodlike rather than disklike. In the SANS detection mode experiments are made over a few hundred seconds.

Egelhaaf et al.[112] have studied the kinetics of the micelle-to-vesicle transition when lecithin/cholesterol/bile salt micelles are diluted in water to form vesicles. A suitable reference to this procedure has been provided by Ollivon and coworkers.[113] The scope and applicability for time-resolved studies was previously discussed in detail by Egelhaaf some years ago.[114] For the lecithin system given earlier, the transition to vesicles was followed by time-resolved SANS (instrument D22 at ILL). Changes in structure were occurring over a period of hours, with a structural reorganization taking place at the start of the reaction that was too fast to be resolved. This was thought to indicate the formation of flat bilayer sheets with a thickness of about 5 nm. At the end state the neutron data revealed that the final vesicles form with a narrow size distribution and that their number increases during the course of the reaction. Additional experiments using static and dynamic light scattering have faster time resolution than SANS but cannot provide the same detailed information. However, it is likely that, in the intermediate state, the lecithin and cholesterol will be present in the middle of the disk with the bile salt located at the edges, where it is in pseudo-equilibrium with bile salt in the external water medium. There is a large increase in intensity associated with the transition. A summary of the stages in the micelle-to-vesicle transition is as follows:

	fast (s)		slow (hours)		fast (s)	
Micelles	$\rightarrow$	Small Disks	$\rightarrow$	Large Disks	$\rightarrow$	Vesicles

It is clear, however, from the range of studies discussed above, that a combination of different techniques is likely to offer the best way to study this sort of system at the present time.

Grillo, Kats, and Muratov[115] have recently used the new dedicated stopped-flow facility, located on the SANS instrument D22 at ILL, to investigate the micelle-to-vesicle transformation of AOT in aqueous solution upon direct mixing with solutions of salts such as NaCl, NaBr, KCl, and KBr. It was shown that the overall process is quite slow but intensity changes can be monitored over a total period of more than 5 h. The dead time of the stopped-flow instrument used is a few milliseconds, but the main limiting factor is the time needed to obtain a transient spectrum. This effectively puts a time limit of the order of 1 sec to obtain a data point, although by analyzing multiple spectra from a series of experiments, the signal-to-noise ratio can of course be improved. The region analyzed is essentially that of vesicle growth, which the authors identify with incorporation of micelles into the growing vesicles. They report that the vesicle radius increases as $t^{1/6}$ (t = time after mixing).

Vesicle formation in aqueous mixtures of lecithin and bile salt that takes place upon dilution with water has been studied intensively by means of time-resolved scattering methods (static and dynamic light scattering, SANS), and it has been concluded that this transition takes place via elongated micelles and disklike micelles as intermediate structures.[116,117]

The transformation of lipid vesicles into micelles upon addition of sodium cholate[118] or surfactants[119] is slow enough that the structural changes can be followed by cryo-TEM, but this is generally not the case as the reactions are taking place on too fast a time scale. Walter et al.[118] showed that the intermediate structures involved in the vesicle-to-micelle transition of eggPC upon addition of sodium cholate are open vesicles, large bilayer sheets, and long flexible cylindrical micelles, which evolved from the edges of the bilayer sheets. Silvander et al.[119] showed that the transformation of lecithin vesicles into micelles upon addition of sodium alkylsulfates involves intermediate structures that depend on the surfactant chain length. Likewise, the transformation of the doubly lamellar vesicles of the gemini surfactant 12-20-12 into micelles upon addition of the threadlike micelle-forming gemini surfactant 12-2-12 was slow enough to be followed by cryo-TEM.[120] This study evidenced numerous intermediate structures: disklike micelles, ringlike micelles, short threadlike micelles, and metastable irregular ribbonlike aggregates.

A slow reorganization was reported to occur upon addition of amphiphilic polymers to a system of small vesicles of DPPC/DPPA (dipalmitoylphosphatidic acid) that resulted in mixed aggregates of micellar size.[121] The added polymers were poly(sodium acrylate/isopropylacrylamide) bearing some pendant hydrophobic alkyl chains with 8 or 18 carbon atoms. The change of vesicle hydrodynamic radius with time was biphasic

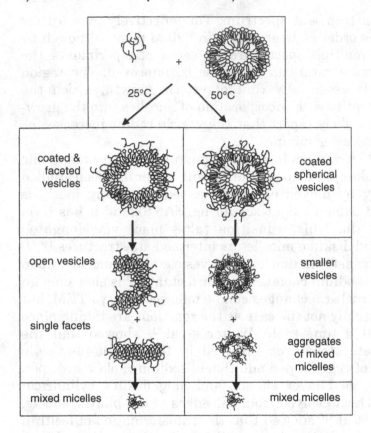

Figure 6.15 Tentative mechanism of vesicle reorganization in the presence of hydrophobically modified poly(sodium acrylate-isopropylacrylamide). Path 50°C: disruption resulting from the presence of intermediate species having a broad range of size or a size continuously decreasing with time up to the final size. Path 25°C: direct formation of small mixed globules in the absence of intermediate size between the initial faceted vesicles or sheets of membranes and the final mixed complexes. Reproduced from Reference 121 with permission of Elsevier.

(or more) and stretched over hours to months depending on the length and amount of hydrophobic groups. Freeze fracture electron microscopy (FFEM) was used to identify an intermediate structure. Figure 6.15 shows the tentative mechanism of vesicle disruption postulated on the basis of the FFEM and light scattering observations.

These results illustrate the variety of behaviors observed when studying the vesicle-to-micelle and micelle-to-vesicle transformations. This diversity reflects the different intermediate structures that can be encountered during the transformations, depending on the way the transformation is achieved and the nature of the lipids or surfactants involved. When the additives that induce the transformation can enter or leave the aggregates very rapidly, such as the hydrotropes that do not take part in the formed vesicles, and when the aggregates are nonionic with weakly repulsive or even attractive interactions, the transformation can be very rapid and characterized by a single relaxation time. Indeed, collisions between aggregates can then occur and can completely dominate the transformation. This may be the case in the studies of Friberg at al.[92–95] It would be interesting to examine these systems by cryo-TEM to check if indeed intermediate structures are absent between micelles and vesicles. Most of the other studies of the vesicle-to-micelle transformation involved electrically charged aggregates and surfactants that enter and leave aggregates much more slowly than hydrotropes. This effect results in both slowing down the transformation and making transient and slowly evolving intermediate structures more likely to occur. More generally cryo-TEM studies should always be performed in parallel to kinetic studies whenever possible. This permits a surer assignment of the observed relaxation processes.

IV. FATTY-ACID AND MIXED PHOSPHOLIPID/FATTY ACID VESICLE SYSTEMS

Fatty acid vesicles were first characterized by Gebicki and Hicks in 1973.[122] At pH values greater than 10, the carboxylic acid head groups are fully ionized, and micelles are formed above the cmc. As the pH is reduced to the region around pH 7–9, depending on the particular fatty acid, partial protonation of the anionic head takes place, and there is a micelle-to-vesicle transition. It is usually assumed that weakening of the head group repulsion through formation of hydrogen-bonded surface interactions is a major factor in driving the structural transition, although structural rearrangement of the hydrocarbon region is also likely to be important. In the region where vesicles exist, there remains some uncertainty as to whether micelles and/or monomers coexist in the system, as evidenced by a recent esr study in which the spin-label probe molecule 16-doxylstearic acid was incorporated into oleic acid/oleate vesicles. However, particularly stable vesicles were found at pH = 8.5.[123]

Equilibration of a fatty acid vesicle system across a semi-permeable dialysis membrane which is impermeable to vesicles has recently been studied.[124] Extensive dialysis (> 24 h) apparently leads to very slow formation of vesicles in the coexisting buffer solution; in contrast, equilibration in the corresponding micellar system, at higher pH, is achieved much more rapidly, on the time scale of minutes.

Phospholipids can be readily combined with fatty acids to form mixed vesicle assemblies. The permeability of the membranes can be influenced by the proportion and type of fatty acid added. Muranushi and coworkers[125] incorporated the fatty acids oleic, linoleic, and stearic acid into egg phosphocholine lipid membranes and looked at the effect on permeability of the drugs sulfanilic acid and procainaimide ethobromide. They showed that the fatty acids interacted with the polar region of the membrane, and also increased disorder in the membrane interior, facilitating release of both drugs, with unsaturated acids having a greater effect. Kleinfeld and coworkers[87] measured partition coefficients of oleic, linoleic, and palmitic acids into phospholipid bilayers and transport across the membrane. The rate-limiting step in transfer was flip-flop across the membrane; first-order rate constants in the 0.3-1 s^{-1} region were measured. Doody and coworkers[65] investigated the rate of migration of the fluorescent fatty acid 9-(3-pyrenyl)-nonanoic acid across phospholipid bilayers. A value of 5 s^{-1} was determined, suggesting the rate-determining step was the exit of the fatty acid from a vesicle. Treyer et al.[126] measured the permeability of the nucleotides AMP,

UMP, UDP, and UTP in POPC/sodium cholate vesicle systems and found a maximum of permeability in systems where the cholate concentration within the vesicles approached the saturation limit before destabilization of vesicles occurred. Kraske and Mountcastle[127] studied the effect of temperature and added cholesterol on the permeability of DMPC bilayers. For the leakage of glucose, they showed that maximum permeability was achieved at 24°C, the main phase transition temperature of DMPC, and was due to lateral area and volume changes in the lipid domains as they fluctuate between gel and fluid states.

V. DYNAMIC CHANGES ASSOCIATED WITH VESICLE PHASE TRANSITIONS

The temperature T_C at which the gel-to-liquid crystalline phase transition of lipid and surfactant bilayers occurs is an important characteristic of these systems. Maybrey and Sturtevant[128] observed an increase in the phase transition temperature when palmitic acid was added to DMPC, but both increases and decreases can be observed with different fatty acids.[129–131] It has been shown in Section II.A that the lateral and transverse motions of lipids in vesicle bilayers become more rapid as the temperature is increased from below to above T_C. This section reviews dynamic processes that occur in bilayers at around T_C.

The ultrasonic relaxation method has been used a great deal to study very fast processes detected at the approach to T_C. At a given ultrasonic frequency, the ultrasonic absorption goes through a well-defined maximum at T_C indicating the occurrence of relaxation processes associated with this transition.[132,133] The measured relaxation time was around 10^{-8} s for dipalmitoyl lecithin vesicles, and it was attributed by Gamble and Schimmel to the isomerization of the lipid chain (kink formation).[132] This finding and assignment were confirmed in studies of other vesicle systems, DPPC,[134] DODDMAB (dioctadecyldimethylammonium chloride)[135] and DMPC,[136] that used the ultrasonic absorption relaxation technique[134,135] and also p-jump[135] and T-jump with Joule[134] or laser[136] heating (see Chapter 2, Section III.D.1). These studies also evidenced

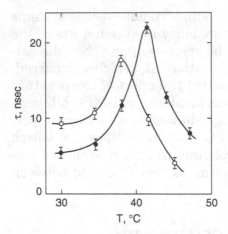

Figure 6.16 Temperature dependence of the relaxation time associated with isomerization of the lipid alkyl chain in DODDMAB vesicles as measured by the ultrasonic relaxation method for small vesicles (o) and for large vesicles (•). Reproduced from Reference 134 with permission of the American Chemical Society.

several other relaxation processes occurring on longer time scales and showed that all the associated relaxation times and relaxation amplitudes go through a maximum at T_C (see Figure 6.16).

Holzwarth and coworkers performed extensive studies of the dynamics in the phase transition region of DMPC, DPPC, DML (dimyristyl-1-α-lecithin), and DPL (dipalmitoyl-1-α-lecithin) vesicles using a laser T-jump apparatus.[136–139] One of the key advantages of the laser heating method is that it avoids the strong electric fields (15–40 kV cm^{-1}) used in the more conventional Joule-heating T-jump method, which can distort vesicles and lead to pore generation (see Section VI). The relaxation signals (turbidity and fluorescence intensity changes) were characterized by up to five relaxation times in the 4 ns–10 ms time range. Figure 6.17 shows the model used for the interpretation of the observed relaxation behavior of vesicles at around T_C. The fastest processes are associated with kink formation in the alkyl chains in the membrane (1 → 2) as discussed earlier and membrane expansion and increased mobility of the lipid head groups (2 → 3). More

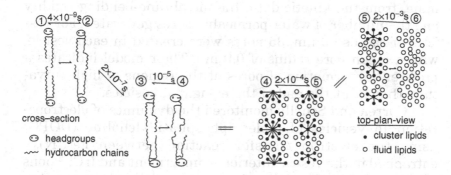

Figure 6.17 Model for the five relaxation processes associated with the phase transition of vesicle bilayers. Reproduced from Reference 136 with permission of Springer-Verlag.

complex isomers are then created and lateral diffusion increases $(3 \rightarrow 4)$. This allows clusters of different order to be formed and the fluid clusters grow $(4 \rightarrow 5)$. In the last step, the rigid lipid clusters are completely dissolved and the membrane is in the fluid state $(5 \rightarrow 6)$. The cooperative nature of the process was indicated by the dramatic increase in the slow relaxation times at the transition temperature itself.

The dynamics of the slow processes had been discussed earlier by Kanehisa and Tsong.[140]

VI. EFFECT OF APPLIED ELECTRIC FIELDS ON THE PROPERTIES OF VESICLE BILAYERS

When biological cells and lipid vesicles are exposed to an electric field, they are mechanically deformed and transient pores can be induced to form in the membrane (electroporation).[141–145] The field has been recently reviewed.[146] Only two recent examples are presented.

Kakorin and Neumann[147] studied the deformation of salt-filled phosphatidylcholine/phosphatidylglycerol vesicles, when electrical field pulses of 10 kV cm^{-1} were applied over a period of some milliseconds. The change in electrical conductivity and vesicle volume were determined. The theoretical analysis is quite complex, but the authors were able to determine, from the kinetic data, the membrane-bending rigidity and the number of water-permeable pores generated. For vesicles of radius 50 nm, 35 pores were created in each vesicle, with a mean pore radius of 0.9 nm. Their model indicates a preferential formation of pores at the regions of high curvature, that is, at the end of the elongated vesicles.

Correa and Schelly monitored the dynamics of electroporation in vesicles of dioleoylphosphatidylcholine (DOPC) using the electron transfer reaction between $Fe(CN)_6^{4-}$ entrapped in the vesicle interior compartment and $IrCl_6^{2-}$ ions in the outside[148] or the chemical reaction between Ag^+ (entrapped) and Br^- (outside).[149] In both instances birefringence measurements showed that the applied field induced an elongation of the vesicles in the direction of the field. The increased pressure inside the vesicle results in the ejection of part of its content when pores are formed. The electron transfer reaction was detected, indicating electroporation, only for pulse lengths longer than 200 μs for an applied field strength of 8 kV cm^{-1}. The Ag^+/Br^- reaction was detected after a shorter time of about 30 μs because the entrapped Ag^+ is of smaller dimension than $Fe(CN)_6^{4-}$ and requires smaller pores to be ejected. This permitted to evaluate as 4 μm/s the average rate of growth of the pores.[148] The reaction between Ag^+ and Br^- takes place outside the vesicles.

VII. CONCLUSIONS

The study of vesicle dynamics is now a mature area of research. The main processes involved in bilayer mobility and solute transport are quite well understood, but this is not so with the spontaneous formation and breakdown of vesicles. There is a significant difference between the dynamics of micelle formation/breakdown and vesicle formation/breakdown. Micelle phenomena occur on a very short time scale, with processes for most micellar systems taking place in time scales less than 1 sec. For vesicles of synthetic surfactants like the alkylbenzenesulfonates, the relevant processes are in the second to minute time range, although surfactant monomer exchange between vesicles and aqueous solution may well take place in the millisecond time range. The critical vesicle concentration is an important parameter, which can be measured for systems like sodium 6-phenyltridecanesulfonate, but not in the case of phospholipids, where the monomer concentration external to the bilayer is likely to be of the order of nanomolar. This means that dynamics of micelle-to-vesicle transformations can be very slow, and often what is studied in the classical time window (seconds to days) is the evolution of vesicle structures. Sometimes this evolution is clear from experimental measurements, but too often the picture is still confusing. There is a need to define some appropriate model systems for study.

Again, it would be useful to observe directly the evolution of intermediate morphologies by cryo-TEM for instance, but this is difficult in systems where the transition from/to vesicles is highly cooperative as the intermediate concentrations may then be very low. Another problem is that of hysteresis. Sometimes this is observed, but often it is ignored or un recorded. This is linked to the possibility that the system under any given set of conditions may not always be at a true thermodynamic equilibrium.

REFERENCES

1. Bangham, A.D., Horne, R.W. *J. Mol. Biol.* 1964, 8, 660–668.

2. Evans, E., Needham, D. *J. Phys. Chem.* 1987, 91, 4219.

3. Menger, F.M., Keiper, J.S. *Adv. Materials* 1998, 10, 888–890.

4. Lasic, D.D. *American Scientist* 1992, 80, 20–31.

5. Mayer, L.D., Hope, M.J., Cullis, P.R. *Biochim. Biophys. Acta* 1986, 858, 161–168.

6. Hunter, D.G., Frisken, B.J. *Biophys. J.* 1998, 74, 2996–3002.

7. Kunitake, T., Okahata, Y. *J. Am. Chem. Soc.* 1977, 99, 3860–3861.

8. Danino, D., Kaplun, A., Talmon, Y., Zana, R. In *Structure and Flow in Surfactant Solutions*, Herb, C.A., Prud'homme, R.K. Eds., ACS, Washington DC, 1994, p. 105–119.

9. Engberts, J.B.F.N., Hoekstra, D. *Biochim Biophys Acta* 1995, 1241, 323–340.

10. Krafft, M.P., Reiss, J.G. *Biochemie* 1998, 80, 489–514.

11. Ferro, Y., Krafft, M.P. *Biochim. Biophys. Acta* 2002, 1581, 11–20.

12. Lipowsky, R., Sackmann, E. *Structure and Dynamics of Membranes*, Vol. 1, Elsevier, Amsterdam, 1995.

13. Barenholz, Y., Lasic, D.D. *Handbook of Non-Medical Applications of Liposomes*, CRC Press, Boca Raton, FL, 1996.

14. Hackel, W., Seifert, U. Sackmann, E. *J. Phys. II* 1997, 7, 1141–1147.

15. New, R.R.C. In *Liposomes: A Practical Approach*, Oxford University Press, 1990.

16. Wilschut, J., Duguenes, N., Fraley, R., Papahadjopoulos, D. *Biochemistry* 1980, 19, 6011–6020.

17. Zellmer, S., Cevc, G., Risse, P. *Biochim. Biophys. Acta* 1994, 1196, 101–113.

18. Garcia, R.A., Pantazatos, S.P., Pantazatos, D.P., McDonald, R.C. *Biochim. Biophys. Acta* 2001, 1511, 264–270.

19. Jaycock, M.J., Ottewill, R. *Fourth International Congress on Surface Active Substances*, Brussels 1964, Section B, paper 8.

20. Zana, R., Lang, J. *C. R. Acad. Sci. (Paris) C* 1968, 266, 893–896 and 1347–1350.

21. Sams, P.J., Wyn-Jones, E., Rassing, J. *Chem. Phys Lett.*, 1972, 13, 233–236.

22. Aniansson, E.A.G., Wall, S.N., Almgren, M., Hoffmann, H., Kielmann, I., Ulbricht, W., Zana, R., Lang, J., Tondre, C. *J. Phys. Chem.* 1976, 80, 905–922.

23. Aniansson, E.A.G., Wall, S.N. *J. Phys. Chem.* 1974, 78, 1024–1030 and 1975, 79, 857–863.

24. Kahlweit, M., Teubner, M. *Adv. Colloid Interface Sci.* 1980, 13, 1–20.

25. Kahlweit, M. *J. Colloid Interface Sci.* 1980, 90, 92–110.

26. Herrmann, C.-U., Kahlweit, M. *J. Phys. Chem.* 1980, 84, 1536–1544.

27. Patist, A., Kanicky, J.R., Shulka, P.K., Shah, D.O. *J. Colloid Interface Sci.* 2002, 245, 1–15.

28. Luisi, P.L. *J. Chem. Ed.* 2001, 78, 380–384.

29. Laughlin, R.G. *Colloids Surf. A* 1997, 128, 27–38.

30. Laughlin, R.G. *The Aqueous Phase Behaviour of Surfactants*, Academic Press, New York, 1994.

31. Lasic, D.D. *J. Liposome Res.* 1999, 91, 43–52.

32. Yatcilla, M., Herrington, K.L., Brasher, L.L., Kaler, E.W., Chiruvolu, S., Zasadzinski, J.A. *J. Phys. Chem.* 1996, 100, 5874–5879.

33. Lampietro, D.J., Kaler, E.W. *Langmuir* 1999, 15, 8590–8601.

34. Israelachvili, J., Mitchell D.J., Ninham B.W. *J. Chem. Soc. Faraday Trans. 2* 1976, 72, 1525–1535.

35. Nagarajan, R. *Langmuir* 2002, 18, 31–38.

36. Maeda, N., Senden, T.J., di Meglio J.M. *Biochim. Biophys. Acta* 2002, 1564, 165–172.

37. Lasic, D.D. *Biochim. Biophys. Acta* 1982, 692, 501–512; *J. Theor. Biol.* 1987, 124, 35–45; *Biochem. J.* 1988, 256, 1–18; *Liposomes: From Physics to Applications*, Elsevier, Amsterdam, 1993.

38. Ishikawa, Y., Kuwahara, H., Kunitake, T. *J. Am. Chem. Soc.* 1994, 116, 5579–5591.

39. Kornberg, R.D., McConnell, H.M. *Biochemistry* 1971, 10, 1111–1120.

40. Bai, J., Pagano, R.E. *Biochemistry* 1997, 36, 8840–8848, and references cited.

41. Galla, J.J., Theilen, U., Hartmann, W. *Chem. Phys. Lipids* 1979, 23, 239–251.

42. Almgren, M. *J. Phys. Chem.* 1981, 85, 3599–3603.

43. Cocera, M., Lopez, O., Estelrich, J., Parra, J.L., de la Maza, A. *Langmuir* 1999, 15, 6609–6612.

44. Moss, R.A., Fujita, T., Ganguli, S. *Langmuir* 1990, 6, 1197–1199.

45. Moss, R.A., Ganguli, S., Okumura, Y., Fujita, T. *J. Am. Chem. Soc.* 1990, 112, 6391–6392.

46. Moss, R.A., Li, J.M., Kotchevar, A. *Langmuir* 1994, 10, 3380–3382.

47. Devaux, P., Mc Connell, H.M. *J. Am. Chem. Soc.* 1972, 94, 4475–4481.

48. Wu, E.–S., Jacobson, K., Pahadjopoulos, D. *Biochemistry* 1977, 16, 3936–3941.

49. Gaub, H., Sackmann, E., Büschl, R., Ringsdorf, H. *Biophys. J.* 1984, 45, 725–731.

50. Kano, K., Kawaizumi, H., Ogawa, T., Sunamoto, J. *J. Phys. Chem.* 1981, 85, 2204–2209.

51. Van den Zegel, M., Boens, N., De Schryver, F.C. *Biophys. Chem.* 1984, 20, 333–345.

52. Dobretsov, G., Borchevskaya, T., Petrov, V. *Biofizika* 1980, 25, 960–966.

53. Papahadjopoulos, D., Kimelberg, H.K. *Phospholipid Vesicles (Liposomes) as Models for Biological Membranes*, Pergamon Press, Oxford, 1973.

54. Eisinger, J., Scarlata, S. *Biophys. Chem.* 1987, 28, 273–280.

55. Tachiya, M., Almgren, M. *J. Chem. Phys.* 1981, 75, 865–870.

56. Tachiya, M. *Chem. Phys. Lett.* 1975, 33, 289–293.

57. Almgren, M. *Chem. Phys. Lett.* 1980, 71, 539–543.

58. Almgren, M. *J. Amer. Chem. Soc.* 1980, 102, 7882–7888.

59. Ducwitz–Peterlein, G., Moraal, H. *Biophys. Struct. Mechanism* 1978, 4, 315–326.

60. Marchi–Artzner, V., Jullien, L., Belloni, L., Raison, D., Lacombe, L., Lehn, J.M. *J. Phys. Chem.* 1996, 100, 13844–13856.

61. Sengupta, P., Sackmann, E., Kuhnle, W., Scholz, H.P. *Biochim. Biophys. Acta* 1976, 436, 869–878.

62. Almgren, M., Swarup, S. *Chem. Phys. Lipids* 1982, 31, 13–22.

63. Kano, X., Yamaguchi, Y., Matsuo, Z. *J. Phys. Chem.* 1980, 84, 72–76.

64. Charlton, S., Olsson, J., Wong, K.Y., Pownall, H., Louie, D., Smith, L. *J. Biol. Chem.* 1976, 251, 7952–7955.

65. Doody, M., Pownall, Kao, Y., Smith, L. *Biochemistry* 1980, 19, 108–116.

66. Roseman, M., Thompson, T. *Biochemistry* 1980, 19, 439–444.

67. De Cuyper, M., Joniau, M., Engberts, J.B.F.N., Sudholter, E.J.R. *Colloids Surf.* 1984, 10, 313–319.

68. McLean, L.R., Phillips, M.C. *Biochemistry* 1984, 23, 4624–30.

69. Bisby, R.H., Mead, C., Mitchell, A.C. Morgan, C.G. *Biochem. Biophys. Res. Comm.* 1999, 262, 406–410.

70. Morgan, C.G. Bisby, R.H., Johnson, S.A., Mitchell, A.C. *FEBS Lett.* 1995, 375, 113–116.

71. Storch, J., Kleinfeld, A.M. *Biochemistry* 1986, 25, 1717–1726.

72. Kleinfeld, A.M., Storch, J. *Biochemistry* 1993, 32, 2053–2061.

73. Kamp, F., Zakim, D., Zhang, F., Noy, N., Hamilton, J.A. *Biochemistry* 1995, 34, 11928–11937.

74. Kleinfeld, A.M., Chu, P., Storch, J. *Biochemistry* 1997, 36, 5702–5711.

75. Disalvo, E.A. *Adv. Colloid Interface Sci.* 1988, 29, 141–170.

76. Olson, U., Nakamura, K., Kunieda, H., Strey, R. *Langmuir* 1996, 12, 3045–3051.

77. Brock, W., Stark, G., Jordan, P.C. *Biophys. Chem.* 1981, 13, 329–348.

78. Knoll, W., Stark, G. *J. Membrane Biol.* 1979, 37, 13–21, and 1975, 25, 249–260.

79. Jordan, P.C., Stark, G. *Biophys. Chem.* 1979, 10, 273–290.

80. Clement, N.J., Gould, J.M. *Biochemistry* 1981, 20, 1534–1538.

81. Biegel, C.M., Gould, J.M. *Biochemistry* 1981, 20, 3474–3479.

82. Hamilton, R., Kaler, E. *J. Colloid Interface Sci.* 1987, 116, 248–255.

83. Hamilton, R., Kaler, E. *J. Phys. Chem.* 1990, 94, 2560–2566.

84. Sato, T., Kijima, M., Shiga, Y., Yonezawa, Y. *Langmuir* 1991, 7, 2330–2335.

85. Edwards, K., Almgren, M. *Langmuir* 1992, 8, 824–832.

86. Ohno, H., Ukaji, K., Tsuchida, E. *J. Colloid Interface Sci.* 1987, 120, 486–494.

87. Kleinfeld, A.M., Chu, P., Romero, C. *Biochemistry* 1997, 36, 14146–14158.

88. Viscio, D.B., Prestegard, J.H. *Proc. Nat. Acad. Sci. USA* 1981, 78, 1638–1642.

89. Vinson, P.K., Talmon, Y., Walter, A. *Biophys. J.* 1989, 56, 669–681.

90. Danino, D., Talmon, Y., Zana, R. *J. Colloid Interface Sci.* 1997, 185, 84–93.

91. Sun, C., Ueno, M. *Colloid Polym. Sci.* 2000, 278, 855–863.

92. Campbell, S.E., Yang, H., Patel, R., Friberg, S.E., Aikens, P.A. *Colloid Polym. Sci.* 1997, 275, 303–306.

93. Friberg, S.E., Campbell, S.E., Fei, L., Yang, H., Patel, R., Aikens, P.A. *Colloid Surf. A* 1997, 129–130, 167–173.

94. Zhang, Z., Ganzuo, L., Patel, R., Friberg, S.E., Aikens, P.A. *J. Surfactants Deterg.* 1998, 1, 393–398.

95. Campbell, S.E., Zhang, Z., Friberg, S.E., Patel, R. *Langmuir* 1998, 14, 590–594.

96. Zasadzinski, J.A., Jung, H.T., Coldren, B., Kaler, E.W. In *Self-Assembly*, Robinson, B.H., Ed., IOS Press, Amsterdam, 2003, pp. 432–442.

97. Farquhar, K.D., Misran, M., Robinson, B.H., Steytler, D.C., Moroni, P., Garrett, P.R., Holzwarth, J.F. *J. Phys. Condens. Matter* 1996, 8, 9397–9404.

98. Brinkmann, U., Neumann, E., Robinson, B.H. *J. Chem. Soc. Faraday Trans.* 1998, 94, 1281–1286.

99. Robinson, B.H., Bucak, S., Fontana, A. *Langmuir* 2000, 16, 8231–8237.

100. Robinson, B.H., Bucak, S., Fontana, A. *Langmuir* 2002, 18, 8288–8294.

101. Fontana, A., de Maria, P., Siani, G., Robinson, B.H. *Colloid. Surf. A*, 2003. 32. 365–374.

102. Eastoe, J., Heenan, R., Grillo, I., Steytler, D., Rogerson, M., Robinson, B.H., to be published.

103. O'Connor, A.J., Hatton, T.A., Bose, A. *Langmuir* 1997, 13, 6931–6935.

104. Xia, Y., Goldmints, I., Johnson, P.W., Hatton, T.A., Bose, A. *Langmuir* 2002, 18, 3822–3828.

105. Shioi, A., Hatton, T.A. *Langmuir* 2002, 18, 7341–7348.

106. Lipowsky, R. *J. Phys. II France* 1992, 2, 1825–1832, *Physica A* 1993, 194, 114–121.

107. Wan, M., O'Connor, A., Bose, A., Grieser, F. In *Self-Assembly* Robinson, B.H., Ed., IOS Press, Amsterdam, 2003, pp. 454–463.

108. Drummond, C.J., Warr, G.G., Grieser, F. *J. Phys. Chem.* 1985, 89, 2103–2109.

109. Drummond, C.J., Grieser, F., Healy, T.W. *J. Phys. Chem.* 1988, 92, 5580–5585.

110. Blume, A. In *Self-Assembly* Robinson, B.H., Ed., IOS Press, Amsterdam, 2003, pp. 389–409.

111. Gradzielski, M., Grillo, I., Narayanan, T. In *Self-Assembly*, Robinson, B.H., Ed., IOS Press, Amsterdam, 2003, pp. 410–421.

112. Egelhaaf, S.U., Leng, J., Sazlonen, A., Schurtenberger, P., Cates, M.E. In *Self Assembly*, Robinson, B.H., Ed., IOS Press, Amsterdam, 2003, pp. 422–431.

113. Ollivon, M., Lesieur, S., Grabielle–Mademont, C., Paternostre, M. *Biochim. Biophys. Acta* 2000, 1508, 34–48.

114. Egelhaaf, S.U. *Curr. Opin. Colloid Interface Sci.* 1998, 3, 608–613.

115. Grillo, I., Kats, E.I., Muratov, A.R. *Langmuir* 2003, 19, 4573–4581.

116. Egelhaaf, S.U., Olsson, U., Schurtenberger, P. *Physica B* 2000, 276, 326–340.

117. Egelhaaf, S.U., Schurtenberger, P. *Phys. Rev. Lett.* 1999, 82, 2804–2814.

118. Walter, A., Vinson, P.K., Kaplun, A., Talmon, Y. *Biophys. J.* 1991, 60, 1315–1325,

119. Silvander, M., Karlsson, G., Edwards, K. *J. Colloid Interface Sci.* 1996, 179, 104–115.

120. Bernheim–Groswasser, A., Zana, R., Talmon, Y. *J. Phys. Chem. B* 2000, 104, 12192–12201.

121. Ladavière, C., Toustou, M., Gulik–Krzywicki, T., Tribet, C. *J. Colloid Interface Sci.* 2001, 241, 178–187.

122. Gebicki, J.M., Hicks, M. *Nature* 1973, 243, 232–234,

123. Walde, P., Morigaki, K. In *Self–Assembly*, Robinson, B.H., Ed., IOS Press, Amsterdam, 2003, pp. 443–453.

124. Morigaki, K., Walde, P., Misran, M., Robinson, B.H. *Colloids. Surf. A* 2003, 213, 37–44.

125. Muranushi, N., Takagi, N., Muranushi, S., Sezaki, H. *Chem. Phys. Lipids* 1981, 28, 269–279.

126. Treyer, M., Walde, P., Oberholzer, T. *Langmuir* 2002, 18, 1043–1050.

127. Kraske, W.V., Mountcastle, D.B. *Biochim. Biophys. Acta* 2001, 1514, 159–164.

128. Mabrey, S., Sturtevant, J.M. *Biochim. Biophys. Acta* 1977, 486, 444–450.

129. Inoue, T., Yanagihara, S., Misono, Y., Suzuki, M. *Chem. Phys. Lipids* 2001, 109, 117–133.

130. Cevc, G., Seddon, J.M., Hartung, R., Eggert, W. *Biochim. Biophys. Acta* 1988, 940, 219–240.

131. Seddon, J.M., Templer, R.H., Warrender, N.A., Huang, Z., Cevc, G., Marsh, D. *Biochim. Biophys. Acta* 1997, 1327, 131–147.

132. Gamble, R.C., Schimmel, P.R. *Proc. Nat. Acad. Sci. USA* 1978, 75, 3011–3014.

133. Ma, L.D., Magin, R.L., Dunn, F. *Biochim. Biophys. Acta* 1978, 902, 183–192.

134. Sano, T., Tanaka, J., Yasunaga, T., Toyoshima, Y. *J. Phys. Chem.* 1982, 86, 3013–3016.

135. Harada, S., Takeda, Y., Yasunaga, T. *J. Colloid Interface Sci.* 1984, 101, 524–531.

136. Genz, A., Holzwarth, J.F. *Colloid Polym. Sci.* 1985, 263, 484–493.

137. Genz, A., Holzwarth, J.F. *Eur. Biophys. J.* 1986, 13, 323–330.

138. Gruenwald, B., Frisch, W., Holzwarth, J.F. *Biochim. Biophys. Acta* 1981, 641, 311–320.

139. Groll, R., Böttcher, A., Jager, J., Holzwarth, J.F. *Biophys. Chem.* 1996, 58, 53–65.

140. Kanehisa, M.I., Tsong, T.Y. *J. Am. Chem. Soc.* 1978, 100, 424–434.

141. Harbich, W., Helfrisch, W. *Z. Naturforschung* 1979, 34a, 1063–1065.

142. Kummrow, M., Helfrisch, W. *Phys. Rev. A* 1991, 44, 8356–8360.

143. Winterhalter, M., Helfrisch, W. *J. Colloid Interface Sci.* 1988, 122, 583–586.

144. Hibino, M., Itoh, H., Kinosita, K. *Biophys. J.* 1993, 64, 1789–1800.

145. Hyuga, H., Kinosota, K., Wakabayashi, N. *Biochem. Bioenerg.* 1993, 32, 15–25.

146. Neumann, E., Kakorin, S. *Curr. Opinion Colloid Interface Sci.* 1996, 1, 790–799.

147. Kakorin, S., Neumann, E. *Ber. Bunsenges. Phys. Chem.* 1998, 102, 670–675.

148. Correa, N.M., Schelly, Z.A. *Langmuir* 1998, 14, 5802–5805.

149. Correa, N.M., Schelly, Z.A. *J. Phys. Chem. B.* 1998, 102, 9319–9322.

7

Dynamics of Lyotropic Liquid Crystal Phases of Surfactants and Lipids and of Transitions Between These Phases

RAOUL ZANA

CONTENTS

I. INTRODUCTION

In the presence of water, surfactants and lipids give rise to a variety of phases referred to as lyotropic phases or mesophases.[1,2] The most important of these phases are the lamellar, hexagonal, cubic micellar, and cubic bicontinuous phases denoted by L, H and V, and Q, respectively (see Figure 1.11 in Chapter 1). The subscripts 1 or 2 attached to these phase symbols indicate that the phase is direct (water continuous) or inverse (discontinuous water domains). Many other lyotropic phases have been identified that differ from the main ones by the state of the alkyl chain (crystalline or disordered) and of the head group arrangement (ordered or disordered).[1] In the particular case of the lamellar phase, additional variations come from the possible different orientations adopted by the alkyl chains with respect to the plane of the lamellae (angle of tilt of the chain) and also from the state of the surface of the lamellae that can be planar or rippled.[1] Numerous detailed descriptions have been given for the equilibrium state of the various phases that surfactants and lipids can form in the presence of water.[1-3]

However, relatively few studies have addressed the dynamics of phase transitions of lyotropic phases of surfactants and lipids. The superb book *The Aqueous Phase Behavior of Surfactants*, by R.G. Laughlin,[3] has one chapter dealing with this topic. This chapter is 8 pages long and contains only 25 references, not all dealing with the dynamics of phase transitions, in a book of 558 pages. Likewise, the recent review on the kinetics of phase transitions in surfactant solutions by Egelhaaf[4] includes a rather short paragraph on the kinetics of phase transitions between lyo-tropic phases. In addition, a literature survey shows that the majority of the reported studies on the dynamics of lyotropic phase transitions concern lipids and not surfac-tants. Many of these studies deal with the transformation of lipid lamellar phases into other phases and the reverse processes. The main reason for this situation is that the bilayers making up a lamellar phase are a good model for the cell membrane. Therefore, many studies of the rate of lipid phase transitions were undertaken in order to better understand the behavior of biological cells during replication as well as cell membrane lysis and reconstitution. As Laggner et al.[5] pointed out, studies of dynamics of phase transitions may provide "methods for prolonging the life-times of eventual structural intermediates in the transitions. A benefit from such an achievement could be the better structural description of the intermediates.... Also they are likely to be biomedical benefits from such results since the development of agents modulating the dynamics of mem-brane transformation, such as fusion, is likely to play an important role in many medical applications, e.g., liposome-based gene therapy, fertility modulation, or percutaneous drug applications." Most of the reported studies aimed at bringing information on the time scale of the transitions and on the nature of eventual intermediate phases between the known initial and final phases, and presenting mechanisms that account for the observations.

The various lyotropic phases can transform one into another when changing the temperature T^{1-3} or the pressure p^6 applied to the investigated system, or the surfactant/lipid concentration C^{1-3} in the system. Therefore, transitions between phases will be affected by changes in T, p, or C. This sensitivity provides a way to approach the dynamics of phase transitions in systems containing lipids or surfactants by means of T-jump, p-jump, and stopped-flow methods (see Chapter 2, Section II).

In the most recent studies of the kinetics of phase tran-sitions, the investigated system is enclosed in a dedicated cell to which can be applied a p-jump, a T-jump, or a concentration jump and which can be set on the path of an intense beam of x-rays from a synchrotron source or of neutrons. High-intensity sources of x-rays and neutrons are now available at various facilities in the world. High-radiation flux and collimation of the beam allow a direct identification of the structural intermediates between the initial and final structures with a millisecond time resolution. The cells that contain the investigated systems are of a design similar to those for conventional chemical relaxation studies. The time-resolved small angle x-ray scattering (TR-SAXS) method with pressure, temperature or concentration perturbation has been recently reviewed.[7,8] The time-resolved small-angle neutron scattering (TR-SANS) method where the perturbation is induced by a stopped-flow device has been described.[9] However, conventional p-jump setups[10] and T-jump setups with laser[11] or Joule heating[12] have also been used. In addition to the dynamics of phase transformation, these apparatuses permit the study of the spontaneous fluctuations that occur in mesophases, particularly at the approach of phase transitions. The dynamics of these fluctuations are important for an in-depth understanding of the dynamics of phase transitions and the intermediate structures involved in these transitions.

In this chapter, studies involving surfactants are somewhat arbitrarily separated from those that concern lipids. Similarities or differences of behavior are pointed out whenever necessary. The chapter is organized as follows. Section II reviews the dynamics of the surfactant L_3 phase and of transitions between various lyotropic phases. Section III deals with the dynamics of phase transitions in lipid/water mixtures. Section IV reviews the dynamics of phase transitions induced by shear. Section V concludes this chapter.

Some aspects of the dynamics of phase transitions are dealt with in this book for amphiphilic block copolymers (Chapter 4, Section VIII), microemulsions (Chapter 5, Sections VII.C (rate of dissolution in microemulsions)) and VII.D (electric field-induced dynamics of microemulsions), and vesicles (Chapter 6, Sections V (phase transitions in vesicle bilayers) and VI (electric field effect on vesicles)).

II. DYNAMICS OF SURFACTANT LYOTROPIC PHASES AND TRANSITIONS BETWEEN PHASES

A. Dynamics of the L_3 Phase

The structure of the L_3 phase or sponge phase is now well established[10,13] It is an isotropic phase that consists of a multiconnected bilayer (membrane) that separates the space into two subspaces, each of them self-connected throughout the sample (see Figure 7.1). In phase diagrams the L_3 phase often borders on a lamellar phase. It has been reported to occur in many different systems based on ionic,[13,14] zwitterionic,[13,15] and nonionic surfactants.[10,16] For systems with ionic and zwitterionic surfactants, the L_3 phase is observed in the presence of a cosurfactant, added to tune the curvature of the surfactant bilayer. With nonionic surfactants the cosurfactant is not always needed, as curvature tuning can be achieved by changing the temperature.

The relaxation behavior of the L_3 phase has been investigated using the E-field jump method with electric birefringence detection (Kerr effect),[10,13,15] the Joule effect T-jump with light scattering detection,[10,14,16] the p-jump with conductivity detection,[10] and the dynamic light scattering (DLS).[10,13] The E-field jump studies consistently reported a single relaxation process with a relaxation time τ_{EJ} that was the same when turning the field on and off and that scaled

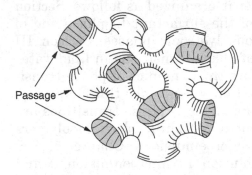

Passage

Figure 7.1 Schematic structure of the L_3 phase. Adapted from Reference 13.

as ϕ^{-3} (ϕ = volume fraction of the bilayer, that is, surfactant + cosurfactant).[10,13,15] This scaling did not depend on whether the surfactant used to generate the L_3 phase was nonionic,[10] ionic,[13] or zwitterionic.[15]. This dependence was accounted for theoretically by assuming for the L_3 phase either the bicontinuous structure defined above[10,13] or a structure where the dispersed phase (surfactant + cosurfactant) is under the form of discrete platelets.[15] This last interpretation is now abandoned. The observed relaxation process was assigned to topology changes in the L_3 phase, more specifically to the spontaneous formation/annihilation of passages.[10,13] This is schematically represented in Figure 7.2. Passages connecting parts of the same subvolume disappear/form. Note that Reference 10 and Reference 13 also reported the observation by DLS of another relaxation with a time constant τ_{DLS} that scaled as $q^{-2}\phi^{-1}$ (q = scattering wave vector).[10,13] The q^{-2} dependence reveals the diffusional nature of this process that was assigned to the relaxation of concentration fluctuations.

Investigations of the L_3 phase by means of T-jump with light-scattering detection revealed a rather complex relaxational behavior and reported differences in the number of observed relaxation processes and their dependencies on the bilayer volume fraction ϕ.[10,14,16] Three relaxation processes were observed for the L_3 phase in the cetylpyridinium chloride/hexanol/brine system.[14] In this study, the fastest relaxation process (relaxation time $\tau_1 \propto q^{-2}\phi^{-1}$, with values very

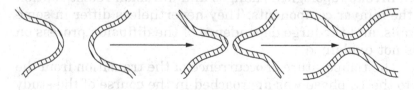

Figure 7.2 L_3 phase: Mechanism for the spontaneous annihilation of a passage connecting two regions of a subvolume. Adapted from Reference 13.

close to those determined by DLS) was assigned to a diffusion process. The intermediate relaxation process (relaxation time $\tau_3 \propto q^{-0}\phi^{-x}$ with x > 5) was assigned to the relaxation of the asymmetry between the two subvolumes of the L_3 phase. The slowest process (relaxation time $\tau_2 \propto q^{-0}\phi^{-3}$, scaling similar to that observed in E-field jump studies[10,13,15]) was assigned to the formation/annihilation of passages connecting two different parts of a subvolume (see above and Figure 7.1 and Figure 7.2). In contrast, a single relaxation process was shown for L_3 phases based on nonionic surfactants of the C_mEO_n type (*m* = carbon number of the alkyl chain; n = number of ethylene oxide (EO) units) in the systems water/$C_{10}EO_4$/decanol[10] and water/$C_{12}EO_5$/decane.[16] The associated relaxation time τ had very different characteristics depending on the value of ϕ.[16] At high volume fraction ($\phi \sim 0.3$) τ scaled as $q^{-2}\phi^{-1}$ just like the time constant measured by DLS. At low volume fraction ($\phi < 0.2$) τ scaled as $q^{0}\phi^{-9}$. This behavior led the authors to conclude that in fact two different relaxation processes are operating in the system, one associated with topology changes that dominates at low ϕ and the other associated with a collective diffusion that dominates at high ϕ. The proposed mechanism postulates that after a T-jump the L_3 phase relaxes first by a topological change, which is followed by a collective diffusion process that redistributes the bilayer material.[16] The interpretations of the relaxation behavior of the L_3 phase in Reference 14 and Reference 16 bear many similarities. In particular they both invoke topological changes and diffusion redistribution of the bilayer components. They nevertheless differ in some details, and the large dependence of the diffusion process on ϕ is not explained.

The temperature of occurrence of the transition from the L_3 to the L_α phase was approached in the course of the study of the dynamics of the L_3 phase in the cetylpyridinium/hexanol/(water+NaCl) system.[14] This was found not to affect the diffusion process but to result in a considerable slowing down of the relaxation associated to the topological change.

B. Dynamics of Phase Transitions

1. Transition from the L_α Phase to the Bicontinuous Cubic Phase (V_1) and $V_1 \rightarrow L_\alpha$ Transition

These studies were all performed using nonionic surfactants of the $C_m EO_n$ type.[12,17–22]

Knight et al.[12] used a Joule effect T-jump with birefringence and turbidity detection to show that the bicontinuous cubic (gyroid) $V_1 \rightarrow$ lamellar L_α phase transition in the system 60% $C_{12}EO_6$ in 1 M CsCl occurs in about 2 s after a short induction time of about 0.5 s (see Figure 7.3). A similar result was found by Clerc et al.[17] for the $V_1 \rightarrow L_\alpha$ transition in aqueous $C_{12}EO_6$ (concentration 62 or 65 wt%) by means of an infrared laser T-jump with TR-SAXS detection. The relaxation signals were fitted to a stretched exponential and yielded

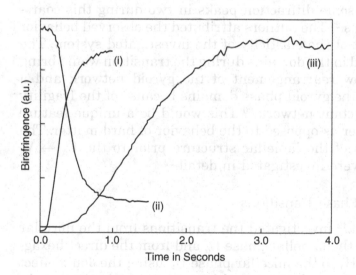

Figure 7.3 Birefringence relaxation signals for the transitions $L_1 \rightarrow L_\alpha$ (42% $C_{12}EO_6$ in 4 M NaCl at 26°C; curve 1), direct hexagonal $H_1 \rightarrow L_1$ (42% $C_{12}EO_6$ in 1 M NaCl at 26°C; curve 2), and bicontinuous micellar $V_1 \rightarrow L_\alpha$ (60% $C_{12}EO_6$ in 1 M CsCl at 26 °C; curve 3). Reproduced from Reference 12 with permission of the American Chemical Society.

relaxation times between 0.3 and 0.6 s. Note that in both studies the observations were restricted to times shorter than about 10–20 s owing to the cooling effect of the thermostatic system that occurs after the T-jump and brings back the system to its initial temperature.

The $L_\alpha \to V_1$ transition in the aqueous $C_{16}EO_7$ system (concentration 45 or 55 wt%) was investigated by Imai et al.[18-21] by TR-SANS and TR-SAXS. In these studies the system was annealed for a long time at a temperature corresponding to the L_α phase and then quenched and maintained at a temperature corresponding the cubic phase, thus avoiding the problem of returning to the initial temperature occurring in the preceding T-jump studies.[12,17] The gyroid phase was observed after 10 min but the kinetics of the $L_\alpha \to V_1$ transformation was not investigated. However, the authors reported that the diffraction pattern of the gyroid phase continued to evolve for a very long time (many hours) with a splitting of some diffraction peaks in two during this coarsening process.[20] The authors attributed the observed behavior to the effect of the elasticity of the investigated system. The stress stored in the domains during the transition would bring about a slow rearrangement of the gyroid network and a splitting of the gyroid phase domains because of the fragility of the surfactant network.[20] This would be a unique feature of soft matter as opposed to the behavior of hard matter. The fluctuations of the lamellar structure prior to the $L_\alpha \to V_1$ transition were investigated in detail.[22]

2. Other Phase Transitions

Knight et al.[12] investigated the transitions from the micellar phase L_1 to the lamellar phase L_α and from the direct hexagonal phase H_1 to the micellar phase L_1, using the Joule effect T-jump with turbidity or birefringence detection, in $C_{12}EO_6$/(water + salt) systems. All transitions were found to occur very rapidly, in about 0.5 s (see Figure 7.3).

The $H_1 \to V_1$ transition was also shown to occur rapidly with a relaxation time shorter than 1 s.[17]

Simmons et al.[23] studied the kinetics of the phase transition $L_\alpha \rightarrow H_2$ (inverse hexagonal phase) in a lecithin/AOT/water/isooctane system where the isooctane makes up the continuous phase, using TR-SANS. The time constant for the transition was reported to be about 420 s. However, the temperature quench across the L_α/H_2 coexistence region was very large (> 40°C) and relatively slow (about 100 s). Both effects may have much affected the reported time constant.

III. DYNAMICS OF PHASE TRANSITIONS IN LIPID LYOTROPIC PHASES

A. Dynamics of Lamellar-to-Lamellar Phase Transitions

The lamellar phase is one of the most common lipid mesophase. Several studies deal with the dynamics of the lamellar phase because such studies may have relevance in cell fusion and cell replication. Numerous types of lamellar phases have been described depending on the packing of the lipid alkyl chains and head groups.[11,24] The phase stable at very low temperature is the crystalline subgel lamellar phase L_C where the alkyl chains are perpendicular to the bilayer surface and the head groups are regularly arranged. In the gel phase, the hydrocarbon chains are arranged in a hexagonal subcell with the chains either perpendicular to the bilayer surface, as in the L_β phase, or tilted with respect to this surface, as in the $L_{\beta'}$ phase. A variant of the $L_{\beta'}$ phase is the $P_{\beta'}$ phase where the bilayer surface is rippled. At higher temperature, in the liquid crystalline state L_α the alkyl chains are disordered. Figure 7.4 gives schematic representations of the phases L_C, $P_{\beta'}$, $L_{\beta'}$, and L_α. Each of these phases is characterized by a specific x-ray diffraction pattern that permits its identification. Transitions between these phases can be induced in different manners and the kinetics of these transitions has been investigated. Before we review these studies, recall that surfactants also give rise to similar lamellar phases. In fact, Section V in Chapter 6 deals with the dynamics of the lamellar-to-lamellar $L_\beta \rightarrow L_\alpha$ phase transition in

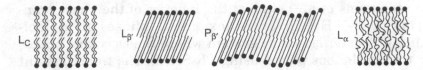

Figure 7.4 Schematic representation of the lamellar phases L_C, $L_{\beta'}$, $P_{\beta'}$ and L_{α}. Reproduced from Reference 6 with permission of R. Oldenburg Verlag, GmbH.

vesicle bilayers. Such studies are therefore not reviewed in this chapter.

One of the first study of lamellar-to-lamellar phase transition dynamics was performed on the lamellar phase formed by lipid extracts from E-coli membranes and on the membrane itself by TR-SAXS using a cell with a built-in Peltier element that allowed large T-jumps and rapid cooling of the cell.[25] The time required to achieve the disorder-to-order transition was found to be 1–2 s for the lipid lamellar phases and the membrane. The transformation involved no induction period after the T-jump. Some results suggested that the time course of the transformation involved two relaxation times.

Later studies that used higher flux TR-SAXS apparatus and pure lipids (DHPE = dihexadecylphosphatidylethanolamine[26] or monoacylglycerides such as monoolein and monoelaidin[27]) confirmed that the lamellar-to-lamellar transitions are very rapid. Thus the $L_{\beta'} \rightarrow L_{\alpha}$ transition for the DHPE system was complete in less than 2 s while the $L_{\alpha} \rightarrow L_{\beta'}$ transition required about 6 s.[26] For the monoglycerides the $L_{\beta} \rightarrow L_{\alpha}$ transition was complete in less than 3 s.[27] Within the sensitivity limit of the method used, no intermediate structure was detected between the initial and final phases.[26] The two-state nature of the transitions between lamellar phases was confirmed in other studies involving 1-stearoyl-2-oleoyl-3-phosphatidylethanolamine (SOPE)[28] and DHPE.[29] In Reference 29 the DHPE lamellar phase was perturbed by a p-jump and the equilibrium pressure was reached in 2 s, that is, much more rapidly that in the previous studies by the same group where the perturbation of the system was achieved by a relatively slow T-jump generated by a circulation of fluid.[26,27] The $L_{\beta'} \to L_\alpha$ transition was found to be completed within 37 s with a nonexponential course, while the $L_\alpha \to L_{\beta'}$ transition was complete in less than 13 s. These times are much longer than those measured in T-jump studies and according to the authors the difference probably arose from a "disparate experimental protocol."[29] Cheng et al.[30] reexamined the kinetics of the $L_{\beta'} \to L_\alpha$ and $L_\alpha \to L_{\beta'}$ transitions by p-jump coupled to TR-SAXS. They reported still shorter times for the transitions, less than 50 ms and about 1 s, respectively, and concluded that the transitions were of the two-state type. Note that the perturbation of the system in the T-jump studies[26,27] was of much larger amplitude than in the p-jump studies. This probably affected the absolute values of the reported transition times. The effect of the amplitude of the perturbation on the measured times is further discussed below.

The transitions $P_{\beta'} \to L_\alpha$ in DPPC (1,2-dipalmitoyl-glycero-3-phosphatidylcholine)[11] and $L_C \to L_\alpha$ in DHLG (1,2-di-O-hexadecyl-3-O-lactosylglycerol)[31] were investigated by laser T-jump coupled to TR-SAXS. The transitions were found to be extremely rapid, with characteristic times of less than 2 ms for the first transition and 57 ms for the second. No intermediate state was observed. Note that the characteristic times found for phase transitions induced by heating the system with laser pulses were always shorter by as much as two orders of magnitude than when the transition was induced by T-jumps obtained by the circulation of fluids or the Peltier effect (see Table 7.1).

However, not all lamellar-to-lamellar transitions are of the two-state type. Tenchov and Quinn[24] used TR-SAXS to show that whereas the $L_C \to L_\alpha$ transition is often of the two-state type, the $L_C \to L_{\beta'}$ transition is often continuous, i.e., there is a progressive shift of some diffraction peaks, which corresponds to progressive changes in the packing of the lipid alkyl chains as the system proceeds from one state to the other. The authors also reported that some systems show a behavior that is neither two-state nor continuous.[24]

Two more studies are cited to illustrate the observation of intermediate states. The first study involved multilamellar liposomes of various lipids that were incubated for a suffi-

TABLE 7.1 Characteristic Time τ for Lamellar-to-Lamellar Phase Transitions

Lipid	Type of Perturbation	Transition	τ (s)	Reference
Lipid extract from *E-coli*	Peltier effect T-jump	$L_\beta \rightleftarrows L_\alpha$	1–2	25
DHPE	Fluid flow T-jump	$L_{\beta'} \rightleftarrows L_\alpha$	1–6	26
DPPE	Fluid flow T-jump	$L_\beta \rightleftarrows L_\alpha$	2	35
DPPC + trehalose	Fluid flow T-jump	$L_\beta \rightleftarrows L_\alpha$	2	36
DMPC alone	Fluid flow T-jump	$L_{\beta'} \rightleftarrows L_\alpha$	1	34
DMPC + butanol			4	
DMPC + octanol			4	
DMPC + dodecanol			4	
DMPC + tetradecanol			1	
SOPE	Laser T-jump	$L_\beta \rightarrow L_\alpha$	< 0.002	28
DMPC[a]	p-jump	$P_{\beta'} \rightleftarrows L_\alpha$	50–100	37
		$L_\alpha \rightleftarrows P_{\beta'}$	10–150	
DHPE	Microwave T-jump	$L_{\beta'} \rightleftarrows L_\alpha$	5–10	38
DHPE	p-jump	$L_{\beta'} \rightleftarrows L_\alpha$	37	29
		$L_\alpha \rightleftarrows L_{\beta'}$	13	
DHPE	p-jump	$L_\beta \rightleftarrows L_\alpha$	1	30
		$L_\alpha \rightleftarrows L_{\beta'}$	< 0.05	
1,2-dihexadecyl-3-O-lactosyl-sn-glycerol	Laser T-jump	$L_C \rightarrow L_\alpha$	0.057	31
DEPC[a]	p-jump	$L_\beta \rightarrow L_\gamma \rightarrow L_\alpha$	5	33
		$L_\alpha \rightarrow L_\gamma \rightarrow L_\beta$	15	
POPC	Laser T-jump	$L_\alpha \rightarrow L^*$	< 0.005	5
		$L^* \rightarrow L_\alpha$	2	

[a] The transitions occur in two stages.

ciently long time at a temperature corresponding to the L_α phase and then submitted to a laser T-jump or to a p-jump after which the system was still in the L_α phase.[5] Figure 7.5 shows the observed behavior of the lamellar repeat distance d. This distance first decreases discontinuously by about 0.3 nm and then relaxes back to its initial value after about 15 s. Two time constants were required to fit the relaxation curve, with an average relaxation time of about 3 s. A similar behavior was observed for several other lipids and also when performing the transition on lipid systems with two coexisting lamellar phases, obtained upon addition of salts.[5] This "thin" structural intermediate had been reported earlier by the same group.[32] The change of lamellar repeat distance was found to be close to 0.3 nm, irrespective of the investigated system, and it was attributed to the formation of an intermediate lamellar phase noted L_α^*. Laggner et al.[5] explained the

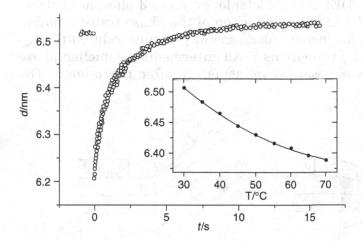

Figure 7.5 Evolution of the interlamellar distance during the transition $L_\alpha \rightarrow L_\alpha^* \rightarrow L_\alpha$ in 20 wt% POPC (1-palmitoyl-2-oleoyl-glycero-3-phosphocholine) induced by a T-jump of 15°C from an initial temperature of 30°C. For comparison the inset represents the temperature dependence of the lamellar repeat distance of POPC under equilibrium condition. Reproduced from Reference 5 with permission of the Royal Society of Chemistry.

observed behavior in terms of the martensitic lattice-disinclination mechanism as shown in Figure 7.6b. The transition is localized in a discontinuous transition plane linking the nascent to the parent phase and moving extremely rapidly (at the speed of sound) through the bilayer. This mechanism is in agreement with the fact that the Bragg peaks remained sharp after the T-jump. The change in d value mostly reflects changes occurring at the level of the water involved in the lamellar structure. The authors considered several possibilities and concluded that the simplest one that still explains the experimental results would be that the water forms localized lentils that do not much perturb the order in the multilamellar liposomes. For the return from the L_α^* to the L_α phase the authors presented the mechanism in Figure 7.6a where the parent thin lattice connects to the nascent thicker one by a zone of disorder.[5]

The second study showing intermediate phases was performed on DEPC (1,2-dielaidoyl-glycero-3-phosphatidylcholine) by TR-SAXS and induction of the phase transformation by p-jump; an increase/decrease of pressure induced the L_α-to-L_β/L_β-to-L_α transitions.[33] An intermediate lamellar phase noted $L_?$ was observed immediately after the p-jump. The

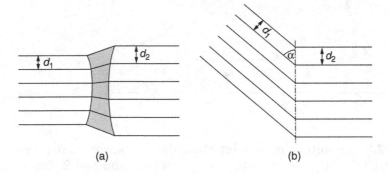

(a)   (b)

Figure 7.6 Scheme of the mechanism of the transitions $L_\alpha \rightarrow L_\alpha^*$ (right, minimum loss or order and coherence) and $L_\alpha^* \rightarrow L_\alpha$ (left, transition with zone of disorder resulting in a loss of coherence). Reproduced from Reference 5 with permission of the Royal Society of Chemistry.

equilibrium phase L_α or L_β started to develop shortly after the appearance of the $L_?$ phase, at its expense (see Figure 7.7). The overall transition time was found to decrease significantly when the pressure difference Δp between the final pressure and the transition pressure increased (see Figure 7.7, left). The overall transformation was achieved in 2–5 s depending on Δp. The same study[33] showed no intermediate state in the course of the $P_{\beta'} \to L_\alpha$ and $L_\alpha \to P_{\beta'}$ transitions of DPPC (1,2-dipalmitoyl-glycero-3-phosphatidylcholine). The transitions were terminated after about 20 and 3 s, respectively. These times increased very much when the amplitude of the p-jump was decreased. These results are in qualitative agreement with the results of another T-jump study of these transitions for DPPC[11] in the sense that no intermediate state was observed, but the values of the times required for the transitions differ in the two studies by up to two orders of magnitude. A literature survey of the lamellar-to-lamellar phase transitions of DPPC[11] showed considerable differences in the reported values of the transition times. However, the results in different studies are difficult to compare owing to differences in the nature of the perturbation (p or T), in the time required for the perturbation to be fully applied, in the values of the pressure (p_i and p_f) and temperature (T_i and T_f) in the initial and final phases, and also in the difference between the value of the transition pressure or temperature and the initial and final values of p and T. It remains that these large qualitative and quantitative differences provide additional evidence that the lipid transition pathways and mechanisms can be rate-dependent[11] and, additionally, dependent on the amplitude of the perturbation.

The effect of alkanols on the rate of the $L_{\beta'} \to L_\alpha$ transition of DMPC (1,2-dimyristoyl-glycero-3-phosphatidylcholine) was found to be relatively small and little dependent on the alcohol chain length.[34]

Table 7.1 summarizes some of the results presented above as well as results reported in other studies.[35–38]

At this stage it is worth recalling the mechanism of the main transition $L_{\beta'} \to L_\alpha$ as detailed by Cevc and Marsh.[39] The first stage in this transition is a nucleation of gauche confor-

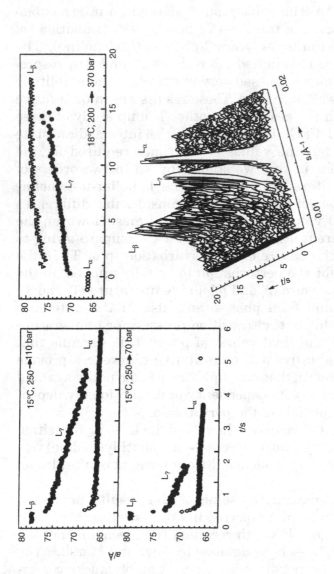

Figure 7.7 Lamellar lattice constant of DEPC (1,2-dielaidoyl-glycero-3-phosphatidylcholine) in excess water after p-jumps inducing the $L_\alpha \rightarrow L_\beta$ transition (left) using different p-jump amplitudes and the $L_\beta \rightarrow L_\alpha$ transition (top right). The series of diffraction patterns shown bottom right, taken at various time intervals after the p-jump, correspond to the same conditions as in the top-right figure. The emergence of the intermediate $L_?$ phase is clearly seen. Reproduced from Reference 33 with permission of the Royal Society of Chemistry.

mations in the lipid chains. These distortions migrate causing density fluctuations in the bilayer and relaxation of the high-density regions via lateral expansion (growth). A redistribution of water in the interbilayer space is coupled to this expansion and is thought to be the rate-limiting step for this postulated nucleation/growth mechanism. The early stages of the transition are those discussed in the phase transition of vesicle bilayers (see Chapter 6, Section V and Figure 6.17). The water redistribution was not directly considered.

A similar mechanism of nucleation and growth was postulated for the gel-to-subgel ($L_\beta \to L_C$) phase transformation provided that the system was kept close to equilibrium.[40]

B.   Dynamics of the Lamellar-to-Inverted Hexagonal (H_2) Phase Transition

The dynamics of this transition has been investigated a great deal.[5,25–28,31,33,41–43] Indeed, as pointed out by Siegel et al.,[42] "the mechanism of the L_α/H_2 transition is of interest for two reasons. First, drastic rearrangements in the topology of the lipid/water interface must occur in this transition. This rearrangement probably requires energy-intensive processes, which would result in slow overall rate. Yet some of these transitions can be quite rapid. Second, many authors have suggested that the first structures to form during this transition are related to the structures that mediate membrane fusion in liposomal systems" (see the review in Reference 44).

Ranck et al.[25] were the first to show that the $H_2 \to L_\alpha$ transition of lipids extracted from E-coli membranes can be extremely rapid, being completed in 1–2 s. This result was confirmed by Caffrey[26], for the $L_\alpha \to H_2$ transition of DHPE. Times shorter than 1 s were observed in some systems. Laggner et al.[28] reported a half-time of about 100 ms for the $L_\alpha \to H_2$ transition of SOPE, induced by means of a laser T-jump and using TR-SAXS for studying the structural changes. Half-times of transition in the range of 5 to 10 s were reported for DOPE and egg-PE (egg yolk phosphatidylethanolamine) in a study where the transition was induced by a p-jump and monitored by TR-SAXS (see Figure 7.8).[33] This study revealed an induction time of a few seconds following the p-jump, after

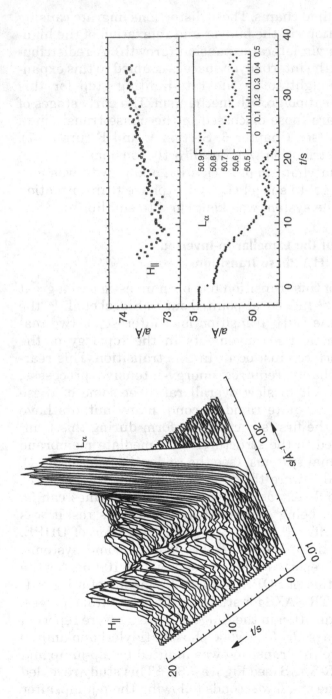

Figure 7.8 Selected diffraction patterns of DOPE after a p-jump from 300 to 120 bar at 20°C that induces the L_α-to-H_2 transition (left) and corresponding changes of lattice constants of the first-order Bragg reflections of the L_α and H_2 phases (right). Reproduced from Reference 33 with permission of the Royal Society of Chemistry.

which the transition occurred (see Figure 7.8). The induction time was attributed to water motion (uptake and release).[33] Tate et al.[41] investigated the $L_\alpha \to H_2$ transition of DOPE and DEPE by T-jump with TR-SAXS detection. The transition also involved an induction period and was clearly more rapid with DEPE than DOPE. A very large slowing down of the kinetics was observed when the final temperature after the T-jump was decreased.[41] The back transition $H_2 \to L_\alpha$ was found to be more rapid. Erbes et al.[43] reported on the $L_\beta \to H_2$ transition in 1:2 DPPC:palmitic acid mixtures. The complete transformation of the L_β phase into the H_2 phase took about 20 s. The transitions investigated in References 25, 26, 28, 33, 41, and 43 were all of the two-state type.

Other studies of the lamellar-to-inverted hexagonal transition evidenced intermediate states. For instance, Laggner et al.[5] showed that the $L_\alpha \to H_2$ transition in phosphatidylethanolamines involves the thin lamellar phase $L_\alpha{}^*$ discussed in the preceding paragraph. Köbert et al.[31] reported the formation of two transient intermediate phases, I' and I, in the course of the lamellar-to-H_2 transition of galactolipids. The first intermediate phase to appear, I', has a structure similar to the thin lamellar phase $L_\alpha{}^*$. Only the intermediate phase I was evidenced in the $L_\alpha \to H_2$ transition of glycolipids.[31] The authors suggested that the intermediate phase I' was not detected because it forms and disappears very rapidly. All these transitions were found to occur on the time scale of a few hundreds of milliseconds.

Siegel et al.[42] attempted to use the T-jump cryo-TEM technique[45] to directly visualize the intermediate states in the course of the $L_\alpha \to H_2$ transition in lipids. The specimen for cryo-TEM was rapidly heated by exposure to the light from a xenon lamp and vitrified to fix possible intermediate states induced by the T-jump. For egg phosphatidylethanolamine, the transition was found to be completed in 9 ms when the specimen was superheated by 20°C. N-monomethylated DOPE showed a slower transition. In both instances no intermediate state could be resolved owing to their lability on the time scale of the method stated to be around 1 ms.

Caffrey[27] reported a very short time for H_2-to-isotropic fluid transition for monoolein.

C. Dynamics of Transitions Involving the Inverse Cubic Phases Q_2

The reasons for the interest in transitions involving cubic phases are the same as for transitions involving the inverse hexagonal phase H_2.[42] Several cubic phases of different symmetry have been found.[46,47] The cubic phase that has attracted the most interest in kinetic studies is the inverse bicontinuous cubic phase Q_2. This phase consists of a single lipid bilayer on either side of which lie two interpenetrating continuous networks of water channels.[47] The primitive, double diamond, and gyroid (body-centered) inverse cubic phases (space groups Im3m, Pn3m and Ia3d, respectively) are denoted Q_2^P, Q_2^D, and Q_2^G (see Figure 7.9).

The first studies involving transitions from and to cubic phases: lamellar-to-cubic, cubic-to-cubic, cubic-to-inverted hexagonal, were performed using a slow T-jump and monitoring the transitions by TR-SAXS.[27] All transitions except the cubic-to-cubic ones were completed in less than 3 s. The body-centered cubic-to-primitive cubic phase transitions were much slower and required between 30 and 1800 s. An unidentified stable intermediate phase was evidenced in the transition $L_\alpha \rightarrow Q_2^G$.

The most comprehensive studies of kinetics of transitions involving cubic phases were performed using 1:2 lipid:fatty acid mixtures.[43,47,48] For the 1:2 DMPC:myristic acid mixture a p-jump in the coexistence range of the H_2 and Q_2^P phases showed

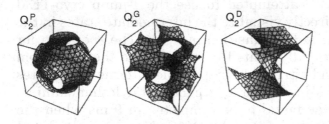

Figure 7.9 Schematic representations of the three inverse bicontinuous cubic phases. Reproduced from Reference 6 with permission of R. Oldenburg Verlag, GmbH.

the occurrence after a time of about 12 s of the Q_2^D phase and the relaxation of the lattice parameters to the values of the original phases in about 6 s.[43] A p-jump that took the system from the L_β phase to the H_2/Q_2^P coexistence range showed the transition to be relatively slow, with changes of the lattice parameters stretching over 1800 s and the occurrence of the L_α phase as a transient intermediate. Figure 7.10 shows the first 35 s of the relaxation of the system, with the rapid disappearance of the L_β phase and the apparition of the H_2 phase. The Q_2^P phase appears after about 120 s.[43] Squires et al.[48] showed by p-jump and TR-SAXS detection that in the 1:2 DLPC:lauric acid mixture the $Q_2^G \rightarrow Q_2^D$ transition is very rapid (< 0.5 s) whereas the transition $Q_2^D \rightarrow Q_2^D + H_2 +$ excess water is much slower, taking 5–10 s depending on the temperature. The reverse transition was much more rapid, taking place in less than 1 s. The appearance of intermediate structures was noted, particularly the Q_2^P phase. The same mixture

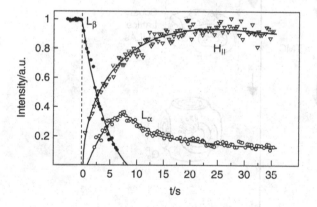

Figure 7.10 Variation of the intensity (in arbitrary units) with time t of the first-order Bragg peaks of the L_α, L_β and H_2 phases of 1:2 DMPC/myristic acid mixture (50 wt% in water) after a p-jump from 500 to 110 bar at 58°C that induces the transition from the L_β to the H_2/Q_2^P coexistence range. Reproduced from Reference 43 with permission of VCH Verlagsgesellchaft mbH.

was used to investigate the $L_\alpha \to Q_2^G$ transition.[47] The transition stretched over a large time scale extending to minutes, and the results revealed the presence of an intermediate phase identified as the Q_2^D phase. The authors proposed for the $L_\alpha \to Q_2^D$ transition a mechanism based on the formation of stalks, transmembrane contacts and interlamellar attachments (ILA) as postulated by Siegel[49] (see Figure 7.11). The Q_2^D phase would arise from a hexagonal array of ILA. Recall

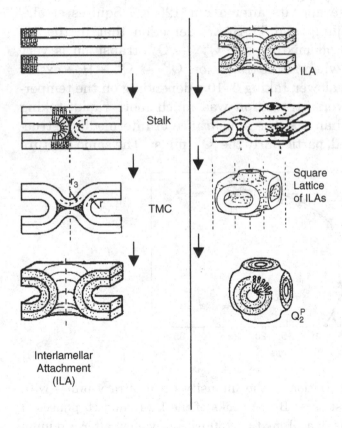

Figure 7.11 Left: Proposed mechanism for the formation of an interlamellar attachment via a stalk and a transmonolayer contact (TMC). Right: Square lattices of ILA can readily form the Q_2^P phase. Reproduced from Reference 47 with permission of the American Chemical Society.

that the $L_\alpha \rightarrow Q_2^G$ transition for the aqueous surfactants $C_{16}EO_7$ and $C_{12}EO_5$ has been investigated.[18-21] The gyroid phase was formed after 10 min but then evolved slowly (see Section II.B).

Additives, such as dextrose or salts, can affect the rate of formation of the cubic phases.[46]

IV. KINETICS OF SHEAR-INDUCED PHASE TRANSITIONS

Several papers have reported that the shearing of surfactant-containing systems for a sufficiently long time and at above a critical shear rate $\dot\gamma$ can result in a phase transition.

The transformation of a lamellar phase L_α into a phase made of tighly packed multilamellar vesicles, referred to as the "onion phase" (see Chapter 1, Section V) is probably the most investigated phase transition induced by shear.[50-52] Zipfel et al.[51] used rheology and TR-SANS and TR-small angle light scattering to show that an intermediate phase occurs between the L_α and the onion phases. The time scale for the overall transition is of the order of 2000 s, while the intermediate phase forms after about 300 s in the experimental conditions used ($\dot\gamma = 10$ s^{-1}, see Figure 7.12). The scattering results suggest that the intermediate phase is made up of parallel multilamellar cylinders (tubuli) or lamellar cylinders where the bilayers are rolled up around the cylinder axis. A very different interpretation of similar results was presented by Escalante and Hoffmann,[53] and one cannot discard the possibility that different systems correspond to different mechanisms of transition and different time scales. Escalante et al.[52] noted that the formed onion phase does not revert to the L_α phase when the shear is stopped.

The effect of shear on the L_3 phase in the cetylpyridinium chloride/hexanol/(water+NaCl) system with a significant amount of added dextrose has been examined.[54] Additions of of dextrose slow down the diffusional process discussed above. At low shear rate, the sponge phase structure is retained but starting at a shear rate above 500 s^{-1}, shear thinning reveals the occurrence of a transition to a lamellar phase. Unfortu-

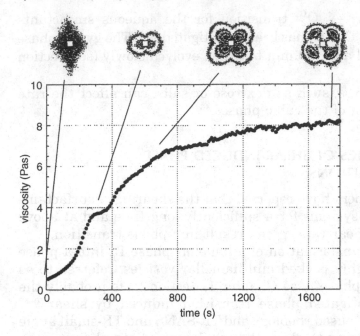

Figure 7.12 Time-dependence of the viscosity in the course of the formation of multilamellar vesicles from a lamellar phase of $C_{10}EO_3$ submitted to the shear gradient 10 s^{-1}. The depolarized small-angle light scattering (SALS) patterns obtained at four different times are represented (flow direction in the SALS patterns is vertical). Reproduced from Reference 51 with permission of EDP Sciences.

nately, the time of application of the shear required for the phase transition was not specified, but it was below 1200 s.

Mahjoub et al.[55] showed that the L_3 phase may be transformed into a lamellar phase when shearing the system with a shear rate above a critical value $\dot{\gamma}_c$. This new lamellar phase is formed after 300 s. The transformation is irreversible when performed at $\dot{\gamma} > \dot{\gamma}_c$.

Shear has been reported to induce the transformation of vesicles present in the cetyltrimethylammonium-3-hydroxynaphthalene-2-carboxylate/water system in threadlike micelles.[56] The transformation was finished after a time shorter than 20 min, and this time decreased as the applied shear rate was increased. A similar transformation was investigated by Zheng et al.[57] for the CTAC/sodium 3-methylsalicylate/water by transmission electron microscopy at cryogenic temperature (cryo-TEM). The authors took advantage of the fact that the blotting procedure for preparing specimen for cryo-TEM observation submits the solution to high shear rates. The samples were observed after being allowed to relax on-the-grid for increasing times before vitrifying the sample. The vesicles were transformed into threadlike micelles during the blotting. These micelles were transformed back into vesicles when the specimen was allowed to relax 30 s before vitrification. For a relaxation time of only 15 s an intermediate state was observed, consisting of densely packed, threadlike micelles and referred to as snake-balls. This phase sometimes showed budding vesicles.[57]

Imai et al.[58] showed that a steady shear flow with $\dot\gamma >$ 0.01 s^{-1} suppresses undulations of the lamellar phase in the $C_{16}EO_7$/water system. This effect induces ordering in the lamellar phase with parallel bilayers and hinders the L_α-to- Q_2^G phase transition that this system can undergo. In fact, at $\dot\gamma > 0.3$ s^{-1} the gyroid phase was transformed into a lamellar phase. The authors reported waiting 100 min for full equilibration.

V. CONCLUSIONS

This chapter reviewed the kinetics of phase transitions in systems based on surfactants and lipids. The use of the p-jump and T-jump techniques with a detection of the relaxation by means of TR-SAXS has permitted much progress in the field. The characteristic times for many phase transitions have been determined and found to be relatively short, in most instances in the time range of a few seconds or less. Intermediate phases have been identified. However, work remains to be done in two main directions. First, the effect of the amplitude of the perturbation on the characteristic time of the transition should be investigated more in detail. Indeed, several of the reviewed studies revealed a very large increase of the time characterizing the transition when the amplitude of the p-jump or T-jump was reduced. This may be partly due to the fact that most studies used very large perturbations and that the condition necessary in relaxation studies of very small perturbations was not met. This may affect both the

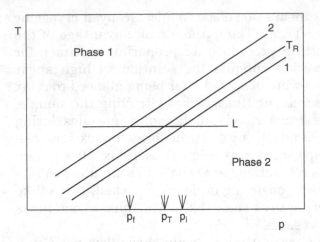

Figure 7.13 Schematic representation of systematic experiments to be performed on a system where a transition from phase 1 to phase 2 can be induced by a change of pressure or temperature.

value of the measured characteristic time and also the shape of the relaxation signal. Also, studies should be performed at constant perturbation amplitude. This is illustrated by considering a system where the equilibrium between the two lipid phases 1 and 2 is sensitive to p and T, as represented in Figure 7.13 (T_R is the line of phase transition). For instance, systematic p-jump experiments should be performed at a given temperature and at constant Δp = final pressure p_f – initial pressure p_i, but with variable $p_i - p_T$, that is, at an initial pressure closer and closer to (or farther and farther from) the pressure p_T at which the transition occurs. In Figure 7.13 this would amount to moving along the horizontal line L. Also experiments should be performed at constant Δp and $p_i - p_T$ and increasing temperature. In Figure 7.13 the initial and final states of the system would be moving on lines 1 and 2, respectively. Second, a systematic study of the effect of the chain length of the lipid or surfactant should be performed on the kinetics of typical phase transitions such as the lamellar-to-lamellar and the lamellar-to-inverted hexagonal transitions. The availability of the phase diagrams of nonionic surfactant C_mEO_n/water mixtures should facilitate the choice of the systems on which such studies should be performed.

REFERENCES

1. Skoulios, A. *Ann. Phys.* 1978, 3, 421.

2. Ekwall, P. In *Adv. Liquid Crystals*, Brown, G.H. Ed., Academic Press, 1975, Vol. 1, p. 1.

3. Laughlin, R.G. *The Aqueous Phase Behavior of Surfactants*, Academic Press, London, 1994.

4. Egelhaaf, S.U. *Current Opinion Colloid Interface Sci.* 1998, 3, 608.

5. Laggner, P., Amenitsch, H., Kriechbaum, M., Pabst, G., Rappolt, M. *Faraday Disc.* 1998, 111, 31.

6. Winter, R., Czeslik, C. *Zeit. Kristall.* 2000, 215, 454.

7. Bras, W., Ryan, A.J. *Adv. Colloid Interface Sci.* 1998, 75, 1.

8. Cunningham, B., Bras, W., Leonard, L., Quinn, P.J. *J. Biochem. Biophys. Methods* 1994, 29, 87.

9. Grillo, I., Kats, E.I., Muratov, A.R. *Langmuir* 2003, 19, 4573.

10. Schwarz, B., Münch, G., Ilgenfritz, G., Strey, R. *Langmuir* 2000, 16, 8643.

11. Rapp, G., Rappolt, M., Laggner, P. *Prog. Colloid Polym. Sci.* 1993, 93, 25.

12. Knight, P., Wyn-Jones, E., Tiddy, G.J.T. *J. Phys. Chem.* 1985, 89, 3447.

13. Porte, G., Delsanti, M., Billard, I., Skouri, M., Appell, J., Marignan, J., Debeauvais, F. *J. Phys. II France* 1991, 1, 1101.

14. Waton, G., Porte, G. *J. Phys. II France* 1993, 3, 515.

15. Miller, C.A., Gradzielski, M., Hoffmann, H., Krämer, U., Thunig, C. *Colloid Polym. Sci.* 1990, 268, 1066 and *Prog. Colloid Polym. Sci.* 1991, 84, 243.

16. Le, T.D., Olsson, U., Wennerstrom, H., Urhmeister, P., Rathke, B., Strey, R. *J. Phys. Chem. B* 2003, 106, 9410.

17. Clerc, M., Laggner, P., Levelut, A.-M., Rapp, G. *J. Phys. II France*, 1995, 5, 901.

18. Imai, M., Kato, T., Schneider, D. *J. Chem. Phys.* 1997, 106, 9362.

19. Imai, M., Kato, T., Schneider, D. *J. Chem. Phys.* 1998, 108, 1710.

20. Imai, M., Nakaya, K., Kato, T. *Phys. Revs. E.* 1999, 60, 734.

21. Imai, M., Saeki, A., Teramoto, T., Kawaguchi, A., Nakaya, K., Kato, T. *J. Chem. Phys.* 2001, 115, 10525.

22. Imai, M., Nakaya, K., Kato, T., Seto, H. *J. Chem. Phys.* 2003, 119, 8103.

23. Simmons, B., Agarwal, V., Singh, M., McPherson, G., John, V., Bose, A. *Langmuir* 2003, 19, 6329.

24. Tenchov, B.G., Quinn, P.J. *Liquid Cryst.* 1989, 5, 1691.

25. Ranck, J.-L, Letellier, L., Shechter, E., Krop, B., Pernot, P., Tardieu, A. *Biochemistry* 1984, 23, 4955.

26. Caffrey, M. *Biochemistry* 1985, 24, 4826.

27. Caffrey, M. *Biochemistry* 1987, 26, 6349.

28. Laggner, P., Kriechbaum, M., Hermetter, A., Paltauf, F., Hendrix, J., Rapp, G. *Prog. Colloid Polym. Sci.* 1989, 79, 33.

29. Mencke, A.P., Caffrey, M. *Biochemistry* 1991, 30, 2453.

30. Cheng, A., Hummel, B., Mencke, A., Caffrey, M. *Biophys. J.* 1994, 67, 293.

31. Köbert, M., Hinz, H.-J., Rappolt, M., Rapp, G. *Ber. Bunsenges. Phys. Chem.* 1997, 101, 789.

32. Laggner, P., Kriechbaum, M., Rapp, G. *J. Appl. Cryst.* 1991, 24, 836.

33. Erbes, J., Gabke, A., Rapp, G., Winter, R. *Phys. Chem. Chem. Phys.* 2000, 2, 151.

34. Phonphok, N., Weterman, P.W., Lis, L.J., Quinn, P.J. *J. Colloid Interface Sci.* 1989, 127, 487.

35. Tenchov, B.G., Lis, L.J., Quinn, P.J. *Biophys. Biochem. Acta* 1988, 942, 315.

36. Quinn, P.J., Koynova, L.D., Lis, L.J., Tenchov, B.G. *Biophys. Biochem. Acta* 1988, 942, 305.

37. Caffrey, M., Hogan, J., Mencke, A.J. *Biophys. J.* 1991, 60, 456.

38. Caffrey, M., Magin, R.L., Hummel, B., Zhang, J. *Biophys. J.* 1990, 58, 21.

39. Cecv, G., Marsh, D. *Phospholipid Bilayers, Physical Principles and Methods*, Wiley, New York, 1987.

40. Tristram-Nagle, S., Suter, R.M., Sun, W.-J., Nagle, J.F. *Biochim. Biophys. Acta* 1994, 1191, 14.

41. Tate, M.W., Shyamsunder, E., Gruner, S.M., D'Amico, K.L. *Biochemistry* 1992, 31, 1081.

42. Siegel, D.P., Green, W.J., Talmon, Y. *Biophys. J.* 1994, 66, 402.

43. Erbes, J., Winter, R., Rapp, G. *Ber. Bunsenges. Phys. Chem.* 1996, 100, 1713.

44. Siegel, D. P. *Biophys. J.* 1993, 65, 2124.

45. Chestnut, M.H., Siegel, D.P., Burns, J.L., Talmon, Y. *Micros. Res. Tech.* 1992, 20, 95.

46. Tenchov, B., Koynova, R., Rapp, G. *Biophys. J.* 1998, 75, 853 and references cited.

47. Squires, A.M., Templer, R.H., Seddon, J.M., Woenckhaus, J., Winter, R., Finet, S., Theyencheri, N. *Langmuir* 2002, 18, 7384.

48. Squires, A.M., Templer, R.H., Ces, O., Gabke, A., Woenckhaus, J., Seddon, J.M., Winter, R. *Langmuir* 2000, 16, 3578.

49. Siegel, D.P., Banschbach, J.L. *Biochemistry* 1990, 29, 5975.

50. Diat, O., Roux, D., Nallet, F. *J. Phys. II France* 1993, 3, 1427.

51. Zipfel, J., Nettlesheim, F., Lindener, P., Le, T.D., Olsson, U., Richterung, W. *Europhys. Lett.* 2001, 53, 335.

52. Escalante, I., Gradzielski, M., Hoffmann, H., Mortensen, K. *Langmuir* 2000, 16, 8653.

53. Escalante, I., Hoffmann, H. *J. Phys. Cond. Matter A* 2000, 12, 438.

54. Porcar, L., Hamilton, W.A., Butler, P.D., Warr, G.G. *Langmuir* 2003, 19, 10779.

55. Mahjoub, H.F., McGrath, K.M., Kléman, M. *Langmuir* 1996, 12, 3131.

56. Mendes, E., Narayanan, J., Oda, R., Kern, F., Candau, S.J., Manohar, C. *J. Phys. Chem. B* 1997, 101, 2256.

57. Zheng, Y., Lin, Z., Zakin, J.L., Talmon, Y., Davis, H.T., Scriven, L.E. *J. Phys. Chem. B* 2000, 104, 5263.

58. Imai, M., Nakaya, K., Kato, T. *Eur. Phys. J. E.* 2001, 5, 391.

8

Dynamics of Adsorption of Cationic Surfactants at Air–Water and Solid–Liquid Interfaces

ROB ATKIN, JULIAN EASTOE,
ERICA J. WANLESS, AND COLIN D. BAIN

CONTENTS

I. INTRODUCTION

The aim of this chapter is to compare and contrast adsorption kinetics of model cationic surfactants at air-water and solid-liquid interfaces, so as to draw general conclusions and identify dominant processes. Recently, strides have been made in understanding surfactant adsorption kinetics, and in this area development and application of new surface selective techniques has been key. Methods of relevance in this chapter are neutron reflectivity (NR), ellipsometry, and optical reflectometry (OR). These techniques are based on scattering and/or interference of neutron radiation or polarized laser light, and hence the principal advantages are that they directly probe surface layer structures and adsorption densities. In the text the terms surface excess, adsorbed amount, and surface density are used interchangeably to express two-dimensional concentrations, either at air-water or solid-liquid surfaces. The main surfactants considered are the family of n-alkyltrimethylammonium bromides: C_mTAB, of alkyl chain carbon number m.

II. AIR–WATER INTERFACES

A. Dynamic Surface Excess and Surface Tension

On generation of a new interface in a surfactant solution, the equilibrium surface tension (γ_{eq}) is not instantly reached. For γ to reach its equilibrium value, surfactant molecules first must diffuse to the surface, then adsorb and orient themselves in the interfacial film. Dynamic surface tension (DST) is critical in many industrial and biological processes, and background can be found in recent reviews.[1-3] The stabilization of alveoli by lung surfactant is perhaps the best-known biological application of DST, and this is central to gas transport across the pulmonary membrane. Industrially, in petrochemicals, the DST of aqueous foams contributes to the efficiency of enhanced oil-recovery processes. The manufacture of photographic film and paper utilizes slide coating of multiple layers of thin gelatin films with high coating speeds and flow velocities. This technology relies on careful control of dynamic surface tension in each of the layers to prevent film dewetting and air entrainment, and to promote long-term stability of the final dried film products. DST is also essential in agricultural sprays, which must spread rapidly over hydrophobic leafy surfaces, and in a range of other wetting, foaming, and emulsification processes. For all surfactant-based technologies in which the characteristic time is similar to, or shorter than, that for equilibrium adsorption, dynamic surface tension will be of interest.

There is a now an array of experimental techniques that can be used to measure dynamic surface tensions, $\gamma(t)$, including maximum bubble pressure (MBP), oscillating jet, inclined plate, drop volume, drop shape, and overflowing cylinder (OFC).[1-8] With the aid of an appropriate equation of state, it is possible to infer the dynamic surface excess, $\Gamma(t)$. Uncertainty in the adsorption isotherm can lead to problems in the interpretation of DST data and incorrect conclusions as to the adsorption mechanisms. A more direct approach is to measure $\Gamma(t)$ itself by neutron reflection (NR),[9-12] or ellipsometry.[7,11,12] Here we review the state of the art, with particular attention to recent results on model single-chain cationic surfactants

C_mTAB and direct measurements of $\Gamma(t)$ by these two methods. These measurements allow the unambiguous determination of the adsorption mechanism.

B. Experimental Design

1. Selection of Surfactants

The simplest case to treat theoretically is that of a pure, chemically stable, monomeric surfactant below its critical micelle concentration (cmc), in which adsorption is rapid and reversible. It should be noted that many surfactants do not conform to this ideal model. Commercial surfactants are generally contaminated with other surface-active impurities. While trace impurities do not have a major effect on dynamic interfacial behavior on short time scales, they can have a major influence on the equilibrium behavior, and knowledge of the equilibrium adsorption isotherm is essential for interpreting DST data. Many surfactants are mixtures of isomers (such as alkylbenzene sulphonates) or chain lengths (such as nonylphenylethoxylates) or are hydrolytically unstable (such as sodium dodecylsulfate (SDS) or trisiloxane "superspreaders"). Proteins and other biomolecules may denature at the air–water interface. For surfactant concentration C above the cmc, surfactant aggregates also contribute to the dynamic interfacial behavior. The presence of micelles can complicate the adsorption kinetics in three ways. First, micelles diffuse at a different rate from monomers. Second, micelles and monomers interconvert on timescales ranging from microseconds to seconds[13] (see Chapter 3, Sections III.B and III.C). The addition or loss of single monomers from micelles occurs at the shorter time scales and complete micellar breakdown/formation at the longer time scales. A micelle can thus be considered as an active reservoir of surfactant molecules, both as a source of and as a sink for monomer. Third, micelles may adsorb by a different mechanism and at a different rate from monomers, if they adsorb at all. In this section we will consider surfactant solutions both above and below the cmc.

The cationic surfactants, C_mTAB, are appropriate model compounds for a number of reasons:

- Dilute aqueous phase behavior, cmc values, and micellar properties are well known in the literature. The cmc can be tuned over a large range by addition of salt.
- The surfactants are commercially available and readily purified. The C_mTAB surfactants are chemically stable to hydrolysis.
- The synthesis of deuterated analogues needed for neutron reflection is straightforward.[9-11]

2. Experimental Techniques

Many techniques have been developed for measuring $\gamma(t)$,[3] and of late maximum bubble pressure (MBP) has been the most widely used.[2] However, as outlined below, the overflowing cylinder (OFC)[7-9, 11,12] has a number of advantages over other dynamic methods. For details of the noninvasive measurement of dynamic surface tension $\gamma(t)$ by surface light scattering in the OFC, see Reference 8. To elucidate the adsorption mechanism, in addition to measuring DST, it is also desirable to determine the dynamic adsorbed amount $\Gamma(t)$, since $\gamma(t)$ is merely a response to changes in the surface excess. Although $\Gamma(t)$ is the fundamental quantity of interest, it is much more difficult to measure directly than the surface tension. $\Gamma(t)$ can be inferred from $\gamma(t)$, but only if the equation of state $\Gamma(\gamma)$ is known accurately and the assumption that the same relationship holds for both dynamic and equilibrium adsorption is correct. The main experimental techniques relevant to this review are briefly described below.

a. The Overflowing Cylinder (OFC)

The OFC, shown in Figure 8.1 and Figure 8.2, is suitable for measurements on the 0.1–1 s time scale. The surface properties of this cell, in particular the relationship between the surface excess Γ, surface expansion rate, $\theta = (1/r)d(ru_s)/dr$ (r = radial distance; u_s = radial surface velocity), and C, depend solely on adsorption kinetics at that particular C and surface excess. The OFC has been designed to provide a large (~50 cm²), almost flat surface for analysis by ellipsometric, spectroscopic, and other scattering techniques. The cell consists

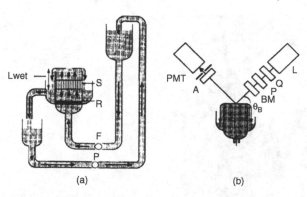

Figure 8.1 (a) Schematic illustration of the overflowing cylinder (OFC): S – flow straightener; R – resistance plate; F – flowmeter; P – magnetic drive pump; Lwet – wetting length. (b) Schematic illustration of the ellipsometer: L – He-Ne laser; P – polarizer; Q – quarterwave plate; BM – birefringence modulator; θ_B – Brewster angle; A – analyser; PMT – photomultiplier tube. Reprinted with permission from [7a]. © (1997) Elsevier.

of a stainless steel cylinder, 80 mm in diameter and 140 mm tall. Liquid is pumped vertically upward through a baffle that reduces the height required to ensure plug flow, and the solution then flows over the top rim of the cylinder, achieving a pure steady state at the horizontal surface. The surface flows radially from the center toward the rim. For surfactant solutions, surface tension gradients can increase the surface velocity by an order of magnitude compared to that for pure water. The surface expansion rate is typically in the range 1 to 10 s^{-1}, which for dilute surfactant solutions ($\leq$ a few mM) leads to a surface that is far from equilibrium. Since the surface is at a steady state, experiments lasting up to several hours are feasible. This feature is especially useful to improve signal to noise, for example for neutron reflection at low surface coverage. The OFC requires a solution volume of approximately 1.5 liters and thus when working with deuterated materials for neutron reflectivity studies, the amount and cost of surfactant required becomes a consideration. Technical details and specifications can be found elsewhere.[7]

(a)

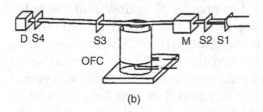

(b)

Figure 8.2 (a) The overflowing cylinder (OFC). (b) Schematic set-up of the OFC with neutron reflection. Slits - S1, S2, S3, S4. M - incident beam monitor. D - final detector. Reprinted with permission from [16]. © (2003) American Chemical Society.

b. Ellipsometry

Ellipsometry measures the change in polarization of light reflected from a liquid surface and is dependent on the thickness of the adsorbed layer and its refractive index. The technique is extremely sensitive to the presence of surfactant on the surface of water, and measurements can be made with a precision of < 1% of a monolayer, in just a few seconds. Recently it has been shown that the measured coefficients of ellipticity, $\bar{\rho}$, for monolayers of $C_m TAB$ (m = 12–18) and for a nonionic surfactant are approximately linear functions of the surface excess.[11,12] This simple relationship between $\bar{\rho}$ and Γ is a significant factor in the use of ellipsometry for dynamic surface studies, since it obviates the need to determine a calibration curve, $\Gamma(\bar{\rho})$, for each and every surfactant.

A typical experimental configuration is shown in Figure 8.1b. The polarization of a He-Ne laser is modulated photoelastically at 50 kHz by a quartz plate, and the laser is then directed onto the surface of the liquid in the OFC at the Brewster angle, θ_B. The reflected beam is detected by a photomultiplier tube and lock-in amplifiers extract the signals at 50 and 100 kHz. $\bar{\rho}$ is recorded every second, and an average over 50 readings is calculated. Technical details and examples of the application of ellipsometry to dynamic liquid surfaces can be found in the literature.[7,11,12]

c. Neutron Reflection

Neutron reflection (NR) provides an accurate method for quantifying the adsorption of surfactants at liquid interfaces. It is an important, direct method for determining Γ_{eq} and Γ_{dyn}, and an ideal complement to lab-based ellipsometry. Figure 8.2 shows an image of an OFC interfaced to the SURF reflectometer at the ISIS pulsed neutron source in the U.K., and a schematic of the experimental configuration. The application of neutron reflectivity to equilibrium adsorbed layers has been reviewed recently.[10] Technical details of the OFC-NR setup and experimental protocols are covered elsewhere.[9,11,12] We simply note here that the NR experiment measures an average surface excess over the central 4 cm^2 of the surface of the OFC with a typical precision of $\pm$ 0.2 $\times$ 10^{-6} mol m^{-2} and an accuracy of 5%.

The key approach in NR is to match the refractive index of the aqueous subphase to that of air, so that the signal arises from the adsorbed surfactant layer only. The scattering length of H_2O is negative while that of D_2O is positive: an 8 mol% solution of D_2O in H_2O gives rise to a null reflecting substrate (called null reflecting water or contrast-matched water). To determine the adsorbed amount, measured reflectivity profiles, $R(Q)$ (where Q is the scattering vector) are fitted to an optical matrix model to obtain values of the layer thickness, τ, and scattering length density, ρ. The area per molecule, A_s, and adsorbed amount, Γ, can be determined from

$$A_s = \frac{\sum_i b_i}{\rho \tau} = \frac{1}{\Gamma N_a} \qquad (8.1)$$

where Σb_i is the sum of atomic scattering lengths in the molecule and N_a is the Avogadro Number. More detail on data analysis, and limitations, can be found in the literature.[9,10]

d. Laser Doppler Velocimetry

The surface velocity on the OFC depends on the surfactant concentration and the radial distance, r, from the center of the cell, but is independent of flow rate.[7] The technique of laser Doppler velocimetry (LDV)[14] can be used to determine the radial surface velocities, u_s. Details of the applications of LDV to surfactants in an OFC are given elsewhere.[7] Plots of $u_s(r)$ are analyzed to obtain the surface expansion rate, θ, and θ is found to be approximately constant across the surface of the cylinder. For the C_mTAB surfactants, θ lies in the range 0.5 to 6 s^{-1}.

C. Dynamic Surface Properties of C_mTAB Solutions

The dynamic and equilibrium surface tensions are shown in Figure 8.3a as a function of C for the model surfactant $C_{16}TAB$.[7,8] There is a large difference between γ_{dyn} and γ_{eq} for C = 0.1–1 mM, with a maximum value of $\Delta\gamma = \gamma_{dyn} - \gamma_{eq} = 17$ mN.m^{-1} at C = 0.4 mM. At lower concentration, both curves tend toward the value of pure water. At higher C they con-

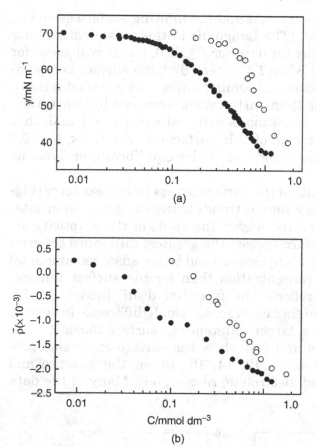

Figure 8.3 Equilibrium (●) and dynamic (○) surface tensions (a) and ellipticities (b) for C_{16}TAB solutions. Measurement technique: du Nöuy ring for equilibrium solutions (●) and surface light scattering from the surface of the OFC for dynamic tensions (○). Reprinted with permission from [7b]. © (2000) Elsevier.

verge to the limiting value of γ above the cmc. The concentration regime in which this "dynamic envelope" occurs can be understood qualitatively by considering the ratio of the depletion length, Γ/C and the diffusion length, $(D/\theta)^{0.5}$, where D is the diffusion coefficient of the surfactant. For C_{16}TAB in the OFC, $(D/\theta)^{0.5} = 10\text{--}30$ μm and Γ/C~ $\Gamma_\infty/(K+C)$ where the Langmuir constant $K \sim 0.1$ mM and the limiting surface excess Γ_∞ $\sim 4 \times 10^{-6}$ mol.m^{-3}. (The Langmuir isotherm is a rather poor fit to the experimental data for C_{16}TAB, but it will serve for this illustration.) When $\Gamma/C \ll (D/\theta)^{0.5}$, the surface is replenished with surfactant (assuming diffusion-controlled adsorption) much faster than surfactant is removed by convection. For C_{16}TAB, this condition is satisfied when $C > 1$ mM, that is, just above the cmc of this surfactant. For $C \ll K \sim 0.1$ mM, there is little adsorption under equilibrium or dynamic conditions.

Measurements of the surface excess by ellipsometry (Figure 8.3b) show very similar trends to the surface tension data. As explained later, the *higher* the value of the ellipticity $\bar{\rho}$, the lower the surface excess. The greatest difference between dynamic and static ellipticities (and hence adsorbed amounts) occurs at lower concentration than for the surface tension. This difference reflects the fact that $d\gamma/d\Gamma$ increases with increasing Γ: At higher coverages, a small difference in surface excess results in a larger difference in surface tension.

Figures 8.4a and 8.4b show the surface expansion rate $\theta(C)$ for C_mTAB (m = 12, 14, 16, 18) in the absence and presence of added electrolyte, respectively. Many of the data

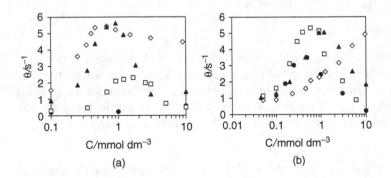

Figure 8.4 Surface expansion rate, θ, in an overflowing cylinder as a function of the concentration, C, of C_mTAB (a) in the absence of added electrolyte, and (b) in 0.1 M NaBr: ($\bullet$) m = 12, ($\square$) m = 14, ($\blacktriangle$) m = 16, ($\diamond$) m = 18.

sets have a characteristic bell shape. Comparison of the data for $C_{16}TAB$ in Figures 8.3a and 8.4a shows that the maximum surface expansion occurs when the $\Delta\gamma$ is large. This correlation can be understood because the surface expansion is driven by Marangoni effects (surface-tension gradients). At low C, Γ_{dyn} is small and γ is close to its value in pure water everywhere on the surface of the OFC. At sufficiently high C, the surface is in equilibrium and the surface tension gradients again vanish.

Not all surfactants conform to this bell-shaped behavior. $C_{12}TAB$ solutions show no surface acceleration caused by the surfactant (Figure 8.4a). For $C_{12}TAB$, the Langmuir constant K is sufficiently high (~ 1 mM) that the condition $\Gamma/C \ll (D/\theta)^{0.5}$ is satisfied at all C and the surface is in equilibrium with the bulk (as confirmed by ellipsometry measurements). Addition of salt lowers K and the effects of DST are observed once more (Figure 8.4b). For $C_{18}TAB$, the values of θ do not return to the pure water values at high C, with or without salt. For $C_{18}TAB$, most of the surfactant is present as micelles, and it is likely that either slow diffusion or slow micelle breakdown limits the rate at which surfactant can adsorb to the surface.

Differences between the equilibrium and dynamic surface excess values must be the underlying reason for the differences in dynamic and equilibrium surface tensions and ellipticities shown in Figure 8.3a and 8.3b. For quantitative analysis of the dynamic data, the values of $\bar{\rho}$ need to be converted to values of Γ. To perform this conversion, Γ_{dyn} was measured directly at selected surfactant concentrations by NR for deuterated $C_{14}TAB$, $C_{16}TAB$ and $C_{18}TAB$ (Figure 8.5). The equilibrium surface excess, measured by Lu et al.,[15] for $C_{16}TAB$ is also indicated by a dotted line. As expected, the dynamic surface excess is lower than the equilibrium value at C below 1 mM. In all three cases, the limiting surface coverage is approached at bulk C of 1–2 mM. This value is determined by the balance of convection and diffusion to the surface and is not strongly correlated to the cmc values (shown by arrows in Figure 8.5). Two fluorocarbon surfactants (based on a dialkyl sulphosuccinate structure) showed similar

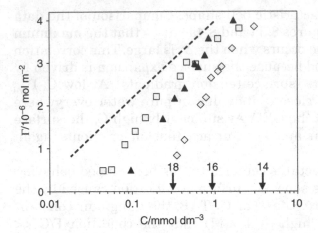

Figure 8.5 Dynamic surface excess of $C_m TAB$ surfactants in the OFC without added electrolyte: ($\square$) $m = 14$, ($\blacktriangle$) $m = 16$,($\Diamond$) $m = 18$. Arrows mark the cmc's of the three surfactants. Dotted line indicates the equilibrium surface excess for $C_{16}TAB$ [15]. Data have been amalgamated and re-plotted from references 9 and 12.

behavior,[16] as do the nonionic surfactants $C_{10}E_8$ and $C_{12}E_8$.[11] The "dynamic envelope" is thus determined principally by the rate of surface expansion (the strain rate) not by the properties of the individual surfactant. There are two general exceptions to this rule:

- Surfactants with cmc values sufficiently high that the surface is always in equilibrium (cf. $C_{12}TAB$ without salt).
- Surfactants that form large aggregates in solution (e.g., $C_{12}E_3$). In the latter case, mass transport to the surface by diffusion is very slow and the dynamic envelope is extended to higher C.

Figure 8.6 correlates the coefficients of ellipticity measured on the OFC to the surface excess values determined by NR or, for $C_{12}TAB$, by tensiometry.[9,12] The significance of the near-linear response is that this calibrated ellipsometric measurement can be used to determine dynamic surface excess values for this family of surfactants. Ellipsometry has signif-

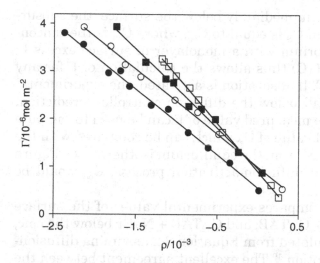

Figure 8.6 Relationship between dynamic ellipticity and dynamic surface excess for $C_{18}TAB$ (●), $C_{16}TAB$ (○) and $C_{14}TAB$ (■) and for equilibrium ellipticity and equilibrium surface excess for $C_{12}TAB$ (□). Reprinted with permission from [12]. © (2003) American Chemical Society.

icant advantages over neutron reflection, in terms of experimental convenience, time, and cost.

D. Adsorption Mechanism for C_mTAB at the Air-Water Surface

1. Surfactant Solutions Below the cmc

Solution of the convection-diffusion equation provides a relationship between the surface excess and the surface expansion rate[7b]:

$$\theta = \frac{2D(C - C_s)^2}{\pi \Gamma^2} \tag{8.2}$$

where the diffusion coefficient D may be obtained from the literature or determined by NMR.[17,18] If the adsorption rated is diffusion-controlled, then the monolayer is in equilibrium with the solution immediately below the surface: the subsurface concentration C_S is equal to C_{eq}, where C_{eq} is the concentration in equilibrium with a monolayer of surface excess Γ. A knowledge of $\Gamma(C)$ thus allows the calculation of Γ for any values of θ and C. If adsorption is activated, the experimental value of Γ will fall below the diffusion-controlled prediction. Alternatively, the measured value of Γ can be used to compute the experimental value of C_s, which can be compared with the value, C_{eq}, expected from the equilibrium isotherm (see Figure 8.3). For a mixed diffusion-activation process, C_{eq} would be less than C_s.

Figure 8.7 compares experimental values of the surface excess of $C_{14}TAB$, $C_{16}TAB$, and $C_{12}TAB$ + NaBr below the cmc, with values calculated from Equation 8.2 assuming diffusion-controlled adsorption.[19] The excellent agreement between the calculated and measured values demonstrates the absence of an activation barrier in this system. The same conclusion was drawn for the fluorinated anionic surfactant sodium bis(1H,1H nonafluoropentyl)-2-sulfosuccinate (di-CF4). A

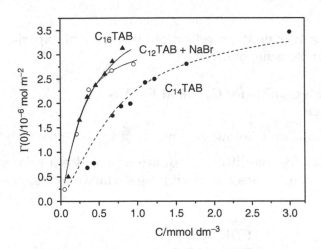

Figure 8.7 Surface excess in the center of the cylinder as a function of surfactant concentration: $C_{16}TAB$ (▲); $C_{14}TAB$ (●); $C_{12}TAB$ + 0.1 M NaBr (○). Reprinted with permission from [19]. © (2004) American Chemical Society.

detailed analysis of the mass transport through the electrical double layer[19] shows that an activation barrier arising from the electrical double layer is unlikely to limit the rate of adsorption except at strain rates much higher than those encountered in the OFC.

The type of isotherm used to model the equilibrium adsorption of ionic surfactants is crucial for interpreting dynamic data. For the analysis in Figure 8.7, we employed a model that includes lateral interactions between surfactant molecules and accounts for counterion binding. Kralchevsky et al.[20] have argued that the most appropriate isotherms for these kinds of systems are the van der Waals isotherm for the surface-active ions and the Stern isotherm for counterions. An earlier analysis,[7b] in which counterion binding was neglected, suggested that the deviations from diffusion control were observed at low C of C_{16}TAB. Counterion binding reduces the surface potential and therefore reduces the height of the activation barrier. The inclusion of counterion binding in the model restores agreement with experiment.

2.   Surfactant Solutions above the cmc

The simplest model for adsorption kinetics from micellar solutions assumes that equilibration of the monomers and micelles in solution is fast on the time scale of θ^{-1}. With this assumption, the solution can be divided into two regions separated by the surface $C(Z) = \text{cmc}$. For axial positions $z < Z$ (where $z = 0$ defines the surface plane), the micellar concentration is zero and mass transport is due only to monomers. For $z > Z$, the monomer concentration is constant and equal to the cmc and transport is due only to the micelles. The convection-diffusion equation can be solved in each region and the fluxes of monomers and micelles matched at $z = Z$. With a knowledge of the monomer and micellar diffusion coefficients and the adsorption isotherm, the surface excess can then be predicted as in the case for surfactants below the cmc.[19] Figure 8.8a compares the experimental and calculated values of Γ for C_{14}TAB + salt above and below the cmc. There is excellent agreement with a diffusion-controlled model with rapid micellar breakdown. Figure 8.8b shows similar data for C_{16}TAB + salt at C above the cmc. Here the experimental

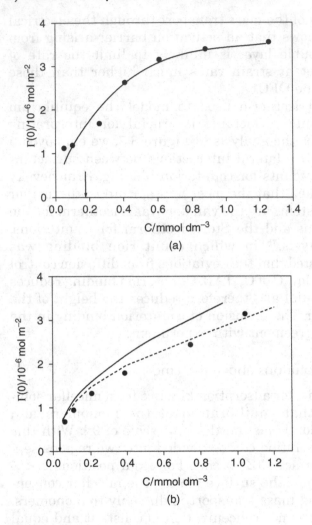

Figure 8.8 Comparison between experimental and theoretical values for the surface excess, $\Gamma(0)$, at the center of the OFC. ($\bullet$) experimental data, ($\square$) diffusion-controlled adsorption with rapid micelle breakdown, (- - -) diffusion-controlled adsorption with finite breakdown kinetics (see text for model). Arrows mark the cmc's. (a) $C_{14}TAB$ + 0.1 M NaBr, (b) $C_{16}TAB$ + 0.1 M NaBr. Reprinted with permission from [19]. © (2004) American Chemical Society.

surface excess is less than that calculated. Recently, asymptotic solutions have been developed for a model that incorporates simple micellar kinetics explicitly, and these closed-form solutions have been tested against numerical simulations.[21] The monomer-micelle equilibrium is treated as a single reaction between N monomers and a micelle, where N is the aggregation number of the micelle. This model has been applied to the data for $C_{16}TAB$ + salt in Figure 8.8b. Excellent agreement with experiment is found for a micellar breakdown rate $k_{-1} = 20$ s^{-1} for an assumed value $N = 90$ (dashed line in Figure 8.8b). This rate constant is reasonable based on relaxation measurements on related systems,[22] though it should be noted that the rate constants for micellar breakdown are found experimentally to be a function of C[22b] and only an average value is derived from fits to the OFC data.

While a need remains to validate the mathematical model for adsorption in micellar solutions against additional experimental data sets, indications are that the kinetics of adsorption of C_mTAB surfactants both above and below the cmc can be explained quantitatively on a diffusion-controlled model if finite micelle breakdown kinetics are allowed for.

E. General Observations for Other Ionic Surfactants

Limitations on neutron beam time mean that only selected surfactants can be investigated by OFC-NR. However, parametric and molecular structure studies have been possible with the laboratory-based method maximum bubble pressure tensiometry (MBP). This method has been shown to be reliable for C > 1 mM.[23] Details of the data analysis methods and limitations of this approach have been covered in the literature.[2,23] Briefly, the monomer diffusion coefficient below the cmc, D, can be measured independently by pulsed-field gradient spin-echo NMR measurements. Next, $\gamma(t)$ is determined by MBP and converted to $\Gamma(t)$ with the aid of an equilibrium equation of state determined from a combination of equilibrium surface tensiometry and neutron reflection. The values of $\Gamma(t)$ are then fitted to a diffusion-controlled adsorption model with an effective diffusion coefficient D_{eff}, which is sensitive to the dominant adsorption mechanism: $D_{eff}/D = 1$ for pure diffusion control, $D_{eff}/D < 1$ for mixed activation-diffusion, or in extreme cases, a pure activation mechanism. For both the anionic surfactant di-CF$_4$ (by MBP and OFC studies[16]) and the catanionic surfactant n-hexylammonium dodecylsulfate (MBP only), the MBP data were consistent with a diffusion-controlled mechanism for C > 1 mM.[23] For lower C, the MBP data suggested there might be a small activation barrier. We note, however, that the OFC data for di-CF4 did not show evidence for an activation barrier at strain rates of 10 s^{-1}.

III. SOLID–LIQUID INTERFACES

Until recently, the fast rate at which a surfactant layer forms at the solid-liquid interface has prevented accurate investigation of the adsorption process. As a result, the mechanism of surfactant adsorption has been inferred from thermodynamic data. Such explanations have been further confused by misinterpretation of the equilibrium morphology of the adsorbed surfactant as either monolayers or bilayers, rather than the discrete surface aggregates that form in many surfactant-substrate systems.[24–47] However, the recent development of techniques with high temporal resolution has made possible studies of the adsorption, desorption,[25,38,41,48–60] and exchange[54] rates of surfactants. In this section, we describe the adsorption kinetics of C$_m$TAB surfactants at the silica-aqueous solution interface, elucidated by optical reflectometry in a wall-jet flow cell. The adsorption of C$_m$TAB surfactants to silica is the most widely studied system[25,38,41,48,50–54] and hence the adsorption kinetics can be related to the adsorption process with great clarity. For a more thorough review of adsorptions isotherms, the types of surfactant structures that form at the solid-liquid interface, and the influence of these factors on adsorption, the reader is directed to Reference 24.

A. Optical Reflectometry and the Wall-Jet Cell

In optical reflectometry (OR)[61] a linearly polarized light beam is reflected from a surface and the reflectivity of s- and p-polarized components is measured (Figure 8.9). The intensity

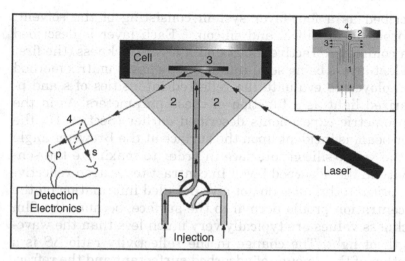

Figure 8.9 Schematic diagram of the reflectometer (after ref 61). A linearly polarised beam from the He-Ne laser (1) enters the cell through a 45-degree glass prism (2) and passes through the solution striking the surface under investigation (3). The reflected beam is split into its p and s components (4). Both components are detected by photodiodes and recorded separately. Solution is passed into the cell by means of an injection system. This consists of two flasks, one containing only solvent and the other surfactant solution, and a valve (5) that directs one solution to the cell and the other to waste. Solutions are passed into the cell by teflon tubing via a cylindrical hole in the glass prism (2). Inset: Schematic diagram of stagnation point flow. The solution is perpendicularly injected towards the flat surface. As the jet stream spreads over the surface, a stagnation point forms at the spot where the axis of the impinging jet intersects with the surface. Reprinted with permission from [48]. © (2000) American Chemical Society.

of the s and p polarizations can be measured with a temporal resolution of < 0.1 s. As material adsorbs to the silica-water interface the intensities of the reflected beams are altered. We note the difference between ellipsometry (Section II) in which the (complex) ratio, ρ, of the amplitudes of the reflected p and s-polarized electric fields is measured and reflectometry, in which the (real) ratio, S, of the intensities of p- to s-polarized light is determined. To interpret S, the interface is described as a four-layer system consisting of the solvent, adsorbed layer, silica, and silicon.[62] Each layer is described by a complex refractive index and a layer thickness (the first and last layers being semi-infinite). An optical matrix method is employed to evaluate the reflected intensities of s and p-polarized light as a function of these parameters. As in the ellipsometric experiments described earlier (Section II), the laser beam is incident upon the surface at the Brewster angle for the water-silicon interface in order to maximize the sensitivity to the adsorbed layer. In contrast to neutron reflectivity, optical techniques do not give detailed information on the concentration profile normal to the surface, because the film thickness values are typically very much less than the wavelength of light. The change in the reflectivity ratio ΔS is a function of the amount of adsorbed surfactant and the refractive index increment, dn/dc, of the surfactant solution. For thin films, there is a linear relationship between ΔS and Γ:

$$\Gamma = \frac{\Delta S}{S_0} \frac{1}{A_s} \tag{8.3}$$

where S_0 is the initial value of S prior to adsorption and the sensitivity factor, A_s, can be calculated from the optical matrix method and the known optical properties of the surfactant and the bulk phases.[61].

The experimental setup employs stagnant point flow hydrodynamics in a wall jet, shown in the inset of Figure 8.9. There is no convection at the stagnation point, so transport of surfactant to and from the surface is by diffusion only. The hydrodynamics associated with stagnation point flow have been investigated by Dabros and Van de Ven,[62] who showed that the diffusion-limited flux, J, of an adsorbate to the surface is given by

$$J = 0.766 \mu^{1/3} R^{-1} D^{2/3} (\alpha \, \text{Re})^{1/3} C \tag{8.4}$$

where v is the kinematic viscosity, Re is the Reynolds number, and the geometric parameter α is determined by the ratio h/R, where h is the distance between the surface and the inlet tube and R is the radius of the tube.

B. Kinetics of Cationic Surfactant Adsorption

There have been a number of investigations of the adsorption kinetics of C_mTAB surfactants to the silica-aqueous solution interface.[38,41,48,50-54] It might be expected that a positively charged head group would provide the driving force for adsorption to the negatively charged silica substrate. However, the native charge associated with the silica surface is neutralized at low surface excess values.[63] For surfactant concentration in excess of the surface charge neutralization condition, adsorption proceeds against a repulsive electrostatic barrier and is driven by hydrophobic interactions.

1. Surfactant Chain Length Effects

The adsorption kinetics for C_{12}TAB, C_{14}TAB, and C_{16}TAB in 10 mM KBr is presented in Figure 8.10. The initial adsorption rate is equal to the gradient of the linearly increasing region of an adsorption experiment.[50] For $\Gamma < 0.25 \times 10^{-6}$ mol m^{-2}, electrostatic interactions control the adsorption rate.[50] For

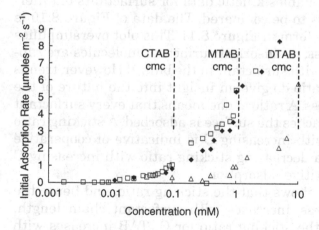

Figure 8.10 Initial Adsorption Rate for C_{16}TAB (CTAB, open squares), C_{14}TAB (MTAB, closed diamonds) and C_{12}TAB (DTAB, open triangles) in the presence of 10 mM KBr. The dashed vertical lines represent the solution cmc for each surfactant in 10 mM KBr. Reprinted with permission from [50]. © (2003) Elsevier.

higher values of Γ, both electrostatic and hydrophobic interactions influence the kinetics. Since the electrostatic contribution will not be very sensitive to the nature of the surfactant for the same value of Γ, the differences in measured adsorption rates for the three surfactants are primarily a consequence of different hydrophobic interactions associated with the different lengths of the hydrocarbon chains.

Generally, the adsorption rate increases with surfactant concentration due to the increased flux of surfactant to the surface. For a quantitative comparison of adsorption rates it is necessary (a) to compare the rates of adsorption not at the same C, but in the same region of the adsorption isotherm, so that the surface excess and adsorbed layer morphologies are similar, and (b) to account for the rate at which the surfactant is being delivered to the surface. The first of these conditions is largely accomplished by normalizing the C axis by the cmc. The second is achieved by comparing the actual initial rate of adsorption to the theoretically determined diffusion limited flux to the surface, a ratio that we call the sticking ratio.[48] A plot of sticking ratio versus the C normalized by the cmc enables kinetic data for surfactants of different chain lengths to be compared. The data of Figure 8.10 is presented in this form in Figure 8.11. This plot oversimplifies the actual process, as adsorbed surfactant molecules are rapidly exchanging with surfactant in the bulk.[54] However, trends in the sticking ratio do give an insight into the nature of the adsorption process. A ratio of one means that every surfactant molecule that reaches the surface is adsorbed. A sticking ratio that increases with increasing C is indicative of cooperative adsorption and a decreasing sticking ratio with increasing C indicates competitive adsorption.

Figure 8.11 shows that the sticking ratio, and hence the adsorption success, increases with surfactant chain length. Below the cmc, the sticking ratio for C_{16}TAB increases with C, indicating that the process is predominantly cooperative, and therefore hydrophobically driven. Conversely, the sticking ratio for C_{14}TAB and C_{12}TAB decreases with increasing C, consistent with a competitive and predominantly electrostatically driven adsorption process.

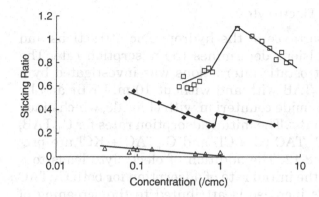

Figure 8.11 Sticking Ratio versus concentration normalized by the cmc for $C_{16}TAB$ (open squares), $C_{14}TAB$ (closed diamonds) and $C_{12}TAB$ (open triangles) in the presence of 10 mM KBr. The lines are drawn to guide the eye. Reprinted with permission from [50]. © (2003) Elsevier.

The most striking feature in Figure 8.11 is the jump in the sticking ratio for $C_{16}TAB$ at the cmc. This increase in adsorption efficiency may be attributed to direct adsorption of micelles to the surface. As the bulk surfactant concentration, and hence the concentration of micelles in solution, rises, the sticking ratio gradually falls. This decline is indicative of competitive adsorption between micelles. Micelles only adsorb as a single layer, so previously adsorbed micelles inhibit adsorption of subsequent micelles. A smaller jump in sticking ratio at the cmc can be detected for $C_{14}TAB$. This reduced effect is attributed to the greater concentration of monomers in solution. Above the cmc there is competition between monomers and micelles for adsorption. As the chain length decreases, the cmc and the relative number of monomers present in solution increases. Thus, the magnitude of the change in sticking ratio at the cmc decreases with decreasing chain length. It was not possible to obtain data above the cmc for $C_{12}TAB$ due to optical artifacts associated with mixing, induced by the large concentration of surfactant in the solution.[50]

2.   Influence of Electrolytes

The relative importance of the hydrophobic attractions and electrostatic repulsions determines the adsorption rate. The effect of the electrostatic interactions was investigated by a comparison of $C_{16}TAB$ with and without 10 mM KBr and by replacement of bromide counterion with chloride, which binds less strongly than Br^-. The initial adsorption rates for $C_{16}TAB$, $C_{16}TAB$ + KBr, $C_{16}TAC$ (C ≡ Cl) and $C_{16}TAC$ + KCl are presented in Figure 8.12. The addition of electrolyte leads to a large increase in the initial rate of adsorption for both $C_{16}TAC$ and $C_{16}TAB$. This increase is attributed to the screening of the electrostatic repulsion between the monomers and the surface (note that the surface charge is reversed early in the adsorption process due to surfactant adsorption). The raw adsorption rates for $C_{16}TAB$ and $C_{16}TAC$ are very different in the absence of added salt, but very similar with added electrolyte. Both the nature of the counterion and the electrolyte

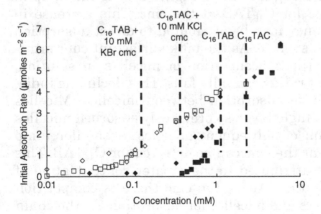

Figure 8.12 Initial Adsorption Rate for $C_{16}TAB$ in the presence (open squares) and absence (closed squares) of 10 mM KBr and for $C_{16}TAC$ in the presence (open diamonds) and absence (closed diamonds) of 10 mM KCl. Reprinted with permission from [50]. © (2003) Elsevier.

concentration affect the cmc so, as above, it is useful to present the sticking ratio as a function of normalized concentration (Figure 8.13).

For C_{16}TAB in the absence of KBr, the sticking ratio increases up to the cmc, but is much lower than in the presence of salt. In both cases, hydrophobic interactions influence the adsorption kinetics, but the screening of repulsions by the electrolyte increases the sticking ratio in the presence of salt. Above the cmc, the sticking ratio for C_{16}TAB without salt is independent of C, showing that micelles do adsorb directly but that there is less competition for adsorption sites. This is partly because the maximum surface excess is lower than with added KBr and partly because the monomer concentration is higher and therefore the contribution of micellar adsorption is less significant.

C_{16}TAC behaves very differently from C_{16}TAB. There is no increase in sticking ratio at the cmc, either with or without KCl. A lack of direct involvement of micelles in the adsorption process may be a consequence of the decreased level of ion

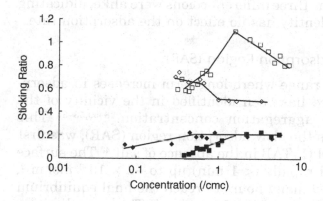

Figure 8.13 Sticking Ratio versus concentration normalized by the cmc for C_{16}TAB in the presence (open squares) and absence (closed squares) of 10 mM KBr and for C_{16}TAC in the presence (open diamonds) and absence (closed diamonds) of 10 mM KCl. The lines are drawn to guide the eye. Reprinted with permission from [50]. © (2003) Elsevier.

binding of chloride, which means that the C_{16}TAC micelles carry a larger surface charge than C_{16}TAB micelles and hence are repelled more strongly from the surface. In the absence of KCl, the expected increase in sticking ratio with C was observed below the cmc. This is a result of hydrophobic interactions. For C_{16}TAC + KCl, however, the sticking ratio decreases with increasing C. The most likely explanation is that the higher cmc in the C_{16}TAC + KCl system compared to the C_{16}TAB + KBr system leads to a greater flux of monomers and hence a transition to a situation where electrostatic interactions dominate at a concentration C that is still below the cmc. Similar behavior was observed with C_{14}TAB and C_{12}TAB in KBr, which have shorter alkyl chains and hence higher cmc values.

The influence of the coion identity on adsorption behavior has also been investigated.[50] The C_{16}TAC system was chosen to investigate this effect, as the presence of added chloride has little influence on the maximum surface excess or the aggregate morphology. The rates of adsorption for C_{16}TAC in the presence of 10 mM KCl, NaCl, and LiCl were determined. The results for the three different coions were alike, indicating that the coion identity has no effect on the adsorption rate.

3. The Slow Adsorption Region (SAR)

A concentration range where long-term increases in adsorption are observed has been identified in the vicinity of the critical surface aggregation concentration.[48-50,53-54] This region, known as the slow adsorption region (SAR), was first noted for 0.6 mM C_{16}TAB in the absence of salt.[48] The surface excess increased rapidly (< 1 min) up to ~2×10^{-6} mol.m^{-2}, but then required many hours to reach the final equilibrium adsorbed amount of ~4.7×10^{-6} mol.m^{-2}. The same limiting surface excess was reached at higher concentration of C_{16}TAB in less than 1 min. A more detailed investigation of these results was conducted using cetylpyridinium bromide CPBr.[49] The UV activity of CPBr allowed solution surfactant C to be determined spectrophotometrically with high precision, permitting the effect of small increases in surfactant concentration on slow adsorption phenomena to be elucidated. Figure 8.14 shows that solutions of CPBr of concentration of 0.202 mM and 0.554 mM lead to equilibrium coverages of 1×10^{-6} mol.m^{-2} and 4×10^{-6} mol.m^{-2}, respectively. Only between these two concentrations were long-term increases in CPBr adsorption noted. AFM imaging was used to investigate the structural changes that accompany these slow increases in adsorption. No structure was discernable in the adsorbed layer 20 min after surfactant solution was passed into the AFM cell. After an additional 22 h, elongated admicelles were observed.[49] The SAR appears to be a consequence of kinetic barriers to the formation of the thermodynamically stable arrangement of adsorbed surfactant.[24] The boundaries of the SAR are determined by the adsorbed layer structure and coverage, and the morphology of the surfactant aggregates in solution. Below the SAR, the concentration in bulk is not sufficient to raise the chemical potential of the monomer to a

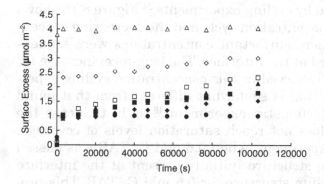

Figure 8.14 Measured surface excess of CPBr at the hydroxylated silica-solution interface versus time for CPBr concentrations of 0.202 (filled circles), 0.274 (filled diamonds), 0.284 (filled squares), 0.306 (filled triangles), 0.315 (open squares), 0.336 (open diamonds) and 0.554 mM (open triangles) mM with no added electrolyte. Adsorption occurs over a much greater time period than for other concentrations. The surface excess initially increases rapidly to 1×10^{-6} mol m^{-2} (or 2.4×10^{-6} mol m^{-2} for 0.336 mM CPBr) then continues to increase over many hours. Reprinted with permission from [49]. © (2003) Elsevier.

level where surface aggregation is favorable. At C above the SAR, surface aggregates are clearly forming and are doing so rapidly. It may be that at concentration above the SAR, but below the cmc, a number of transient semi-formed aggregates are present in solution, and these premature aggregates adsorb to the surface. This removes the requirement for monomer units to reorganize slowly into aggregates on the surface.

A SAR for $C_{16}TAC$ in the presence of 10 mM LiCl has been observed.[50] This important result shows that long-term increases in adsorption can occur in the presence of electrolyte, provided that the electrolyte does not influence the surface excess[50] or structure.[40] This indicates that it is the structural barrier to adsorption that is critical for the evolution of slow adsorption effects.

4. Evidence for Kinetic Trapping

Optical reflectometry enables different surfactant concentrations to be analyzed consecutively. This allowed the SAR to be further probed by cycling experiments.[24] Figure 8.15 shows the increasing concentration cycle and the decreasing concentration cycle, where surfactant concentrations were sequentially equilibrated at the interface. The long-term increase for 0.6 mM $C_{16}TAB$ observed in this concentration cycling experiment (Figure 8.15a) is somewhat different from that found in a normal reflectometry experiment,[48] as in this case the surface excess does not reach saturation levels of coverage. In the cycling experiment, when 0.6 mM $C_{16}TAB$ was passed into the cell, the structure initially present at the interface was the equilibrium structure for 0.5 mM $C_{16}TAB$. This preadsorbed structure leads to a different adsorption path that becomes kinetically trapped at a surface excess below the equilibrium surface excess. In the cycling experiment, saturation coverage was not reached until a bulk concentration of 0.8 mM.

For C below 0.5 mM, the surface coverages observed in the desorption cycle were higher than in the adsorption cycle (Figure 8.15b). Only some of this discrepancy can be attributed to the drift in the baseline of $\sim 0.2 \times 10^{-6}$ mol.m^{-2}. This

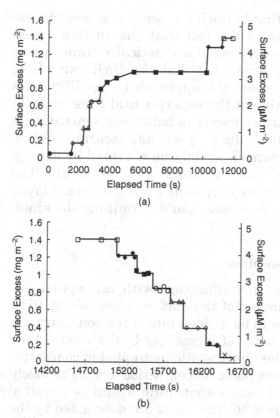

(a)

(b)

Figure 8.15 Surface excess of $C_{16}TAB$ on hydroxylated silica versus time for (a) sequentially increasing and (b) sequentially decreasing $C_{16}TAB$ concentrations in the absence of electrolyte. A stable baseline was obtained in water; the surfactant solutions were passed into the cell in series. Surfactant concentrations used were 0.05 mM (filled circles), 0.15 mM (open diamonds), 0.3 mM (open triangles), 0.5 mM (open circles), 0.6 mM (closed squares), 0.7 mM (closed diamonds), and 0.8 mM (open squares). Note the long-term increase observed for 0.6 mM. Surfactant concentrations of 1, 3 and 5 mM were also used. These concentrations all produced an equilibrium surface excess of 4.6×10^{-6} mol m^{-2} (within experimental limitations) and are not shown here for clarity. Water (crosses) is passed into the cell after 16,500 seconds and the non-zero surface excess obtained is due to baseline drift. Only representative data points are provided. However, the line to guide the eye provides a reasonable indication of the form of the experiment. Reprinted with permission from [24]. © (2003) Elsevier.

observation suggests that in the increasing cycle equilibrium was not achieved despite the fact that the surface excess was stable — an extreme case of a kinetically trapped surface conformation. It is possible that the SAR extends to lower concentrations where the approach to equilibrium is so slow as to fall outside of the experimental time scale, or that the change in surface excess is below the sensitivity of the instrument. These results suggest that accurate equilibrium adsorption isotherms may only be obtained by desorption from higher concentrations and highlight the importance of the structure of the adsorbed surfactant layer, and not merely the surface excess, in determining the kinetics of adsorption.

C. Mechanism of Adsorption

Detailed kinetic data, in conjunction with adsorption isotherms and direct imaging of the surface morphology, allow the adsorption process to be divided into three concentration spans.[24] They are, in order of increasing C, the electrostatic concentration span, the electrostatic-hydrophobic concentration span, and the hydrophobic concentration span. The mechanism of adsorption in each span differs and is depicted schematically in Figure 8.16. The spans are delineated by the adsorption mechanism that is newly available in that concentration range. Thus, in the hydrophobic span all three mechanisms may be operating simultaneously but at different rates. The hydrophobic concentration span may be subdivided into above-cmc and below-cmc spans to reflect the direct adsorption of micelles.

1. The Electrostatic Concentration Span

Here the surfactant ions are electrostatically adsorbed to the oppositely charged surface sites. The presence of a positively charged head group at the interface renders nearby hydroxyl groups more acidic, which induces more negatively charged sites in the vicinity of the initial negative site.

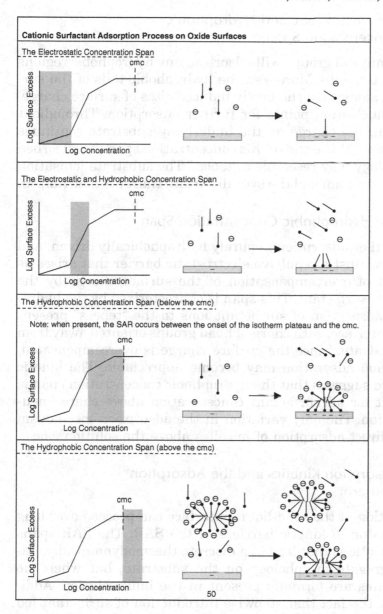

Figure 8.16 Adsorption process for cationic surfactants at the silica-aqueous solution interface. Each span is described in the text. Reprinted with permission from [24]. © (2003) Elsevier.

2. The Electrostatic and Hydrophobic Concentration Span

Surfactant tail groups will adsorb on any hydrophobic regions on the substrate. Moreover, the hydrophobic tails of the surfactant, along with the newly induced sites of surface charge, act as nucleation points for further adsorption. Throughout this span the charge on the underlying substrate continues to increase. At the end of this concentration span the adsorbed morphology may resemble a "tepee." The substrate ionisation is at a maximum and the overall surface charge is neutralized.

3. The Hydrophobic Concentration Span

Any further adsorption is purely hydrophobically driven, and will be against a repulsive electrostatic barrier that arises as a result of overcompensation of the surface charge by the adsorbed surfactant. This span is characterized by the hydrophobic adsorption of surfactant ions to the "tepees" present at the interface, with charged head groups oriented away from the substrate. Since the surface charge is overcompensated, counterion adsorption may become appreciable. The kinetic evidence suggests that the hydrophobic concentration span is relevant for all surfactant concentration above charge neutralization. The only variation in the adsorption mechanism is the direct adsorption of micelles above the solution cmc.

4. Adsorption Kinetics and the Adsorption Isotherm

Adsorption at the solid-liquid interface can proceed over long periods due to kinetic barriers in the SAR. The SAR spans concentrations that lead to a discrete thermodynamically stable aggregate morphology on the substrate, but where no aggregates are formally present in the bulk solution. Additionally, the fact that stepwise introduction of surfactant led to values of surface excess less than those seen at the same concentration for stepwise reductions in surfactant concentration (as shown in Figure 8.15) can be interpreted as evidence that equilibrium has not been reached in the former case. In this case, the kinetic trap is so strong that no further adsorption occurs. Thus, equilibrium adsorption isotherms that are determined by stepwise changes in surfactant concentration should always be conducted under a dilution regime to prevent false equilibrium values from being obtained.

IV. CONCLUSIONS AND OUTLOOK

A. Air–Water Interfaces

The accumulated evidence from studies on the OFC suggests that the adsorption of C_mTABs is diffusion-controlled below the cmc. Above the cmc, there are deviations from a diffusion-controlled model for C_{16}TAB + NaBr, which can be quantitatively explained by slow micellar breakdown kinetics. The alternative of an adsorption barrier cannot be ruled out, though there is as yet no evidence of structures at the air–water interface akin to those observed in the SAR at the solid–liquid interface. More limited studies on other families of ionic surfactants in the OFC and MBP apparatus do not show large deviations from diffusion control. The importance of well-defined hydrodynamics and accurate equilibrium adsorption isotherms cannot be overstressed in quantitative studies of adsorption mechanisms. There is still a need for measurements at higher strain rates, such as occur in turbulent foams, jet breakup and impacting drops, and for additional studies with micellar systems to establish quantitatively the connection between micellar breakdown kinetics and rates of adsorption.

B. Solid–Liquid Interfaces

In contrast to air-water interfaces, the situation at solid-liquid surfaces is complicated by the formation of discrete surface morphologies (hemimicelles, surface aggregates). These structures can be slow to develop; hence the equilibration time scales can be much longer than for ideal air-water interfaces. For this reason it is much more difficult to accurately model adsorption kinetics at the solid-liquid interface: the limiting adsorbed amounts are not simply determined by formation of a complete monolayer (as for air-water surfaces), and material may continue to adsorb as the interfacial structures equilibrate by packing rearrangements.

Analysis of detailed kinetic data in conjunction with adsorption isotherms and surface aggregate morphologies has provided new insight into the adsorption process. The adsorption isotherm is divided into three concentration spans, indicating dominant mechanistic processes in each region. The hydrophobic concentration span can be divided into above-cmc and below-cmc spans to reflect the direct adsorption of micelles, and in this region it is likely that all the processes may be operating simultaneously and at different rates. This fundamental scheme can be used as a foundation to explore surfactant mixtures, surfactant-polymer, and surfactant-polyelectrolyte complexes. Moreover, it is now possible to further understanding of a variety of solid-liquid interfaces of practical interest, such as mineral ore surfaces in froth flotation processes, adsorption onto dental enamel, antiwear layers on lubricating metal surfaces, and bio-deposition onto bone.

V. ACKNOWLEDGMENTS

We would like to thank Richard Darton, Turgut Battal, Dmitrii Strykas, Dimitrina Valkovska and Gemma Shearman (from Oxford), Alex Rankin, James Dalton, Philippe Rogueda, Donal Sharpe, Simon Stebbing, Adrian Downer, Sandrine Nave and Alison Paul (from Bristol), Simon Biggs (from Leeds), Vince Craig (from Australian National University), Jeff Penfold and John Webster (ISIS, Rutherford Appleton Laboratory, U.K.), and Alan Pitt (Kodak U.K.) for stimulating discussions and experimental assistance.

REFERENCES

1. Chang, C-H., Franses, E.I. *Colloid Surf. A* 1995, 100, 1.

2. (a) Eastoe, J., Dalton, J.S. *Adv. Colloid Interface Sci.* 2000, 85, 103. (b) Eastoe, J., Rankin, A., Wat, R., Bain, C.D. *Int. Rev. Phys. Chem.* 2001, 20, 1.

3. Dukhin, S.S., Kretzschmar, G., Miller, R. *Dynamics of Adsorption at Liquid Interfaces*, Elsevier, Amsterdam, 1995.

4. Battal, T., Bain, C.D., Weiss, M., Darton, R.C. *J. Colloid Interface Sci.* 2003, 263, 250.

5. Horozov, T., Arnaudov, L. *J. Colloid Interface Sci.* 1999, 219, 99.

6. Hsu, C.T., Shao, M.J., Lee, Y.C., Lin, S.Y. *Langmuir* 2000, 16, 4846.

7. (a) Manning-Benson, S., Bain, C.D., Darton, R.C. *J. Colloid Interface Sci.* 1997, 189, 109. (b) Bain, C.D., Manning-Benson, S., Darton, R.C. *J. Colloid Interface Sci.* 2000, 229, 247.

8. Manning-Benson, S., Bain, C.D., Darton, R.C., Sharpe, D., Eastoe, J., Reynolds, P. *Langmuir* 1997, 13, 5808.

9. Manning-Benson, S., Parker, S.R.W., Bain, C.D., Penfold. J. *Langmuir* 1998, 14, 990.

10. Lu, J.R., Thomas, R.K., Penfold, J. *Adv. Colloid Interface Sci.* 2000, 84, 143.

11. Valkovska, D., Wilkinson, K.M., Campbell, R.A., Bain, C.D., Wat R., Eastoe J. *Langmuir* 2003, 19, 5960.

12. Battal, T., Shearman, G.C., Valkovska, D., Bain, C.D., Darton, R.C., Eastoe, J. *Langmuir* 2003, 19, 1244.

13. Aniansson, E.G., Wall, S.N., Almgren, M., Hoffmann, H., Kielmann, I., Ulbricht, W., Zana, R., Lang, J., Tondre, C. *J. Phys. Chem.* 1976, 80, 905.

14. (a) Drain, L.E. *The Laser Doppler Technique*, Wiley: Chichester, 1980. (b) Durst, F., Melling, A., Whitelaw, H.J. *Principles and Practice of Laser-Doppler Anemometery*, 2nd Ed, Academic Press, London, 1981.

15. Lu, J.R., Hromadovam M., Simister, E.A., Thomas, R.K., Penfold, J. *J. Phys. Chem.* 1994, 98, 11519.

16. Eastoe, J., Rankin, A., Wat, R., Bain, C.D., Strykas, D., Penfold, J. *Langmuir* 2003, 19, 7734.

17. Griffiths, P.C., Stilbs, P., Paulsen, K., Howe, A.M., Pitt, A.R. *J. Phys. Chem.* 1997, 101, 915.

18. Lide, D.R. (Ed.) *Handbook of Chemistry and Physics*, 76th Ed., pp. 5–91. CRC Press, Boca Raton, FL, 1993.

19. Valkovska, D., Shearman, G.C., Bain, C.D., Darton, R.C., Eastoe, J. submitted for publication in *Langmuir*: Adsorption of ionic surfactants at an expanding air-water interface.

20. Kralchevsky, P.A., Danov, K,D., Broze, G., Mehreteab, A. *Langmuir*, 1999, 15, 2351.

21. Breward, C.J.W., Howell, P.D., submitted for publication in *Euro. J. Appl. Math.*: Straining flow of a micellar surfactant solution.

22. (a) Michels, B., Waton, G. *J. Phys. Chem. A*. 2003, 107, 1133, (b) Tondre, C., Zana, R. *J. Colloid Interface Sci.* 1978, 66, 544.

23. (a) Eastoe, J., Dalton, J.S., Rogueda, P.G.A., Crooks, E.R., Pitt, A.R., Simister, E.A. *J. Colloid Interface Sci.* 1997, 188, 423, (b) Eastoe, J., Rogueda, P., Dalton, J., Dong, J., Sharpe, D., Webster, J.R.P. *Langmuir* 1996, 12, 2706.

24. Atkin, R., Craig, V.S.J., Wanless E.J., Biggs, S. *Adv. Colloid Interface Sci.* 2003, 103, 219.

25. Manne, S., Warr, G.G. In *Supramolecular Structure in Confined Geometries*, Manne, S., Warr, G.G. Eds., ACS, Washington, DC 1999, 2.

26. Manne, S., Cleveland, J.P., Gaub, H.E., Stucky, G.D., Hansma, P.K. *Langmuir* 1994, 10, 4409.

27. Patrick, H.N., Warr, G.G., Manne, S., Aksay, I.A. *Langmuir* 1997, 13, 4349.

28. Manne, S., Gaub, H.E. *Science* 1995, 270, 1480.

29. Manne, S., Schaffer, T.E., Huo, Q., Hansma, P.K., Morse, D.E., Stucky, G.D., Aksay, I.A. *Langmuir* 1997, 13, 6382.

30. Koganovkii, A.M. *Colloid J. USSR* 1962, 24, 702.

31. Wanless, E.J., Ducker, W.A. *J. Phys. Chem. B* 1996, 100, 3207.

32. Wanless, E.J., Ducker, W.A. *Langmuir* 1997, 13, 1463.

33. Grant, L.M., Tiberg, F., Ducker, W.A. *J. Phys. Chem. B* 1998, 102, 4288.

34. Holland, N.B., Ruegsegger, M., Marchant, R.E. *Langmuir* 1998, 14, 2790.

35. Ducker, W.A., Grant, L.M. *J. Phys. Chem. B* 1996, 100, 11507.

36. Ducker, W.A., Wanless, E.J. *Langmuir* 1996, 12, 5915.

37. Ducker, W.A., Wanless, E.J. *Langmuir* 1999, 15, 160.

38. Velegol, S.B., Fleming, B.D., Biggs, S., Wanless, E.J., Tilton, R.D. *Langmuir* 2000, 16, 2548.

39. Liu J.-F., Min, G., Ducker, W.A. *Langmuir* 2001, 17, 4895.

40. Subramanian, V., Ducker, W.A. *Langmuir* 2000, 16, 4447.

41. Atkin, R., Craig, V.S.J., Wanless E.J., Biggs, S. *J. Phys. Chem. B* 2003, 107, 2978.

42. Subramanian, V., Ducker W.A. *Phys. Chem. B* 2001, 105, 1389.

43. Lokar, W.J., Ducker, W.A. *Langmuir* 2002, 18, 3167.

44. Fragneto, G., Thomas, R.K. *Langmuir* 1996, 12, 6036.

45. McDermott, D.C., Lu, J.R., Lee, E.M., Thomas, R.K., Rennie, A.R. *Langmuir* 1992, 8, 1204.

46. McDermott, D.C., McCarney, J., Thomas, R.K., Rennie, A.R. *J. Colloid Interface Sci.* 1994, 162, 304.

47. Ström, C., Hansson, P., Jönsson, B., Söderman, O. *Langmuir* 2000, 16, 2469.

48. Atkin, R., Craig, V.S.J., Biggs, S. *Langmuir* 2000, 16, 9374.

49. Atkin, R., Craig, V.S.J., Biggs, S. *Langmuir* 2001, 17, 6155.

50. Atkin, R., Craig, V.S.J., Wanless E.J., Biggs, S. *J. Colloid Interface Sci.* 2003, 266, 236.

51. Eskilsson, K., Yaminsky, V.V. *Langmuir* 1998, 14, 2444.

52. Pagac, E.S., Prieve, D.C., Tilton, R.D. *Langmuir* 1998, 14, 2333.

53. Fleming, B.D., Biggs, S., Wanless, E.J. *J. Phys. Chem. B.* 2001, 105, 9537.

54. Clark, S.C., Ducker, W.A. *J. Phys. Chem. B.* 2003, 107, 9011.

55. Atkin, R., Craig, V.S.J., Wanless E.J., Hartley, P.G., Biggs, S. *Langmuir* 2003, 19, 4222.

56. Frantz, P., Granick, S. *Phys. Rev. Lett.* 1992, 66, 899.

57. Tiberg, F., Jönsson, B., Lindman, B. *Langmuir* 1994, 10, 3714.

58. Brinck, J., Tiberg, F. *Langmuir* 1996, 12, 5042.

59. Tiberg, F., Landgreen, M. *Langmuir* 1993, 9, 927.

60. Eskilsson, K., Tiberg, F. *Macromolecules* 1997, 30, 6323.

61. Dijt, J.C., Cohen Stuart, M.A., Fleer, G.J. *Adv. Colloid Interface Sci.* 1994, 50, 79.

62. Dabros, T., van de Ven, T.G.M. *Colloid Polym. Sci.* 1983, 261, 694.

63. Goloub, T.P., Koopal, L.K. *Langmuir* 1997, 13, 663.

9

Rheological Properties of Viscoelastic Surfactant Solutions: Relationship with Micelle Dynamics

HEINZ REHAGE

CONTENTS

I.   INTRODUCTION

"παντα ρει" — All things are in a state of flux and as a function
of time everything tends to flow; everything changes. This
famous statement of the ancient Greek philosopher Heracli-
tus is still valid and expresses the basic ideas of rheological
research. Many substances in daily life cannot be simply
classified as solids or liquids, but show more complicated
intermediate properties. Such systems are often composed of
super molecular, colloidal microstructures. For many industrial applications, it is of special interest to explore the basic
relationships between molecular structures of complex fluids
and their mechanical properties. Such experience is valuable
in the optimal design of products, including foams, emulsions,
suspensions, gels, cosmetics, or cleaning liquids. Surfactant
solutions are usually pumped, extruded, stirred, or mixed
during their processing. Typical products based on advanced
rheological techniques are drag-reducing liquids, which are
used to transport fluids through elongated tubes or pipelines
at relatively low energy costs. Shampoos or hair conditioners
should exhibit gel-like properties in order to form stable
foams. Emulsions or suspensions often have yield values,
which will protect these systems against sedimentation or
coagulation. Well-defined flow properties are often required
in industry and research, and consequently there are increasing demands to study these phenomena. This chapter is an
attempt to summarize recent knowledge on complex flow processes, with special emphasis drawn on nonlinear phenomena. The materials, as we will discuss in the next paragraphs,

are viscoelastic surfactant solutions. It turns out that many rheological properties of these solutions are controlled by micellar kinetics. Given this special situation, viscoelastic surfactant solutions can sometimes be characterized by simple theoretical laws. This holds, at least, in certain concentration regimes or certain temperature intervals. These special features lead to ideal conditions, which are difficult to realize in other types of colloidal systems. Viscoelastic surfactant solutions may, therefore, be used as simple model liquids in order to gain a deeper insight into fundamental principles of flow.

II. VISCOELASTIC SURFACTANT SOLUTIONS

A. Dilute Solutions of Spherical Micelles

Viscoelasticity is a general phenomenon that can be observed in many colloidal systems. Typical examples are gels, liquid crystalline phases, or concentrated emulsions or dispersions. Elastic response can always arise when a mechanical force leads to reversible sample deformation. In this case, internal structures are deformed or changed, and they recover their quiescent states after removal of stress. During the entire deformation process, elastic energy is reversibly stored in the sample. This work is instantaneously released after attaining the original state. Alternatively, elastic properties occur if supermolecular structures are oriented or changed during flow. Typical examples are shear-induced phase transitions, for instance, the transformation of a lamellar phase into a dispersion of onionlike vesicles. In contrast to these phenomena, an external force can also be dissipated as heat due to the action of flow processes. In this case, we observe viscous properties, and the deformation that has taken place is irreversible. Most materials show both processes at the same time, and this leads to striking viscoelastic properties. The occurrence of such phenomena depends in a sensitive way on the molecular structures of the samples. Surfactants dissolved in aqueous solutions generally show both rheological contributions: a viscous resistance, resulting from liquid flow and an elastic response that is caused by the deformation, orientation, or change of supermolecular, micellar structures. A rather simple, limiting case is a highly dilute aqueous solution of spherical micelles, slightly above the critical micelle concentration (cmc). In this case, we should expect mainly viscous forces that are caused by distortion processes of streamlines around the micelles. The viscosity increase, induced by the presence of the spherical aggregates, can approximately be represented by Einstein's equation[1]:

$$\eta = \eta_{cmc}(1+2.5\varphi_m) \tag{9.1}$$

Here, η_{cmc} denotes the viscosity of the surfactant solution at the cmc and φ_m is the volume fraction of the globular aggregates. Einstein's equation does not include interactions, and this law is strictly limited to the regime of low concentrations. This equation can be used to calculate the volume occupied by the globular micelles from viscosity measurements. φ_m represents the hydrodynamic effective volume of all micelles, including solvent layers and, in case of ionic surfactants, surrounding shells of adsorbed counter ions. Besides a certain increase of the viscous resistance, which is predicted from Equation 9.1, we will not expect considerable elastic effects in a highly dilute solution of spherical micelles. This holds if the micelles do not change their size or shape during flow; that means they behave as solid particles. It is also important that they cannot form supermolecular structures in the quiescent state or during flow. In dilute solutions of ionic surfactants, the occurring phenomena are generally more complicated, and purely viscous behavior is only expected when the counter ion distribution is not disturbed during flow (electroviscous effects). In most solutions of spherical micelles, elastic effects are very small, but a few exceptions exists that do not obey this rule. In this case, spherical micelles or dilute solutions of rod-shaped aggregates exhibit striking viscoelastic effects, even at concentrations slightly above the cmc.[2-4] These phenomena are induced by structural changes, which occur during flow and they are called *shear-induced phase transitions*. Due to enhanced collision processes, which occur during flow, these solutions can form new micellar structures under the action of a velocity gradient. This effect, denoted as *orthokinetic coagulation*, was first discovered by Smoluchowski.[5-7] In this theoretical approach, coagulation occurs due to the relative velocity of the micelles. The resulting longer micelles have a larger coagulation area and a higher relative velocity. This effect tends to increase the coagulation rate. When the flow stops, the solution recovers its original state and the metastable structures, formed in the streaming solutions, decay. In this case, elastic response is caused by structural changes.

B.   Viscoelastic Solutions of Rod-Shaped Micelles

Whereas dilute surfactant solutions usually exhibit only small elastic effects, this phenomenon is more pronounced at elevated concentrations. Typical, gel-like properties are observed in dilute solutions of entangled rod-shaped micelles. This effect was already known for a long time and is described in the classical books on colloid chemistry by Freundlich[8] and Bungenberg de Jong.[9] Later on, extensive studies were done by Gravsholt,[10] Candau et al.,[11-18] Talmon et al.,[19-29] Porte et al.,[30-36] Shikata et al.,[37-56] Cates et al.,[17,18,57-66] Decruppe et al.,[67-73] Hoffmann et al.,[74-78] and by our group.[74,79-86] In the last years, interest on viscoelastic surfactant solutions has even grown, and many data were published from different experimental groups.[68,87-116] These studies include reverse micelles in organic solvents and also aqueous solutions of ionic surfactants. The aggregates that are present in these solutions were characterized by different experimental techniques, such as static and dynamic light scattering, electric birefringence, NMR, rheological measurements, kinetic experiments (temperature-jump, pressure-jump), small-angle neutron scattering, and transmission electron microscopy. The experimental results of these different experimental techniques all point to the existence of elongated, threadlike aggregates. These micelles are often described as *living polymers*, whose chains are subject to reversible scission and recombination processes. In analogy to polymer solutions, these anisometric aggregates are denoted as *pseudo-polymers* or *wormlike micelles*. The lengths of these particles depend in a complicated way on the surfactant concentration, the temperature, and the ionic strength of the solution.[18,19,25,33,38,39,46,62,96,101,117–131] In viscoelastic surfactant solutions, anisometric micelles can reach lengths of a few micrometers. This results in rather low entanglement thresholds. Low surfactant concentrations of the order of 0.01 mol/L are sometimes sufficient to reach the overlap threshold. As long as the lengths of the rods are still smaller than their mean distance of separation, the solutions are not very elastic. Above the critical overlap concentration, however, gel-like properties appear. Elastic features can arise from mechanical contacts between the anisometric micelles (entanglements)[24,38,39,41,47,54, 56,87,132–134] or from intermicellar branching processes.[16,62,98] Cross-linking phenomena between anisometric micelles can lead to foamlike structures, and these processes were often observed in regimes of high ionic strengths or at elevated surfactant concentrations.[12,16,62,98,135–138] In this context, it is interesting to note that branching processes were observed in aqueous solutions of alkylamineoxide and alkylethoxylatesulfate mixtures by means of cryo-transmission electron microscopy (cryo-TEM).[139] Depending on the salt concentration or other parameters, anisometric micelles might be completely stiff or semi-flexible. By analogy with solutions of charged macromolecules, one expects conformational changes when adding solvent, organic additives, or excess salt.[140] Two limiting cases are well understood. A strong polyelectrolyte is highly extended in a solvent of low ionic strength, on grounds of repulsive forces between the surface charges. Under these conditions, the threadlike micelles should behave as rigid particles. The other extreme case will be reached at high concentrations of excess salt. Here, electric forces are weak, because they are screened by the surrounding atmosphere of counterions. In this case, we may assume that the anisometric micelles are rather flexible, and attain wormlike properties. The stiffness of these structures can be characterized by the persistence length, which describes the critical dimension where stiffness turns into flexibility. Whenever the persistence length l_p is much smaller than the total size of the micelle, we may state that the micelle is still flexible at large scales. On the other hand, if l_p is much larger than the total length of the chain, we obtain a rigid rod. In surfactant solutions, the persistence length is a complicated function of the surfactant concentration and the interactions between the micellar species.[39,46,47,69,140–143] Typical values are of the order of 10–100 nm. In the regime of high ionic strengths and elevated surfactant concentrations one usually observes the formation of very flexible, anisometric micelles that do behave similar to entangled macromolecules.[39,46,47,69,140–143]

C. Flow Properties of Surfactant Mesophases

In the regime of elevated concentration, detergents tend to form supermolecular structures with pronounced viscoelastic properties. Lyotropic liquid crystals are multicomponent systems formed in mixtures of amphiphilic molecules and a polar solvent. Some of these phases are capable of storing elastic energy. Typical phases showing this behavior are nematic, hexagonal, or lamellar liquid crystals and cubic phases. In contrast to semi-dilute solutions of wormlike micelles, liquid crystals are anisotropic materials, even in the quiescent state. The rheological properties of these solutions depend, therefore, on the direction of measurement. Because of these anisotropic properties, flow regimes are more complex and more difficult to study experimentally than in isotropic liquids. Generally, translation motions of liquid crystals are coupled to orientation processes, and velocity gradients tends to disturb the alignment. In addition, boundary conditions at the surface of the measuring cells are sometimes unknown, and the state of orientation in the vicinity of solid surfaces is often not well understood. Another problem arises from the poly-domain texture of liquid crystalline samples, which might be changed during flow. In order to compensate for these effects, defined alignment is often induced by external electric or magnetic fields. Since the rheological properties of liquid crystalline materials may also depend on history and aging of the sample, it is of special interest to insert samples into measuring instruments without applying or exerting mechanical forces, which might finally lead to undesired pre-orientations or phase transitions. In order to avoid these problems, Hoffmann et al. have developed special techniques, where liquid crystalline lamellar phases were formed by means of chemical protonation reactions.[144–147] The action of shear can also change phase boundaries, deform microstructures, or induce the formation of entirely new types of structures. Given these problems, many of the existing data in the literature cannot be used for quantitative interpretations.

In conclusion, there exists a broad range of different phenomena in viscoelastic solutions of liquid crystalline surfactants. Non-Newtonian viscous shear flow, viscoelastic properties, thixotropic and rheopectic phenomena, differences of normal stresses and stress relaxation processes are caused by deformation, orientation, coagulation, bursting processes, or shear-induced phase transitions. It is, therefore, of special interest to elucidate the crucial interplay between structure under shear and rheological behavior of self-assembling systems. Complex rheological phenomena may also result from two other problems. In most experiments, flow fields are not completely homogenous, and this might lead to slip, oscillations, plug flow, and phase transitions. The second problem is concerned with self-association phenomena of amphiphilic molecules. At the surface between a solid and water, surfactants tend to form densely packed monomolecular films. These adsorbed monolayers often have a complicated substructure, and they can, for instance, consist of hemi-micelles. The surfactant aggregates in aqueous solutions are different from these adsorbed surfactant layers. The transition zone between surface and bulk solution can lead to slip or flow instabilities (wall effects). Sometimes, this transition zone contains only a small amount of amphiphilic molecules (depletion layer). On the other hand, a solid surface can also induce structure orientation. This holds for liquid crystalline materials, e.g., nematic phases, where this technique is frequently used to produce electronic displays. Because of the anisotropic character of these solutions, rheological properties are complicated and sometimes still unexplored. In contrast to complex features of surfactant mesophases, isotropic solutions of rod-shaped micelles are, nowadays, well understood. These samples can also show complicated processes like shear-induced phase transitions, flow instabilities, or nonlinear rheological effects. Many of these phenomena are equally observed in viscoelastic surfactant mesophases. In the next sections, we will mainly focus on basic properties of entangled solutions of rod-shaped micelles. These liquids can act as simple model systems in order to gain a deeper insight into fundamental principles of complex flow processes.

III. INFLUENCE OF MICELLAR KINETICS ON RELAXATION PROPERTIES OF ENTANGLED SOLUTIONS OF ROD-SHAPED MICELLES

A. Short Reminder on Micellar Kinetics

The micelles formed at concentration above the cmc are fluctuating, dynamic particles, which can change size and shape continuously. These dynamic properties can be described by two different relaxation mechanisms associated with the surfactant monomer exchange between micelles and the micelle formation/breakdown (see Chapter 3, Sections II and III).[148–157] Both processes can be investigated by chemical relaxation techniques (see Chapter 2, Section II).[155–157] The fast relaxation process with a relaxation time τ_1 is related to the association-dissociation exchange of surfactant monomers.[155–157] This process is often diffusion-controlled.[156,157] The slow relaxation process with the time constant τ_2 is attributed to the micellization-dissolution equilibrium.[148–157] In a more general sense, this relaxation time can be denoted as the *average lifetime* or *breaking time* of the aggregates (see Figure 9.1). At least two different mechanisms are known by which micelles are formed and destroyed. In the dilute concentration regime, ionic micelles can vary their aggregation number in a stepwise fashion by reducing the amount of entrapped monomers, whereas at high concentrations breaking and coalescence processes are assumed to be the main factors causing structure breakdown.[155–170] These different processes involve

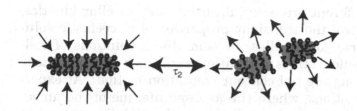

Figure 9.1 Schematic representation of the average lifetime of a surfactant micelle. The breaking process can be described by the time constant τ_2.

complicated molecular rearrangements, which take a long time as compared to the simple diffusion processes of monomeric surfactants.

Typical values of τ_2 range from milliseconds up to hours or days.[155–163,165–170] It turns out that the average lifetime of micellar aggregates is an important parameter, which can vary over many orders of magnitude. Apart from their importance for micellar kinetics, these phenomena have also influence on macroscopic properties of surfactant solutions. As the reader may already suppose, there exists a defined correlation between micellar dynamics and the occurrence of viscoelasticity. This special aspect will be explained in the following section.

B. Relaxation Properties of Entangled Solutions of Rodlike Micelles

The rheological properties of dilute aqueous solutions of flexible, entangled micelles have been extensively investigated within the previous few years.[15,16,19,28–30,38,39,41,43,45,53,54,56,76,87, 92–94,98,111,120,121,143,171–179] Viscoelastic properties usually occur at conditions, where the lengths of the rod-shaped particles are larger than their mean distance of separation. The overlap threshold of anisometric micelles can be very low, and it is often observed in the millimolar range. Semi-dilute solutions of entangled micelles show rather simple scaling laws; the osmotic pressure π and the equilibrium shear modulus G_0 increase as a function of the volume fraction[59,62,180,181]:

$$G_0 \sim \phi^{2.25} \tag{9.2}$$

Two different processes, diffusion and micellar kinetics, usually control the transient properties of viscoelastic solutions of wormlike micelles. In semi-dilute solutions of rod-shaped micelles, stress relaxation can occur by curvilinear diffusion along the rod contour (reptation motion). At experimental conditions, where the average lifetime of the anisometric micelles is much smaller than the diffusion process of the whole aggregate (reptation), there are numerous breaking and reformation processes within the time scale of observation.[59,61-63] Whenever surfactant solutions are subjected to mechanical forces, they develop stresses, which do not immediately fall to zero when the external influence is removed. The time required for the stresses to relax is referred to as *relaxation time*. In semi-dilute solutions of entangled, rod-shaped micelles, any applied shear stress relaxes through chemical pathways. Under conditions of short breaking times, we should observe pure, monoexponential stress decay when investigating a large ensemble of micellar particles.[59,61,62] From a rheological point of view, these are ideal conditions that can be described by simple theories. Monoexponential relaxation processes were observed in different surfactant systems, including polymers as associative thickeners. It is nowadays well established that these phenomena mostly occur in the regime of elevated surfactant concentrations or/and large amounts of excess salt. Under these experimental conditions, the average lifetime of the anisometric aggregates is relatively short in comparison to diffusion processes. The basic dynamic properties of wormlike surfactant micelles were extensively investigated by Cates.[57,59-62] Starting from a modified reptation model, where relaxation occurs by curvilinear diffusion processes, Cates introduced the concept of limited lifetimes, in order to include micelle bursting and stress relaxation. The model predicts the existence of several viscoelastic regimes depending on the competition between diffusive motions and reversible micelle breaking processes.[57,60,62,180] If the monomer exchange rates are extremely slow, one can assume that the lengths of the micelles remain nearly constant within the time scale of experimental observation. This is the limiting regime of pure reptation, where the theory predicts multi-exponential stress decay.[57,59,61–63] In this case, a stretched-exponential relaxation mechanism results with a critical exponent of $1/4$ and a relaxation time τ depending only on reptation processes. This holds for $\tau_2 \gg \tau_{rep}$.

$$\sigma(t) = \sigma_0 \exp\left[-K\left(\frac{t}{\tau_{rep}}\right)^{0,25}\right] \tag{9.3}$$

K is a constant, $\sigma(t)$ denotes the time dependent shear stress, and σ_0 is the initial shear stress at the starting point of the relaxation experiment. In terms of rheological models, such relaxation processes are characterized by a broad distribution or spectrum of different relaxation times. In this regime, the zero shear viscosity η_0 was found to scale with concentration C according to[11,182]

$$\eta_0 \sim C^{5.0-5.3} \tag{9.4}$$

If the micellar kinetics are fast ($\tau_2 \ll \tau_{rep}$), one should, however, observe a purely monoexponential stress decay. In this case, we obtain

$$\sigma(t) = \sigma_0 \exp\left[-\frac{t}{\tau}\right] \tag{9.5}$$

According to Cates's model the rheological relaxation time τ is now a combination of the average lifetime of the micelles and the reptation time.[57,59,61] For the limiting case of rapid breaking micelles one obtains[57,59–63]

$$\tau = \sqrt{\tau_{break}\tau_{rep}} \tag{9.6}$$

The average lifetime of anisometric micelles is theoretically described by τ_2. In this case, the shear stress relaxation time is characterized by the geometric average of the lifetime and the reptation time. The zero shear viscosity and the equilibrium shear modulus G_0 scale in this regime with[11,16,182–184]

$$\eta_0 \sim C^{3.25-3.55}; \quad G_0 \sim kTC^{2.0-2.3} \tag{9.7}$$

A wide range of different reptation and breaking times could be investigated by Cates et al. using computer simulations.[57,185] In these models, the parameter $\zeta = \tau_{break}/\tau_{rep}$ describes the ratio between the average lifetime of the micelles and the reptation time that characterizes the diffusion properties.

For small values of ζ, single-stress relaxation processes are observed. However, for ζ larger than unity, clear departures occur. These deviations can easily be represented in a Cole-Cole plot, where the loss modulus $G''(\omega)$ is plotted as a function of the storage modulus $G'(\omega)$. A pure exponential relaxation process appears in such a diagram as a semicircle passing through the origin:

$$G''(\omega) = \sqrt{G'(\omega)\left(G_0 - G'(\omega)\right)} \tag{9.8}$$

For comparison purposes, experimental data are usually normalized by dividing the storage and loss modulus by G_{osc}, the radius of the osculating circle of the Cole-Cole plot at the origin.[57,135,185] Any departure from monoexponential decay is represented by deviations from the semicircle. This behavior is schematically shown in Figure 9.2, where nonexponential behavior occurs at elevated values of the angular frequency.

Numerically calculated Cole-Cole plots as shown in Figure 9.2 could be successfully compared to experimental data, and this allows us to calculate the parameter ζ and, hence, the ratio τ_{break}/τ_{rep}. From measurements at low angular frequencies, it is possible to determine the zero shear viscosity η_0, which usually coincides with the magnitude of the complex viscous resistance. The plateau modulus G_0 will be obtained by extrapolation of the Cole-Cole data to the horizontal axis. The terminal relaxation time τ_r is then obtained from $\tau_r = \eta_0/G_0$.

The model of Turner and Cates can be used to compare calculated Cole-Cole-plots to experimental data.[57,185] This method allows us to determine the new parameter $\overline{\zeta} = \tau_{break}/\tau_r$. Comparison of $\overline{\zeta}$ and ζ then provides information on the average lifetime and the reptation time of the wormlike micelles. Typical values of the reptation times are of the order

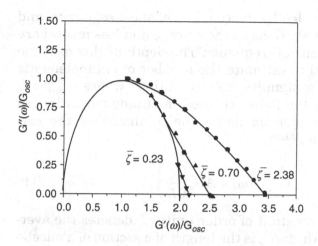

Figure 9.2 Calculated Cole-Cole plots from Ref. 135 for three values of $\bar{\zeta}$.

of a few seconds, and the breaking process usually takes a few hundred milliseconds. These values depend much on the surfactant concentration and the ionic strength of the solutions. It is, in general, not easy to determine the average lifetime of entangled wormlike micelles by means of chemical relaxation methods such as pressure- or temperature-jump. Turner and Cates performed systematic measurements and observed at least in one surfactant system pretty good agreement between τ_{break} determined by kinetic and rheological experiments.[57,185] We can, therefore conclude that this analysis successfully allows determining the breaking time of entangled, rod-shaped micelles.

In the regime of small average lifetimes and/or short time scales, the anisometric micelles do not show reptation motions, but either breathing or local Rouse motions (rotations around the backbone). For such conditions detailed results were obtained from computer simulations by Granek and Cates, using the Poisson renewal model.[59] In this region, the relaxation spectrum is Rouse-like. This results in a minimum in the Cole-Cole diagram followed by a turn-up at the right-hand edge.[11,30,135,181] At these frequencies, aggregate motions are controlled by fluctuations of short segments, and the resulting contributions to the storage and loss moduli are increasing functions of frequency. The depth of this extreme value can be used to estimate the number of entanglements per chain in the system.[181,186] At conditions where $\tau_{break} >> \tau_R$ (where τ_R denotes the Rouse time of an entanglement length l_e), a simple scaling behavior of the minimum of the loss modulus was found[181,186]:

$$G''_{min}(\omega) = AG_0\left(\frac{l_e}{\bar{L}}\right) \qquad (9.9)$$

Here, A is a constant of order unity. $\bar{L}$ denotes the average micellar length, and l_e is the length of a section of a micelle between entanglements. For flexible micelles, the entanglement length can separately be estimated from relation[186]:

$$G_0 = \frac{kT}{l_e^{9/5}} \qquad (9.10)$$

In combination with Equation 9.9 it is now possible to calculate the dimensions of the rod-shaped micelles from the minimum of $G''(\omega)$. Kern et al. performed systematic studies of the average micellar lengths of wormlike aggregates.[186] Typical values are of the order of several micrometers,[186] but extreme values of more than 0.1 mm were also sometimes observed.[187] These data are in fairly good agreement with results of cryo-TEM, which also show the presence of entangled wormlike micelles with dimensions as long as several micrometers.[19–22,24,25,28,173,188,189]

The model of Cates predicts also scaling behavior of the zero shear viscosity, shear modulus, and the relaxation time as a function of the surfactant concentration. Close inspection, however, reveals that these correlations hold only in a limited range of ionic strength. At elevated salt concentrations, rod-shaped micelles can not only form entangled structures but also form networks of branched or multiconnected micelles. The rheological properties of such systems was extensively investigated by Khatory et al.[11,30,135] The general flow properties of these cross-linked networks are similar to those of entangled, wormlike aggregates. The connections between micellar aggregates are very different from the junction points observed in classical polymer solutions. In branched micelles, the situation is similar to foams, and surfactant molecules can flow through the connection points.[135] This leads to a decrease of the shear modulus. It is also possible that in branched micelles the junction points can slide along the aggregates during flow.[11,30,135] From a modified reptation theory, Cates and Turner[190] and later Lequeux[12,91,191] proposed an increasing diffusion motion of the micelle within the reptation tube. This phenomenon leads to more fluid solutions and faster stress relaxation processes as compared to entangled linear micelles of the same size.[11,30,135,137] It is, therefore, possible to detect branching processes on grounds of different scaling properties. Khatory et al. generally observed a lower viscous resistance that one would expect from the reptation theory of linear micelles.[30,135] This phenomenon can be explained by the modified reptation model.[191] Fischer et al. investigated networks of branched surfactant micelles in elongational flow.[98] The results indicate that the presence of branching tends to increase the maximum of the elongational viscosity as it is known for polymer melts.[98] In addition to rheological experiments, multiconnected micelles were also observed by Lin using cryo-TEM.[139] The investigated network structures, formed at high salt concentrations, were similar to bilayer structures of bicontinuous L_3 phases.[139] The aqueous solutions of branched micelles exhibited only weak viscoelastic properties, an apparent yield stress, and at elevated concentrations monoexponential stress relaxation processes occurred.[139] Similar network structures were also discovered by cryo-TEM and light scattering in $C_{12}E_5$/water micellar systems by Talmon and coworkers.[25] In this case, multiconnected network topologies were mainly formed in the vicinity of the critical point.[25]

IV. LINEAR VISCOELASTIC PROPERTIES OF ENTANGLED SOLUTIONS OF ROD-SHAPED MICELLES

A. Linear Viscoelastic Regime

As semi-dilute solutions of wormlike micelles frequently show monoexponential relaxation properties, these surfactant systems can be described by simple equations. These solutions are often used to explore fundamental rheological properties. The flow behavior of these viscoelastic samples can be characterized by time-dependent material functions, such as the relaxation modulus or the transient viscosity. For sufficiently small deformations, the material functions can be expressed by linear differential equations. Under these conditions, the response at any time is proportional to the value of the initiating signal. Doubling the stress will also double the strain (superposition principle). In practice, we observe that most materials show linear time-dependent properties even at finite deformations as long as the shear strain remains below a certain limit. This threshold varies from sample to sample and is thus a basic property of the investigated material. In rubber-elastic systems, the linear stress-strain relation is observed up to high deformations of a few hundred percent, whereas in energy-elastic samples the transition occurs already at very small shear strains of only a few thousandths. On a molecular scale, this effect is caused by reversible structure-breakdown and recovery after removal of stress. In the theory of linear viscoelasticity, all material parameters are functions of the time or, in oscillation experiments, functions of the angular frequency ω (this frequency dependence results from the Fourier transformation of the time domain).

B. Methods of Investigation

At least four different types of rheological experiments are capable of measuring basic material properties. In transient tests, a stepwise transition is used from one equilibrium state to another. In these experiments, a certain shear stress σ, a certain deformation γ or a certain shear rate $\dot{\gamma}$ is suddenly applied at $t = 0$ and then held constant. Due to the initial extension, there is a certain time response of the viscoelastic material that can be measured to get the desired rheological functions. In the start-up flow tests, a step function shear rate is suddenly applied at $t = 0$. The resulting stress is time dependent, and this quantity is measured after the amount of deformation has occurred. From these data, different viscoelastic parameters can be obtained. The relaxation modulus $G(t)$ can be calculated by the relation:

$$G(t) = \frac{\sigma(t)}{\gamma} \tag{9.11}$$

In the limit of linear stress-strain relations, the relaxation modulus does not depend on the initial deformation step and the rheological properties are only described by transient functions. Equation 9.11 suggests that the relaxation modulus describes the stress relaxation after the onset of a step function shear strain. In viscoelastic liquids of entangled solutions of rod-shaped micelles, an applied stress is always relaxing to zero after infinite long periods of time.

We will now consider another type of transient response. Let a shear stress σ be applied at the viscoelastic solution at $t = 0$. In a general case, a time-dependent shear strain is developed that can be measured to get the creep compliance $J(t)$. The creep compliance has the dimensions of a reciprocal modulus, and it is therefore an increasing function of time. From theories of viscoelasticity, it is possible to calculate the creep compliance from the relaxation modulus and inverse:

$$J(t) = \frac{\gamma(t)}{\sigma} \tag{9.12}$$

In a third test, a step function shear rate $\dot{\gamma}$ is suddenly applied at $t = 0$ (start-up flow). In this case, the shear stress σ is measured as a function of time and the shear stress growth coefficient (stressing viscosity) $\eta^+(t)$ can be calculated:

$$\eta^+(t) = \frac{\sigma(t)}{\dot{\gamma}} \tag{9.13}$$

In viscoelastic solutions, this function is exponentially increasing. It is easy to show that the time derivative of $\eta^+(t)$ gives again the relaxation modulus:

$$\frac{d\eta^+(t)}{dt} = G(t) \tag{9.14}$$

In a similar test, the applied shear rate is suddenly reduced to zero. This process is called stress relaxation after cessation of steady-state flow. The decaying shear stress is measured as a function of time:

$$\eta^-(t) = \frac{\sigma(t)}{\dot{\gamma}} \tag{9.15}$$

The shear stress decay coefficient $\eta^-(t)$ evidently describes the relaxation properties of the viscoelastic samples.

Beside these transient methods, there is another test, which can be performed to provide information on the viscoelastic properties of the sample: periodic or dynamic experiments. In this case, the shear strain is varied periodically with a sinusoidal alternation at an angular frequency ω. A periodic experiment at frequency ω is qualitatively equivalent to a transient test at time $t = 1/\omega$. In a general case, a sinusoidal shear strain is applied to the solution. The response of the liquid to the periodic change consists of a sinusoidal shear stress σ, which is out of phase with the strain. The shear stress consists of two different components. The first component is in phase with the deformation and the second one is out of phase with the strain. From the phase angle δ between stress and strain, the amplitude of the shear stress $\hat{\sigma}$ and the amplitude of the shear strain $\hat{\gamma}$ it is possible to calculate the storage modulus G' and the loss modulus G'':

$$G'(\omega) = \frac{\hat{\sigma}}{\hat{\gamma}} \cdot \cos\delta, \quad G''(\omega) = \frac{\hat{\sigma}}{\hat{\gamma}} \cdot \sin\delta \tag{9.16}$$

The storage modulus describes the elastic properties of the sample, and the loss modulus is proportional to the energy dissipated as heat (viscous resistance). It is convenient to express the periodically varying functions as a complex quantity, which is termed the magnitude of the complex viscosity $|\eta^*(\omega)|$. This quantity can be calculated from the following equation:

$$|\eta^*(\omega)| = \frac{\sqrt{G'^2 + G''^2}}{\omega} \qquad (9.17)$$

C.   Maxwell Model for Viscoelastic Solutions

It can be shown that for most viscoelastic solutions there exists a simple correlation between dynamic and steady-state flow characteristics.[192] In a first approximation, the complex viscosity at certain angular frequency ω can be compared with the steady-state value of the shear viscosity $\eta(\dot{\gamma})$ at the corresponding shear rate $\dot{\gamma}$. This empirically observed relationship is called the Cox-Merz rule.[192] Simple mechanical models can represent some of these rheological properties. A purely elastic material deforms under stress, but regains its original shape and size when the load is removed. A practical example of elastic material is a spring working within its limits. For completely elastic materials, stress is directly proportional to strain. However, when a material has viscous as well as elastic properties (as in plastics), deviation from this linear relationship occurs. In this case, deformation is found to be irreversible. This is caused by flow processes, which can be represented by a Newtonian dashpot. The simplest model that can describe viscoelastic surfactant solutions is called the Maxwell material. It consists of a spring and a dashpot connected in series (see Figure 9.3).

The elastic spring corresponds to a shear modulus G_0 and the dashpot represents the constant zero shear viscosity η_0. The spring shown in Figure 9.3 represents the elastic portion of a plastic material's response. When a load is applied to the spring, it instantly deforms by an amount proportional to the load. Moreover, when the load is removed, the spring instantly recovers to its original dimensions. As with all elastic responses, this response is independent of time. The dashpot represents the viscous portion of a material's response. The dashpot consists of a cylinder holding a piston that is

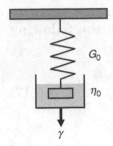

Figure 9.3 Schematic representation of the Maxwell material.

immersed in a viscous liquid. When a load is applied to the dashpot, it does not immediately move. However, if the application of the force continues, the viscous material surrounding the piston is also displaced and the dashpot does deform. This occurs over some period, not instantly, and the length of time depends on both the load and the rate of loading. Viscous response, therefore, has a time-dependent rate of response.

The dynamic properties of the Maxwell-element can be represented by a linear differential equation, and the solutions give the desired material functions. The behavior under harmonic oscillations is characterized by simple formulas:

$$G'(\omega) = G_0 \frac{\omega^2 \tau^2}{1 + \omega^2 \tau^2} \tag{9.18}$$

$$G'(\omega) = G_0 \frac{\omega \tau}{1 + \omega^2 \tau^2} \tag{9.19}$$

$$\left|\eta^*(\omega)\right| = \frac{\eta_0}{\sqrt{1 + \omega^2 \tau^2}} \tag{9.20}$$

In transient experiments, single exponential relaxation processes can describe the Maxwell material:

$$G(t) = G_0 \exp\left[\frac{-t}{\tau}\right] \tag{9.21}$$

$$\eta^{+}(t) = \eta_0 \left[1 - \exp\left(\frac{-t}{\tau}\right)\right], \quad \eta^{-}(t) = \eta_0 \exp\left(\frac{-t}{\tau}\right) \qquad (9.22)$$

$$J(t) = J_0 + \frac{t}{\eta_0} \qquad (9.23)$$

The Maxwell model can also be used in order to characterize the dynamic properties of viscoelastic surfactant solutions. The simplicity of comparing rheological data with the theoretical predictions of a Maxwell material is rather attractive. In this way, it is easy to interpret viscoelastic flow processes.

Before comparing the dynamic characteristics of entangled solutions of rod-shaped micelles with the general properties of the Maxwell material we must first modify the theoretical equations in order to include the effect of solvent friction. In real surfactant solutions, the magnitude of the complex viscosity $|\eta^*(\omega)|$ and the dynamic viscosity $\eta'(\omega) = G''(\omega)/\omega$ cannot decay to zero for $\omega \to \infty$ but they must reach a limiting value in the regime of very high shear rates. Either this threshold value represents the solvent viscosity or, in surfactant solutions, the viscous resistance of the highly aligned micellar structure as, for instance, rods oriented in the direction of flow. This effect can be described by adding the term $\eta(\infty)\omega$ to the corresponding equations of the Maxwell material. From Equation 9.18 and Equation 9.19 we obtain

$$G'(\omega) = G_0 \frac{\omega^2 \tau^2}{1 + \omega^2 \tau^2} \qquad (9.24)$$

$$G''(\omega) = G_0 \frac{\omega \tau}{1 + \omega^2 \tau^2} + \eta(\infty)\omega \qquad (9.25)$$

D. Experimental Results

Monoexponential stress relaxation properties are often observed in cationic detergent solutions at conditions where special counterions are added. Typical examples are aqueous solutions of cetylpyridinium chloride or cetyltrimethylammonium chloride upon addition of salicylate. Combinations of anionic and cationic surfactants may also form entangled solutions of rod-shaped aggregates. Myristyldimethylamine-oxide (MDMAO, $C_{14}H_{29}$-$N(CH_3)_2$-O; trade name AMMONYX MO, from Degussa Goldschmidt AG, Essen) represents an example of neutral surface active compounds, which also may exhibit gel-like properties. Although amine oxides are generally regarded as nonionic surfactants, they can behave in certain aspects similar to amphoteric surfactants. Amine oxides may be partly protonated at low pH-values and their cationic character is then more pronounced. This process can be induced by adding nitric acid. A 15% aqueous solution of MDMAO titrated with HNO_3 to pH = 3.5 leads to a solution showing striking viscoelastic properties. This solution is subsequently denoted as 15%-MDMAO-HNO_3. Typical features of such a solution are summarized in Figure 9.4.

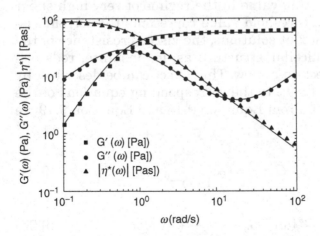

Figure 9.4 Storage modulus G'(ω), loss modulus G''(ω) and the magnitude of the complex viscosity |η*(ω)| as a function of the angular frequency ω compared to the Maxwell model (lines). (Aqueous solution of 15 %wt MDMAO-HNO_3, pH = 3.5, γ = 10 %, T = 20°C).[204, 205]

It is evident that the dynamic properties are represented by a Maxwell-like behavior. Lines represent data calculated from Equation 9.24 and Equation 9.25. In the high frequency regime, elastic properties are dominant. At these conditions, the storage modulus attains a plateau and the loss modulus is much lower than $G'(\omega)$. With decreasing angular velocity, $G''(\omega)$ becomes more important. The intersection point where $G' = G''$ denotes the inverse relaxation time ($\omega_{intsec} = 1/\tau$). From the experimental data one obtains $\lambda = 1.05$ s and an equilibrium shear modulus $G_0 = 549$ Pa. The region of very small angular frequencies is denoted as terminal zone. This regime is controlled by flow processes.

Deviations from the rheological properties of a simple Maxwell material can also be represented in a Cole-Cole plot. In this special diagram, the loss modulus is represented as a function of the storage modulus. For a single-exponential relaxation process, we obtain a semicircle (see Equation 9.8). Relevant data are represented in Figure 9.5.

It is evident that the data follow the theoretical curve in the regime of small angular frequencies. Deviations, however, occur at elevated frequencies. This phenomenon is caused by

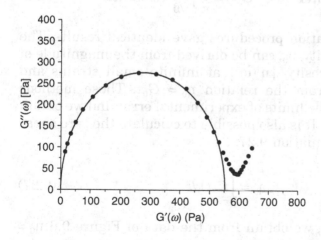

Figure 9.5 Cole-Cole plot for an aqueous solution of 15 %wt MDMAO-HNO$_3$, pH = 3.5, γ = 10%, T = 20°C.[204, 205]

nonvanishing values of $\eta(\infty)\omega$ and the onset of the Rouse regime. Due to the linear presentation, Cole-Cole plots are very sensitive and can be used to detect very small deviations from pure monoexponential stress decay. In double logarithmic scales of $G'(\omega)$ and $G''(\omega)$ small discrepancies are often not visible.

Pure monoexponential stress decrease can also be observed from measurements of the relaxation modulus, $G(t)$, as shown in Figure 9.6. In this experiment the relaxation modulus was measured after applying a step function shear strain. The drawn line corresponds to predictions for a Maxwell model (Equation 9.21) and gives the same relaxation time as obtained from the dynamic measurements $\tau = 1.05$ s $\pm$ 0.14 s. Small deviations, occurring at long times, are sometimes caused by electronic instabilities of the torque transducer and should not be interpreted as real discrepancies to the Maxwell material. For comparison purposes, the relaxation modulus was also calculated from measurements of $G'(\omega)$ and $G''(\omega)$. This can be achieved by solving the following integral equations[193]:

$$G(t) = \frac{2}{\pi} \int_0^\infty \frac{G'}{\omega} \sin \omega t \cdot d\omega, \quad G(t) = \frac{2}{\pi} \int_0^\infty \frac{G''}{\omega} \cos \omega t \cdot d\omega \qquad (9.26)$$

Both integration procedures gave identical results. The zero shear viscosity, η_0, can be derived from the magnitude of the complex viscosity, $|\eta^*(\omega)|$ at infinite small strains and frequencies or from the relation $\eta_0 = G_0\tau$. These methods coincide within the limits of experimental error and we obtain $\eta_0 = 577 \pm 9$ Pa·s. It is also possible to calculate the zero shear viscosity from Equation 9.27:

$$\eta_0 = \int_0^\infty G(t)dt \qquad (9.27)$$

In this case, we obtain from the data of Figure 9.6 $\eta_0 = 571.6$ Pa·s.

It is worthwhile to mention that the dynamic properties of these viscoelastic surfactant solutions can evidently be described by monoexponential relaxation properties. This cor-

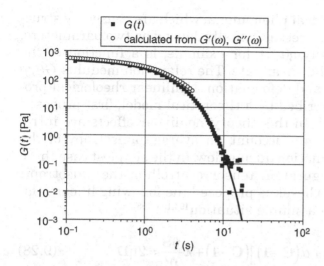

Figure 9.6 Measured values of the relaxation modulus $G(t)$ (filled symbols) and calculated values from $G'(\omega)$ and $G''(\omega)$ (open symbols) as a function of time t, compared to the theoretical predictions of the Maxwell model (line) ($\gamma = 10$ %, $T = 20°C$, 15 %wt MDMAO – HNO_3, pH = 3.5).[204, 205]

responds to the ideal case, which is quantitatively explained by the Cates theory.

V.   NONLINEAR VISCOELASTIC PROPERTIES OF ENTANGLED SOLUTIONS OF ROD-SHAPED MICELLES

The Maxwell model can successfully describe monoexponential stress relaxation properties. This holds for the regime of small deformations, shear rates, or shear stresses. In the region of elevated mechanical forces, severe departures from simple Newtonian and Hookean occur. In viscoelastic surfactant solutions, it is often observed that the shear viscosity decreases markedly with increasing amount of shear. This typical behavior is called shear thinning or pseudo plastic. The non-Newtonian behavior of such solutions is of great practical interest, and it is intimately connected with orientation processes or structural changes that occur during flow.

Nonlinear rheological phenomena, which are caused by structural alignment processes, usually depend on two parameters. The shear viscosity $\eta(t,\dot{\gamma})$ for example, is a function of the shear time and the shear rate. The relaxation modulus $G(t,\gamma)$ depends on time and deformation. Nonlinear rheological processes can be described by a theoretical model, first proposed by Giesekus.[194–200] In this theory, nonlinear effects are introduced by taking into account an average anisotropy of the molecular conformation during flow. In the simplest case there is only one configuration tensor controlling the anisotropic mobility.[194,196–198] Giesekus proposed the following linear relationship for such a simple situation[194,196–198]:

$$\left[1+\alpha\left(\mathbf{C}-\mathbf{1}\right)\right]\left(\mathbf{C}-\mathbf{1}\right)+\lambda\frac{\partial\mathbf{S}}{\partial t}=2\eta\mathbf{D} \qquad (9.28)$$

Here, $\mathbf{C}$ denotes the configuration tensor and $\mathbf{D}$ describes the rate-of-strain tensor of the material continuum. The dimensionless anisotropy factor, α, characterizes the anisotropic character of the particle mobility. It is easy to show that α attains values between zero and one. The limiting case $\alpha = 0$ corresponds to the isotropic motion and ultimately leads to an upper convected Maxwell material. In order to derive a deformation-dependent constitutive equation, a Hookean law connecting the tensor of external stresses $\mathbf{S}$, the configuration tensor $\mathbf{C}$, and the shear modulus G was suggested[195–199]:

$$\mathbf{S}=G(\mathbf{C}-\mathbf{1}) \qquad (9.29)$$

Here, G is the shear modulus and $\mathbf{S}$ the stress tensor. It is easy to see that the Giesekus model gives a simple relation between all rheological parameters by inducing the anisotropy of a streaming solution.

A. Analytical Solutions for Steady-State Shear Flow

For steady shear flow three equations are obtained from Equation 9.28. This holds for the general case, where $0 \le \alpha \le 1$.[79,99,194–197,201]

$$\left[\alpha(N_1(\infty,\dot\gamma)-2N_2(\infty,\dot\gamma))+1\right]\sigma(\infty,\dot\gamma)=\lambda\dot\gamma(1-N_2(\infty,\dot\gamma)) \qquad (9.30)$$

$$\left[\alpha(N_1(\infty,\dot\gamma)-2N_2(\infty,\dot\gamma))+1\right]N_1(\infty,\dot\gamma)=2\lambda\dot\gamma\sigma(\infty,\dot\gamma) \qquad (9.31)$$

$$\left[1-\alpha N_2(\infty,\dot\gamma)\right]N_2(\infty,\dot\gamma)=\alpha\sigma^2(\infty,\dot\gamma) \qquad (9.32)$$

All parameters depend on the time and the shear rate. Steady-state conditions are obtained for $t \to \infty$. Variable $\sigma(\infty,\dot\gamma)$ denotes the steady state values of the shear stress. The anisotropic character of the flowing solutions give rise to additional stress components, which are different in all three principal directions. This phenomenon is called the Weissenberg effect, or the normal stress phenomenon. From a physical point of view, it means that all diagonal elements of the stress tensor deviate from zero. It is convenient to express the mechanical anisotropy of the flowing solutions by the first and second normal stress difference:

$$N_1(\infty,\dot\gamma)=\sigma_{11}(\infty,\dot\gamma)-\sigma_{22}(\infty,\dot\gamma), \quad N_2(\infty,\dot\gamma)=\sigma_{22}(\infty,\dot\gamma)-\sigma_{33}(\infty,\dot\gamma)$$

$$(9.33)$$

N_1 denotes the first normal stress difference and N_2 the second normal stress difference. Both functions depend on the shear rate and the ∞-sign describes the equilibrium, which is reached at steady-state shear conditions.

For steady-state shear flow the set consisting of Equation 9.30, Equation 9.31, and Equation 9.32 can be solved analytically for $\alpha = 0.5$. This special value of the anisotropy factor was often observed in viscoelastic surfactant solutions.[79] The shear stress and first and second normal stress difference can be expressed by the following equations[194–198]:

$$\sigma(\infty,\dot\gamma)=\frac{G_0}{2\tau\dot\gamma}\left(\sqrt{1+4\tau^2\dot\gamma^2}-1\right) \qquad (9.34)$$

$$N_1(\infty,\dot\gamma)=2G_0\left[\frac{1-\Lambda^2}{\Lambda}\right] \qquad (9.35)$$

$$N_2(\infty,\dot{\gamma}) = G_0\left[\Lambda - 1\right] \tag{9.36}$$

where Λ^2 describes the expression

$$\Lambda^2 = \frac{\sqrt{1+4\tau^2\dot{\gamma}^2}-1}{2\tau^2\dot{\gamma}^2} \tag{9.37}$$

Equation 9.34 describes the evolution of shear stress as a function of the shear rate. The steady state shear viscosity can easily be obtained from this equation:

$$\eta(\infty,\dot{\gamma}) = \frac{\sigma(\infty,\dot{\gamma})}{\dot{\gamma}} = \frac{G_0}{2\tau\dot{\gamma}^2}\left(\sqrt{1+4\tau^2\dot{\gamma}^2}-1\right) \tag{9.38}$$

This law gives a decreasing viscous resistance with increasing shear rate and it therefore describes shear thinning or pseudo-plastic behavior. It is often convenient to use dimensionless variables in order to compare experimental data with the predictions of the Giesekus model. This can be achieved by dividing the steady-state shear viscosity by the zero shear viscosity and by multiplying the shear rate with the relaxation time:[79]

$$\chi = \tau\dot{\gamma} \tag{9.39}$$

$$\eta_n(\infty,\chi) = \frac{\eta(\infty,\chi)}{\eta(\infty,0)} \tag{9.40}$$

Here, χ denotes the normalized shear rate and $\eta_n(\infty,\chi)$ the dimensionless steady-state shear viscosity. Relevant data and comparison with Equation 9.38 are summarized in Figure 9.7.

It is evident that the Giesekus model can quantitatively describe the shear thinning behavior of the entangled solutions of rod-shaped micelles. The decrease of the viscous resistance is caused by the alignment of the anisometric aggregates in the streaming solutions. Similar conclusions can be drawn from measurements of the first normal stress difference. This parameter is often represented in terms of the first normal stress coefficient:

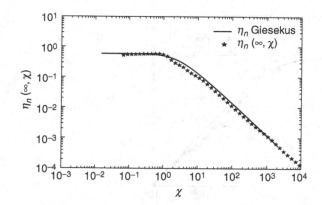

Figure 9.7 Normalized shear viscosity $\eta_n(\infty,\chi)$ as a function of normalized shear rate χ compared to the theoretical predictions of the Giesekus model (lines) (T = 20°C, 15%-MDMAO-HNO$_3$, pH = 3.5).[204, 205]

$$\psi_1(\infty,\dot{\gamma}) = \frac{N_1(\infty,\dot{\gamma})}{2\dot{\gamma}^2} = \frac{\tau_{11}(\infty,\dot{\gamma}) - \tau_{22}(\infty,\dot{\gamma})}{2\dot{\gamma}^2} \qquad (9.41)$$

It is easy to see that in the regime of small shear rates a simple relation holds:

$$\lim_{\dot{\gamma} \to 0}\left(\psi_1(\infty,\dot{\gamma})\right) = \eta(\infty,0)\tau \qquad (9.42)$$

The difference between the first normal stress difference and the zero shear viscosity is just given by the relaxation time τ. Therefore, normalization can be achieved by

$$\psi_{1n}(\infty,\chi) = \frac{\psi_1(\infty,\dot{\gamma})}{\eta(\infty,0)\cdot\tau} \qquad (9.43)$$

In Figure 9.8 we compare measured values of the normalized first normal stress coefficient with Equation 9.35. Again, we observe a pretty good agreement between experimental data and the predictions of the Giesekus model. We can thus conclude that this theory describes very well the mechanical anisotropy of the streaming viscoelastic surfactant solution.

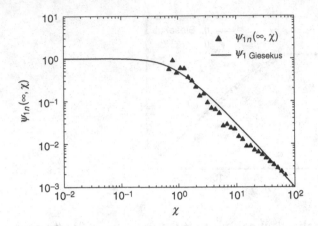

Figure 9.8 Normalized first normal stress coefficient, $\psi_{1n}(\infty,\chi)$ as a function of normalized shear rate χ, compared to the theoretical predictions of the one-mode Giesekus model (lines) (T = 20°C, 15%-MDMAO-HNO$_3$, pH = 3.5).[204, 205]

B. Transient Flow

For transient flow, three coupled differential equations are obtained from Equation 9.28 by using derivations of the configuration tensor, **C**:

$$\dot{\sigma}(t,\dot{\gamma})+\left[\alpha(N_1(t,\dot{\gamma})-2N_2(t,\dot{\gamma}))+1\right]\sigma(t,\dot{\gamma})=\lambda\dot{\gamma}(1-N_2(t,\dot{\gamma}))$$

$$(9.44)$$

$$\dot{N}_1(t,\dot{\gamma})+\left[\alpha(N_1(t,\dot{\gamma})-2N_2(t,\dot{\gamma}))+1\right]N_1(t,\dot{\gamma})=2\lambda\dot{\gamma}\sigma(t,\dot{\gamma}) \quad (9.45)$$

$$\dot{N}_2(t,\dot{\gamma})+\left[1-\alpha N_2(t,\dot{\gamma})\right]N_2(t,\dot{\gamma})=\alpha\sigma^2(t,\dot{\gamma}) \quad (9.46)$$

Here $\dot{\sigma}(t,\dot{\gamma})$ denotes the time derivative of the shear stress, and $\dot{N}_1(t,\dot{\gamma})$ and $\dot{N}_2(t,\dot{\gamma})$ the time-dependent first and second normal stress differences. In order to describe time-dependent properties, one has to solve this set of three equations, which contain three unknown variables. In Equation 9.46 the shear stress enters quadratic and this leads to nonlinear properties. For dynamic processes, such as start-up flow or relaxation after cessation of steady-state flow, one has to solve the entire set of three coupled differential equations given by Equation 9.44, Equation 9.45, and Equation 9.46. Although it is possible to get analytical solutions for some special values of the mobility factor α, the corresponding equations are very complicated and not practical to use. On the other hand, it is rather simple to solve these nonlinear differential equations by numerical computer programs for any arbitrary value of α. This can be done with Runge-Kutta computer programs. We will not go into further details at this point. The interested reader is referred to recent publications, where this method is described in detail.[79,99,201–204] Here, we will briefly discuss two results. A common test in order to determine viscoelastic properties can be performed by suddenly increasing the shear rate from zero to a desired value. The shear stress growth coefficient can be calculated in such an experiment from Equation 9.13. In order to compare experimental results with the predictions of the Giesekus model, it is convenient to use the dimensionless shear time $s = t/\lambda$. The shear stress growth coefficient and the first normal stress growth coefficient can easily be normalized:

$$\eta_n^+(s,\chi) = \frac{\sigma^+(t,\dot{\gamma})}{\chi} = \frac{\sigma^+(t,\dot{\gamma})}{\tau\dot{\gamma}} \tag{9.47}$$

$$\psi_{1n}^+(s,\chi) = \frac{\psi_1^+(t,\dot{\gamma})}{\tau\eta_0} \tag{9.48}$$

Typical examples of experimental values are summarized in Figure 9.9. We observe a striking overshooting behavior of the shear stress growth coefficient and a more gentle increase of the first normal stress growth coefficient. Given the large number of experimental values, it is difficult to see the underlying lines, which represent the numerical solution of the Giesekus model. Close inspection, however, reveals that the data are in perfect agreement with calculated results. Similar results were obtained after suddenly stopping the

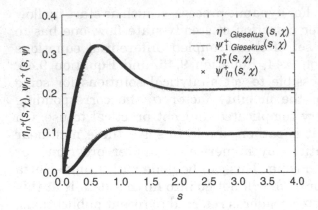

Figure 9.9 Normalized shear stress growth coefficient $\eta_n^+(s,\chi)$ and normalized first normal stress growth coefficient $\psi_{1n}^+(s,\chi)$ as a function of the normalized time s at a shear rate of $2s^{-1}$ (T = 20°C, 15%-MDMAO-HNO$_3$, pH = 3.5). Predictions of the Giesekus model are represented by lines).[204, 205]

shearing action.[79,99,201,203,204] Again, the Giesekus model quantitatively describes the experimental data. It is interesting, to note that the first normal stress difference and the shear stress tend to relax with the same time constants, but this behavior is multi-exponential. This observation is called the Lodge-Meissner criterion.[193,205]

Beside these relaxation experiments, the Giesekus model was also applied for different types of rheological experiments.[79,99,201,203,204] Up to now, no significant deviations between experimental values and theoretical predictions were detected. The Giesekus model thus gives a successful description of the nonlinear rheological properties of viscoelastic surfactant solutions. This holds for an anisotropy factor $\alpha = 0.5$.[79]

VI. SHEAR-INDUCED PHASE TRANSITIONS

Up to now, we have only discussed relatively simple processes where the action of flow leads to the orientation of anisometric, rod-shaped micelles or to the reduction of the entangle-

ment density. At these conditions, one observes characteristic
rheological phenomena resembling semi-dilute solutions of
macromolecules. In addition to these effects, surfactant solutions can change their aggregation structure during flow. In
this case, a new phase is formed that is only stable under the
action of a velocity gradient. This metastable phase, formed
during flow, is called shear-induced structure (SIS).[73,76,206]
When the stirring action stops, the SIS decays and the quiescent state begins reforming again. In the last decades, different types of SIS were observed. This includes flow-induced
formation of vesicles, nematic phases, gel-like structures, liquid crystals, or turbid rings.[19,24,78,101,102,174,207–212] A recent paper
of Mendes et al. even describes a new shear-induced transition
from vesicles into rod-shaped micelles.[14] This special type of
flow-induced phase transition was also observed by Talmon
et al. using cryo-TEM.[213] In this chapter, we mainly focus on
the behavior of rod-shaped micelles, because these solutions
were extensively investigated in the last two decades. Compared to other surfactant systems, entangled solutions of
wormlike micelles show rather simple properties, and we may
use these special samples to understand more complicated
flow-induced phenomena. The large amount of experimental
results gives, at least, qualitative information on the nature
and special features of SIS. It turns out that shear-induced
phases can already be formed in the highly dilute concentration regime, at conditions where the lengths of the rod-shaped
particles are much smaller than their mean distance of separation.[2,3,78,214] Typical examples for complex fluids, showing
these special phase transitions, are aqueous solutions of 2
mM cetylpyridinium or cetyltrimethylammonium salicylate.
The striking effect of SIS can be visualized by simply swirling
the flask and observing the recoil of entrapped air bubbles or
dust particles. In rheological experiments, these solutions
show shear thickening and large first normal stress differences.[2,3] These phenomena occur after a long induction period,
which is of the order of 10^4 strain units. Shear-induced phases
can only be observed after passing a well-defined threshold
value of the velocity gradient.[2,3,78,214] The shear-induced structure behaves like a gel and exhibits strong flow birefrin-

gence.[2,3,76,215] From measurements of the extinction angle, it becomes clear that the flow-induced structure can only grow in the direction of flow.[2-4] Light-scattering experiments suggest that the stirring action forms new micron-sized structures that are much larger than individual wormlike micelles.[216] Similar conclusions were obtained from investigations with an optical Toepler Schlieren technique.[217,218] Patches of the new structure connect to form network superstructures that span hundreds of microns.[219-221] These percolating, gel-like superstructures are responsible for the shear thickening. New investigations of Keller et al. with freeze fracture electron microscopy have revealed large patches of shear-induced structures, which contain a higher surfactant concentration than the surrounding micellar fluid.[222] The new structures are characterized by a submicron stippled or sponge like textures.[222] Images of the shear-induced phase observed with this technique did not show evidence that the new phase still consists of individual, rod-shaped micelles.[222] More likely, the elongated supermolecular structures resembled sponge-like phases or bilayers.[222] With increasing velocity gradient, two different processes can occur in shear-induced structures: it is either possible that more micelles undergo coalescence or the supermolecular gel structure can be stretched in the direction of flow. As soon as the shear-induced gel structure forms, it coexists with aligned and randomly oriented wormlike micelles.[222] By direct imaging of scattered light, Liu and Pine observed the gelation process of anisometric micelles.[216] The initially formed shear-induced gel phase did partly break during flow, and this process induced the formation of extremely elastic, oriented gel bands.[216] SANS studies of Oda et al., which were performed in aqueous solutions of gemini surfactants, were consistent with the shear-induced formation of loosely connected networks of rod-shaped micelles.[223] In this case, the formation of SIS was mainly explained by the attractive forces of the counterions.[223] Cryo-TEM experiments, performed for gemini surfactants with two hydrophobic chains of different lengths, showed that wormlike micelles evolved into a ribbonlike structure (elongated bilayers) after shearing the sample.[224] In the regime of high concentrations, the ribbons transformed into multilayered structures with well-defined diameters.[224]

Systematic experiments, performed by Barentin et al.,[225] showed an instability of flow above a threshold value of the shear rate. At elevated values of the velocity gradient, the micelles aggregated and formed networks of bundles.[225] The magnitude and the temperature dependence of the critical shear rate were in good agreement with experimental data.

Shear-induced phenomena can only occur if the micelles, suspended to flow, acquire enough energy to overcome the repulsive forces between them. According to theories of orthokinetic coagulation, the number of particle collisions increases during flow.[5–7] When the micelles touch, they may coalescence and stick together. These processes can evidently lead to the formation of supermolecular structures during flow. When the stirring action stops, the large aggregates break and return to their initial form. For the mechanisms of formation of the SIS different theories were proposed. Turner and Cates have suggested a simple shear-induced aggregation process, which is based on perikinetic coagulation.[190,226] In this model, anisometric micelles are randomly pushed around by Brownian forces. If they come together with their end caps, they stick and may form elongated structures. The assumption that coagulation will only happen if the micelles approach collinearly is caused by their electric charge.[190,226] Shear flow leads to a partial orientation of the wormlike micelles increasing the coagulation rate. A transient, metastable structure resembling a gel is formed in this way. Bruinsma describes the formation of the shear-induced state employing the idea of orthokinetic coagulation.[227] The different velocities of the micelles induce this process. Collision leads to fusion, and the resulting longer micelles have larger coagulation areas. This effect tends to increase the growth of shear-induced structures. In contrast to other theories, this phenomenon does not require attractive interactions. The Brownian motion and the velocity shear gradient are the dominant transport mechanisms leading to the formation of elongated particles. Employing a mathematical aggregation model, first proposed by Israelachvili, Koch succeeded in calculating hydrodynamic forces with a rigid dumbbell model.[228] In this theoretical approach, it was shown that aggregation forces are weak compared to hydrodynamic forces. SIS formation can then be explained by the flow-induced destruction of rodlike micelles.[228] Wang explained the formation of SIS in the presence of shear and elongational flow by basic thermodynamic principles.[178,219–221,229,230] The free energy of a micelle usually grows if it is compressed or elongated. Long aggregates are better oriented in the direction of flow. Wang showed, that under special conditions, the mean free energy of a surfactant monomer in a longer micelle is smaller than that of shorter micelles.[221,230] This process finally induces micellar growth. Olmsted investigated the behavior of rod-shaped particles on the basis of a modified Doi theory.[107–109,231,232] This work allowed investigating shear-thinning and shear-thickening solutions, and it was performed in order to calculate phase diagrams under condition of constant imposed stress and shear rate. Four fundamental flow curved could be derived, describing different types of shear-induced structures.[107,109,232] New theories of Cates predict that the shear thickening might alternatively be induced by the formation of large, micellar rings.[17] This idea was supported by recent rheological and light scattering results.[18,66] One important consequence of SIS was the detection of nonuniform velocity gradients. This effect is called "shear banding" (see Section VII).

Summarizing the theoretical approaches and experimental studies, we may state that the exact nature of the shear-induced state and the complicated mechanisms involved are not yet completely understood. One important problem, which is still unsolved, concerns the role of electrostatic interactions. Many experiments show that SIS mainly occurs in the regime of very low ionic strengths, where repulsive forces are predominant.[2,3,76,78,215] Addition of salt should lead to enhanced collision frequency between the micellar particles, and this phenomenon should favor SIS. This assumption is, however, in contrast to experimental observations, where shear-induced structures usually disappear upon addition of salt.[2,3,76,78,215] Besides the large number of new results, some phenomena of SIS still remain mysterious. It is interesting to note that shear-induced phases can even have some technical applications. It is well known that the turbulent frictional drag of water can be reduced dramatically by adding small amounts of viscoelastic surfactants, which tend to form shear-induced phases. The amount of drag reduction easily reaches values of 80%. Up to now, several hundred papers were published focusing on this special process. On grounds of these special features, viscoelastic surfactants have numerous drag-reducing applications in petroleum pipelines, heating systems, heat exchangers, bio-reactors, or noise-reduction systems of high-speed vehicles.

VII. INSTABILITIES OF FLOW (SHEAR BANDING)

At least two different types of flow instabilities can occur in complex fluids: the classic Taylor instability and the shear-banding process. In both cases, one observes a banded structure along the vorticity direction in a Couette cell. The internal band microstructure tends to be different for both processes. Whereas the Taylor instability is mainly caused by density and inertia effects, which occur in the regime of elevated shear rates, shear-banding processes are induced by the shear-thinning or -thickening process of complex fluids. The phenomenon of shear banding can be observed in different viscoelastic fluids, including surfactants, polymers, liquid crystals, emulsions, and suspensions.[106,113] The onset of shear banding manifests itself by a discontinuity in the steady-state shear stress of the system.[35,68,71,72,108,113,129,131,174,233–249] Many solutions of complex fluids show shear thinning behavior. In the double logarithmic representation of experimental data, a slope of -1 is often obtained. According to Newtonian's law (Equation 9.49), this means that the stress is constant in certain regimes of the velocity gradient.

$$\sigma(\infty, \dot{\gamma}) = \eta(\infty, \dot{\gamma})\dot{\gamma} \qquad (9.49)$$

The common explanation for the shear stress plateau is slip at the walls of the measuring system or shear banding, which means partitioning of the material into two different zones.[63] In this context, it is interesting to note that many theories, for instance the Giesekus model, predict the onset of a stress plateau. Relevant data are summarized in Figure 9.10.

The shear stress of the Giesekus model is represented by Equation 9.34. This parameter can easily be normalized:

$$\sigma_n = \frac{2\sigma}{G_0} \qquad (9.50)$$

It is easy to recognize that the shear stress σ attains a plateau for $\dot{\gamma} \gg \lambda$:

$$\lim_{\dot{\gamma}\to\infty} \sigma = G_0 \qquad (9.51)$$

The Giesekus model evidently predicts a shear stress plateau that is equal to the shear modulus. Equation 9.51 can easily be verified, and it was found to be in good agreement

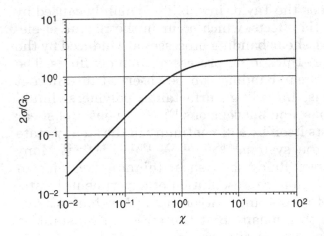

Figure 9.10 Normalized shear stress $2\sigma/G_0$ as a function of the normalized shear rate χ for the Giesekus model.

with experimental data.[79,99,202–204,241] A nonlinear viscoelastic equation based on the reptation-reaction model has been derived by Cates et al.[58,60–62,181] This model also predicts plateau values and a stress maximum where shear banding occurs[58,60–62,181]:

$$\sigma_m = 0.67 G_0 \tag{9.52}$$

This theoretical concept was extensively investigated and found to be in good agreement with experiments.[58,60,62,63,84] Typical examples, obeying these laws, are aqueous solution of cetylpyridinium chloride and sodium salicylate.[58,63,84] Whereas the Cates theory seems to hold in the regime of small and medium salt concentrations, the Giesekus model could, up to now, only be applied in viscoelastic solutions of high ionic strengths. The maximum shear stress, derived from Cates's model, automatically leads to the onset of flow instabilities (the spurt effect). At conditions where the shear stress decreases with increasing velocity gradient, the linear velocity profile becomes unstable. The new state is characterized by two regions (the "bands"). Within this band structure, the shear rate remains constant. In the small region (the "interface"), which connects the two bands, the velocity gradient changes very rapidly with position. The interfaces between the bands can be aligned in the direction of the flow gradient or perpendicular in the direction of the flow vorticity.[174,233,237,245,247,249] The steady-state stress curve of shear banding systems can be qualitatively different, depending on whether the average stress or the average shear rate in the system is held fixed.[63] Each band has its own characteristic features. The phenomenon of flow-induced phase separation is evidently induced by constitutive instabilities (such as isotropic-to-nematic transitions).[63,131,247,250] The transition usually occurs above a critical shear stress or velocity gradient. The flow-induced bands are distinguished by different shear rates or shear stresses, and this may also lead to varying orientations or microstructures.[68,174,233,237,245] In shear-thinning systems, for example, a stress plateau in the flow curve usually indicates gradient banding. This phenomenon has been unambiguously observed in solutions of entangled wormlike micelles.[63,68,72,110,131,174,233,237,239,241,245,247,249] In strain-controlled experiments on shear-thinning solutions, a stress plateau coincides with shear banding in the gradient direction.[235] In shear-thickening solutions, a gel-like phase can be induced during flow.[95] Some surfactant solutions can even show subsequent shear-thinning and shear-thickening effects, depending on the applied shear rate. Typical example are equimolar solutions of 30 or 40 mM cetylpyridinium chloride and sodium salicylate.[3,4,174,212] In the regime of shear thickening, a transient phase separation into turbid and clear ring-like patterns occurred.[174,212] These rings were stacked like pancakes in the direction perpendicular to the vorticity axis. These solutions showed several unusual features as no induction period of the shear-induced phase, no structure formation at the rotating cylinder, a jumping pancake structure, and oscillating shear stresses.[174,212] This example unambiguously shows the existence of different types of shear-banding structures. Despite the large number of experimental results, the exact nature of shear-banding structures and their formation on grounds of shear-induced phase transitions is still mainly unexplored.

VIII. CONCLUSIONS AND PROSPECTS

In this chapter, we discussed the rheological properties of viscoelastic surfactant solutions. We described the complex properties of liquid crystalline phases, and in contrast to these solutions, we have seen that entangled solutions of wormlike micelles exhibit rather simple flow phenomena. It turns out that the kinetic process of micellar breakage and reformation is the main parameter leading to monoexponential stress decay. In situations where the lifetime of elongated micelles is much shorter than reptation motions, we observe very simple scaling laws, which are quantitatively predicted from Cates theory. The linear viscoelastic properties of these solutions can also be described with the phenomenological Maxwell model. Extension of this theoretical approach into the regime of nonlinear phenomena leads to Giesekus model. It is remarkable and surprising that complex properties, such as shear thinning or normal stresses, or transient features, such as stress overshoot and relaxation processes after cessation of steady-state flow, are in excellent agreement with predictions of the Giesekus model. This hold for an anisotropy factor $\alpha = 0.5$. It is interesting to note that the Giesekus model also predicts the existence of different semi-empirical laws, which are well known from experimental observations. Within the limits of experimental errors, this holds for the Cox-Merz rule, the Yamamoto relation, the Laun rule, and both Gleissle mirror relationships.[79,99,201–204] In contrast to these observations, more complex phenomena as shear-induced phase transitions or shear-banding processes are still not completely understood. The rheological properties discussed in this chapter are of general importance, and they are equally observed in different materials, such as polymers and proteins, or inorganic gels. Viscoelastic surfactant solutions can, therefore, be used as simple model systems to study fundamental principles of flow.

REFERENCES

1. Einstein, A. *Ann. Physik* 1906, 19, 289.

2. Rehage, H., Wunderlich, I., Hoffmann, H. *Prog. Colloid Polym. Sci.* 1986, 72, 51.

3. Rehage, H., Hoffmann, H., Wunderlich, I. *Ber. Bunsenges. Phys. Chem.* 1986, 90, 1071.

4. Wunderlich, I., Hoffmann, H., Rehage, H. *Rheol. Acta* 1987, 26, 532.

5. Smoluchowski, M. *Physik Z.* 1916, 17, 585.

6. Smoluchowski, M. *Physik Z.* 1916, 17, 557.

7. Smoluchowski, M. *Z. Phys. Chem.* 1917, 92, 129.

8. Freundlich, H. In *Kapillarchemie*, Akademische Verlagsgesellschaft, Leipzig 1930.

9. Booij, H.L., Bungenberg de Jong, H.G., Heilbrunn, L.V. In *Protoplasmatologia: Handbuch der Protoplasmaforshung*, Weber, F. Ed., Springer Verlag, 1956.

10. Gravsholt, S. *Proc. Intern. Congr. Surf. Activity* 1973, 6, 807.

11. Candau, S.J., Khatory, A., Lequeux, F., Kern, F. *J. Phys. IV* 1993, 3, 197.

12. Lequeux, F., Candau, S.J. In *Structure and Flow in Surfactant Solutions*, Herb, C.A., Prud'homme, R.K., Eds., Washington: ACS Symp. Ser. 578, 1994, 578, Chap. 3.

13. Hassan, P.A., Valaulikar, B.S., Manohar, C., Kern, F., Bourdieu, L., Candau, S.J. *Langmuir* 1996, 12, 4350.

14. Mendes, E., Narayanan, J., Oda, R., Kern, F., Candau, S.J., Manohar, C. *J. Phys. Chem. B* 1997, 101, 2256.

15. Oda, R., Narayanan, J., Hassan, P.A., Manohar, C., Salkar, R.A., Kern, F., Candau, S.J. *Langmuir* 1998, 14, 4364.

16. Candau, S.J., Oda, R. *Colloids Surf. A* 2001, 183, 5.

17. Cates, M.E., Candau, S.J. *Europhys. Lett.* 2001, 55, 887.

18. Oelschlaeger, C., Waton, G., Buhler, E., Candau, S.J., Cates, M.E. *Langmuir* 2002, 18, 3076.

19. Clausen, T.M., Vinson, P.K., Minter, J.R., Davis, H.T., Talmon, Y., Miller, W.G. *J. Phys. Chem.* 1992, 96, 474.

20. Danino, D., Kaplun, A., Talmon, Y., Zana, R. In *Structure and Flow in Surfactant Solutions*, Herb C.A., Prud'homme R.K. Eds., Washington: ACS Symp. Ser. 578, 1994, Chap. 6.

21. Danino, D., Talmon, Y., Zana, R. *J. Colloid Interface Sci.* 1997, 186, 170.

22. Zana, R., Levy, H., Danino, D., Talmon, Y., Kwetkat, K. *Langmuir* 1997, 13, 402.

23. Danino, D., Kaplun, A., Lichtenberg, D., Talmon, Y., Zana, R. *Abstracts of Papers of the American Chemical Society* 1998, 216, 321.

24. Lu, B., Zheng, Y., Davis, H.T., Scriven, L.E., Talmon, Y., Zakin, J.L. *Rheol. Acta* 1998, 37, 528.

25. Bernheim-Groswasser, A., Wachtel, E., Talmon, Y. *Langmuir* 2000, 16, 4131.

26. Konikoff, F.M., Danino, D., Weihs, D., Rubin, M., Talmon, Y. *Hepatology* 2000, 31, 261.

27. Oda, R., Huc, I., Danino, D., Talmon, Y. *Langmuir* 2000, 16, 9759.

28. Lin, Z.Q., Lu, B., Zakin, J.L., Talmon, Y., Zheng, Y., Davis, H.T., Scriven, L.E. *J. Colloid Interface Sci.* 2001, 239, 543.

29. Lin, Z.Q., Zakin, J.L., Zheng, Y., Davis, H.T., Scriven, L.E., Talmon, Y. *J. Rheol.* 2001, 45, 963.

30. Khatory, A., Kern, F., Lequeux, F., Appell, J., Porte, G., Morie, N., Ott, A., Urbach, W. *Langmuir* 1993, 9, 933.

31. Berret, J.F., Roux, D.C., Porte, G., Lindner, P. *Europhys. Lett.* 1995, 32, 137.

32. Roux, D.C., Berret, J.F., Porte, G., Peuvreldisdier, E., Lindner, P. *Macromolecules* 1995, 28, 1681.

33. Oberdisse, J., Regev, O., Porte, G. *J. Phys. Chem. B* 1998, 102, 1102.

34. Berret, J.F., Porte, G. *Phys. Rev. E* 1999, 60, 4268.

35. Radulescu, O., Olmsted, P.D., Berret, J.F., Porte, G., Lerouge, S., Decruppe, J.-P. In *Proceedings of the International Congress on Rheology,* 13th, Binding D.M., Ed., Cambridge, U.K., British Society of Rheology, 2000, 360 p.

36. Filali, M., Michel, E., Mora, S., Molino, F., Porte, G. *Colloids Surf., A,* 2001, 183, 203.

37. Shikata, T., Pearson, D.S. *Abstracts of Papers of the American Chemical Society* 1993, 206, 53.

38. Shikata, T., Pearson, D.S. *Langmuir* 1994, 10, 4027.

39. Shikata, T., Dahmans, S.J., Pearson, D.S. *Langmuir* 1994, 10, 3470.

40. Shikata, T., Pearson, D.S. *Abstracts of Papers of the American Chemical Society* 1994, 207, 13.

41. Shikata, T., Pearson, D.S. In *Structure and Flow in Surfactant Solutions,* Herb C.A., Prod'homme R.K., Eds., Washington: ACS Symp. Ser. 578, 1994, Chap. 8.

42. Shikata, T., Morishima, Y. *Langmuir* 1996, 12, 5307.

43. Shikata, T. *Nihon Reoroji Gakkaishi* 1997, 25, 255.

44. Shikata, T., Niwa, H., Morishima, Y. *Nihon Reoroji Gakkaishi* 1997, 25, 19.

45. Shikata, T., Imai, S., Morishima, Y. *Langmuir* 1997, 13, 5229.

46. Shikata, T., Morishima, Y. *Langmuir* 1997, 13, 1931.

47. Shikata, T., Imai, S., Morishima, Y. *Langmuir* 1998, 14, 2020.

48. Shikata, T., Imai, S. *Langmuir* 1998, 14, 6804.

49. Imai, S., Shikata, T. *Langmuir* 1999, 15, 8388.

50. Imai, S., Shikata, T. *Langmuir* 1999, 15, 7993.

51. Shikata, T., Imai, S. *J. Phys. Chem. B* 1999, 103, 8694.

52. Imai, S., Kunimoto, E., Shikata, T. *Nihon Reoroji Gakkaishi* 2000, 28, 61.

53. Imai, S., Kunimoto, E., Shikata, T. *Nihon Reoroji Gakkaishi* 2000, 28, 67.

54. Imai, S., Shikata, T. *J. Colloid Interface Sci.* 2001, 244, 399.

55. Imai, S., Shiokawa, M., Shikata, T. *J. Phys. Chem. B* 2001, 105, 4495.

56. Shikata, T., Shiokawa, M., Itatani, S., Imai, S. *Korea-Australia Rheol. J.* 2002, 14, 129.

57. Turner, M.S., Marques, C., Cates, M.E. *Langmuir* 1993, 9, 695.

58. Cates, M.E. *Abstracts of Papers of the American Chemical Society* 1993, 206, 10.

59. Cates, M.E. *Physica Scripta* 1993, T49A, 107.

60. Spenley, N.A., Cates, M.E., McLeish, T.C.B. *Phys. Rev. Letters* 1993, 71, 939.

61. Marques, C.M., Turner, M.S., Cates, M.E. *J. Non-Cryst. Solids* 1994, 172, 1168.

62. Cates, M.E. *J. Phys.: Condens. Matter* 1996, 8, 9167.

63. Grand, C., Arrault, J., Cates, M.E. *J. Phys. II* 1997, 7, 1071.

64. Sollich, P., Lequeux, F., Hebraud, P., Cates, M.E. *Phys. Rev. Letters* 1997, 78, 2020.

65. O'Loan, O.J., Evans, M.R., Cates, M.E. *Physica A* 1998, 258, 109.

66. Oelschlaeger, C., Waton, G., Candau, S.J., Cates, M.E. *Langmuir* 2002, 18, 7265.

67. Makhloufi, R., Decruppe, J.P., Cressely, R. *Colloids Surf. A,* 1993, 76, 33.

68. Makhloufi, R., Decruppe, J.P., Sit-Ali, A., Cressely, R. *Europhys. Lett.* 1995, 32, 253.

69. Humbert, C., Decruppe, J.P. *Eur. Phys. J. B* 1998, 6, 511.

70. Humbert, C., Decruppe, J.P. *Colloid. Polym. Sci.* 1998, 276, 160.

71. Lerouge, S., Decruppe, J.P., Humbert, C. *Phys. Rev. Letters* 1998, 81, 5457.

72. Lerouge, S., Decruppe, J.P., Berret, J.F. *Langmuir* 2000, 16, 6464.

73. Berret, J.F., Lerouge, S., Decruppe, J.P. *Langmuir* 2002, 18, 7279.

74. Hoffmann, H., Rehage, H., Rauscher, A. *NATO ASI Ser.* 1992, 369, 493.

75. Hoffmann, II., Ulbricht, W. *Curr. Opin. Colloid Interface Sci.* 1996, 1, 726.

76. Hoffmann, H., Hofmann, S., Kastner, U. *Hydrophilic Polym.* 1996, 248, 219.

77. Bergmeier, M., Hoffmann, H., Thunig, C. *J. Phys. Chem. B* 1997, 101, 5767.

78. Hoffmann, H., Ulbricht, W. *Surf. Sci. Ser.* 1997, 70, 285.

79. Fischer, P., Rehage, H. *Rheol.* 1997, 36, 13.

80. Hoffmann, H., Platz, G., Rehage, H., Schorr, W. *Ber. Bunsenges. Phys. Chem.* 1981, 85, 877.

81. Hoffmann, H., Platz, G., Rehage, H., Schorr, W., Ulbricht, W. *Ber. Bunsen-Ges. Phys. Chem.* 1981, 85, 255.

82. Hoffmann, H., Platz, G., Rehage, H., Schorr, W. *Adv. Colloid Interface Sci.* 1982, 17, 275.

83. Rauscher, A., Rehage, H., Hoffmann, H. *Prog. Colloid Polym. Sci.* 1991, 84, 99.

84. Rehage, H., Hoffmann, H. *Mol. Phys.* 1991, 74, 933.

85. Rehage, H., Platz, G., Struller, B., Thunig, C. *Tenside Surf. Det.* 1996, 33, 242.

86. Fischer, P., Rehage, H., Gruning, B. *J. Phys. Chem. B* 2002, 106, 11041.

87. Lin, Z., Cai, J.J., Scriven, L.E., Davis, H.T. *J. Phys. Chem.* 1994, 98, 5984.

88. Berret, J.F., Roux, D.C. *J. Rheol.* 1995, 39, 725.

89. Schmitt, V., Schosseler, F., Lequeux, F. *Europhys. Lett.* 1995, 30, 31.

90. Kadoma, I.A., van Egmond, J.W. *Phys. Rev. Letters* 1996, 76, 4432.

91. Lequeux, F. *Curr. Opin. Colloid Interface Sci.* 1996, 1, 341.

92. Walker, L.M., Moldenaers, P., Berret, J.F. *Langmuir* 1996, 12, 6309.

93. Wheeler, E.K., Izu, P., Fuller, G.G. *Rheol.* 1996, 35, 139.

94. Berret, J.F. *Langmuir* 1997, 13, 2227.

95. Boltenhagen, P., Hu, Y.T., Matthys, E.F., Pine, D.J. *Phys. Rev. Letters* 1997, 79, 2359.

96. Carl, W., Makhloufi, R., Kroeger, M. *J. Phys. II* 1997, 7, 931.

97. Chen, C.M., Warr, G.G. *Langmuir* 1997, 13, 1374.

98. Fischer, P., Fuller, G.G., Lin, Z.C. *Rheol.* 1997, 36, 632.

99. Fischer, P. *Appl. Rheol* 1997, 7, 58.

100. Gittes, F., Schnurr, B., Olmsted, P.D., MacKintosh, F.C., Schmidt, C.F. *Phys. Rev. Letters* 1997, 79, 3286.

101. Penfold, J., Staples, E., Tucker, I., Cummins, P. *J. Colloid Interface Sci.* 1997, 185, 424.

102. Berret, J.F., Roux, D.C., Lindner, P. *Eur. Phys. J. B* 1998, 5, 67.

103. Ponton, A., Quemada,D. *In Progress and Trends in Rheology V,* *Proceedings of the 5th European Rheology Conference,* Emri I. Ed., Darmstadt: Steinkopff-Verlag, 1998.

104. Watanabe, H., Osaki, K., Matsumoto, M., Bossev, D.P., McNamee, C.E., Nakahara, M., Yao, M.L. *Rheol.* 1998, 37, 470.

105. Callaghan, P.T. *Rep. Prog. Phys.* 1999, 62, 599.

106. Goveas, J.L., Pine, D.J. *Europhys. Lett.* 1999, 48, 706.

107. Olmsted, P.D., Lu, C.Y.D. *Faraday Discuss.* 1999, 112, 183.

108. Olmsted, P.D. *Curr. Opin. Colloid Interface Sci.* 1999, 4, 95.

109. Olmsted, P.D., Lu, C.Y.D. *Phys. Rev. E* 1999, 60, 4397.

110. Soltero, J.F.A., Bautista, F., Puig, J.E., Manero, O. *Langmuir* 1999, 15, 1604.

111. Sood, A.K., Bandyopadhyay, R., Basappa, G. *Pramana* 1999, 53, 223.

112. Kim, W.J., Yang, S.M. *Langmuir* 2000, 16, 6084.

113. Goveas, J.L., Olmsted, P.D. *Eur. Phys. J. E* 2001, 6, 79.

114. Kim, W.J., Yang, S.M. *J. Chem. Eng. Jpn.* 2001, 34, 227.

115. Cardinaux, F., Cipelletti, L., Scheffold, F., Schurtenberger, P. *Europhys. Lett.* 2002, 57, 738.

116. Lee, J.Y., Magda, J.J., Hu, H., Larson, R.G. *J. Rheol.* 2002, 46, 195.

117. Rehage, H., Hoffmann, H. *Faraday Discuss.* 1983, 76, 363.

118. Hoffmann, H., Ulbricht, W. *Tenside Surf. Det.* 1987, 24, 23.

119. Tamori, K., Esumi, K., Meguro, K., Hoffmann, H. *J. Colloid Interface Sci.* 1991, 147, 33.

120. Berret, J.F., Appell, J., Porte, G. *Langmuir* 1993, 9, 2851.

121. Hoffmann, H. In *Structure and Flow in Surfactant Solutions,* Herb C.A., Prud'homme R.K. Eds., Washington: ACS Symp. Ser. 578, 1994, Chap. 1.

122. Panitz, J.C., Gradzielski, M., Hoffmann, H., Wokaun, A. *J. Phys. Chem.* 1994, 98, 6812.

123. Schurtenberger, P., Cavaco, C. *Langmuir* 1994, 10, 100.

124. Hoffmann, H., Thunig, C., Schmiedel, P., Munkert, U. *Faraday Discuss.* 1995 101, 319.

125. Illner, J.C., Hoffmann, H. *Tenside Surf. Det.* 1995, 32, 318.

126. Hoffmann, H., Ulbricht, W. *Tenside Surf. Det.* 1998, 35, 421.

127. Dorfler, H.D., Gorgens, C. *Tenside Surf. Det.* 2000, 37, 17.

128. Truong, M.T., Walker, L.M. *Langmuir* 2000, 16, 7991.

129. Bautista, F., Soltero, J.F.A., Macias, E.R., Puig, J.E., Manero, O. *J. Phys. Chem. B* 2002, 106, 13018.

130. Cannavacciuolo, L., Pedersen, J.S., Schurtenberger, P. *J. Phys.: Condens. Matter* 2002, 14, 2283.

131. Fielding, S.M., Olmsted, P.D. *Los Alamos National Laboratory* 2002, arXiv:cond-mat/0207344 , 15.

132. Shikata, T., Pearson, D.S. *ACS Symp. Ser.* 1994, 578, 129.

133. Shikata, T., Imai, S. *Langmuir* 2000, 16, 4840.

134. Zhang, W.C., Li, G.Z., Ji, K.J., Shen, Q. *Gaodeng Xuexiao Huaxue Xuebao* 2000, 21, 1261.

135. Khatory, A., Lequeux, F., Kern, F., Candau, S.J. *Langmuir* 1993, 9, 1456.

136. Volpert, E., Selb, J., Candau, F. *Macromolecules* 1996, 29, 1452.

137. Hassan, P.A., Candau, S.J., Kern, F., Manohar, C. *Langmuir* 1998, 14, 6025.

138. Shchipunov, Y.A., Hoffmann, H. *Langmuir* 1998, 14, 6350.

139. Lin, Z. *Langmuir* 1996, 12, 1729.

140. Jerke, G., Pedersen, J.S., Egelhaaf, S.U., Schurtenberger, P. *Langmuir* 1998, 14, 6013.

141. Carver, M., Smith, T.L., Gee, J.C., Delichere, A., Caponetti, E., Magid, L.J. *Langmuir* 1996, 12, 691.

142. van der Schoot, P. *J. Chem. Phys.* 1996, 104, 1130.

143. May, S., Bohbot, Y., Ben-Shaul, A. *J. Phys. Chem. B* 1997, 101, 8648.

144. Escalante, J.I., Gradzielski, M., Hoffmann, H., Mortensen, K. *Langmuir* 2000, 16, 8653.

145. Hao, J.C., Hoffmann, H., Horbaschek, K. *J. Phys. Chem. B* 2000, 104, 10144.

146. Hao, J.C., Hoffmann, H., Horbaschek, K. *Langmuir* 2001, 17, 4151.

147. Fischer, A., Hoffmann, H., Medick, P., Rossler, E. *J. Phys. Chem. B* 2002, 106, 6821.

148. Almgren, M., Aniansson, E.A.G.,Wall, S.N., and Holmaker, K. In *Micellization, Solubilization, Microemulsions*, Mittal, K. L. Ed., New York: Plenum 1976. 329 p.

149. Almgren, M., Aniansson, E.A.G., Holmaker, K. *Chem. Phys.* 1977, 19, 1.

150. Aniansson, E.A.G., Wall, S.N. *J. Phys. Chem.* 1974, 78, 1024.

151. Aniansson, E.A.G., Wall, S.N. *NATO Adv. Study Inst. Ser.* 1975, 18, 223.

152. Aniansson, E.A.G., Wall, S.N. *J. Phys. Chem.* 1975, 79, 857.

153. Aniansson, E.A.G., Almgren, M., Wall, S.N. *NATO Adv. Study Inst. Ser.* 1975, 18, 239.

154. Aniansson, E.A.G. *Ber. Bunsenges. Phys. Chem.* 1978, 82, 981.

155. Aniansson, E.A.G., Wall, S.N., Almgren, M., Hoffmann, H., Kielmann, I., Ulbricht, W., Zana, R., Lang, J., Tondre, C. *J. Phys. Chem.* 1976, 80, 905.

156. Hoffmann, H. *GBF Monograph Ser.* 1984, 7, 95.

157. Lang, J., Zana, R. In *Surfactant Solutions: New Methods of Investigation*, Zana, R., Ed., New York, Basel: Marcel Dekker, Inc., 1987, Chap. 8.

158. Nomura, H., Koda, S., Matsuoka, T., Hiyama, T., Shibata, R., Kato, S. *J. Colloid Interface Sci.* 2000, 230, 22.

159. Telgmann, T., Kaatze, U. *J. Phys. Chem. A* 2000, 104, 1085.

160. De Maeyer, L., Trachimow, C., Kaatze, U. *J. Phys. Chem. B* 1998, 102, 8024.

161. von Gottberg, F.K., Smith, K.A., Hatton, T.A. *J. Chem. Phys.* 1998, 108, 2232.

162. Waton, G. *J. Phys. Chem. B* 1997, 101, 9727.

163. Telgmann, T., Kaatze, U. *J. Phys. Chem. B* 1997, 101, 7766.

164. Michels, B., Waton, G., Zana, R. *Langmuir* 1997, 13, 3111.

165. Verrall, R.E., Jobe, D.J., Aicart, E. *J. Mol. Liq.* 1995, 65–66, 195.

166. Kato, S., Harada, S., Sahara, H. *J. Phys. Chem.* 1995, 99, 12570.

167. Hecht, E., Hoffmann, H. *Colloids Surf. A* 1995, 96, 181.

168. Jobe, D.J., Verrall, R.E., Junquera, E., Aicart, E. *J. Phys. Chem.* 1994, 98, 10814.

169. Frindi, M., Michels, B., Zana, R. *J. Phys. Chem.* 1994, 98, 6607.

170. Jobe, D.J., Verrall, R.E., Junquera, E., Aicart, E. *J. Phys. Chem.* 1993, 97, 1243.

171. Berret, J.F., Porte, G., Decruppe, J.P. *Phys. Rev. E* 1997, 55, 1668.

172. Buwalda, R.T., Stuart, M.C.A., Engberts, J.B.F.N. *Langmuir* 2000, 16, 6780.

173. Danino, D., Talmon, Y., Zana, R. *Langmuir* 1995, 11, 1448.

174. Fischer, P., Wheeler, E.K., Fuller, G.G. *Rheol.* 2002, 41, 35.

175. Maeda, H., Yamamoto, A., Souda, M., Kawasaki, H., Hossain, K.S., Nemoto, N., Almgren, M. *J. Phys. Chem. B* 2001, 105, 5411.

176. Mu, J.H., Li, G.Z., Xiao, H.D., Liao, G.H., Liu, Y., Huang, L., Li, B.Q. *Chin. Sci. Bull.* 2001, 46, 1360.

177. Mu, J.H., Li, G.Z., Chen, W.J., Mao, H.Z., Liu, S.J. *Acta Chimica Sinica* 2001, 59, 1384.

178. Mu, J.H., Li, G.Z., Wang, Z.W. *Rheol.* 2002, 41, 493.

179. Schonfelder, E., Hoffmann, H. *Phys. Chem. Chem. Phys.* 1994, 98, 842.

180. Cates, M.E. *Macromolecules* 1987, 20, 2289.

181. Granek, R., Cates, M.E. *J. Chem. Phys.* 1992, 96, 4758.

182. Candau, S.J., Hebraud, P., Schmitt, V., Lequeux, F., Kern, F., Zana, R. *Nuovo Cimento* 1994, 16, 1401.

183. Candau, S.J., Kern, F., Zana, R. *Abstracts of Papers of the American Chemical Society* 1993, 206, 11.

184. Candau, S.J., Lequeux, F. *Curr. Opin. Colloid Interface Sci.* 1997, 2, 420.

185. Turner, M.S., Cates, M.E. *Langmuir* 1991, 7, 1590.

186. Kern, F., Lequeux, F., Zana, R., Candau, S.J. *Langmuir* 1994, 10, 1714.

187. Rehage, H., Hoffmann, H. *J. Phys. Chem.* 1988, 92, 4712.

188. Danino, D., Talmon, Y., Zana, R. *J. Colloid Interface Sci.* 1997, 185, 84.

189. Lin, Z., Hill, R.M., Davis, H.T., Scriven, L.E., Talmon, Y. *Langmuir* 1994, 10, 1008.

190. Cates, M.E., Turner, M.S. *Europhys. Lett.* 1990, 11, 681.

191. Lequeux, F. *Europhys. Lett.* 1992, 19, 675.

192. Cox, W.P., Merz, E.H. *J. Polym. Sci.* 1958, 28, 619.

193. Ferry, J.D. In *Viscoelastic Properties of Polymers*, Wiley, New York, 1980.

194. Giesekus, H. In *Phänomenologische Rheologie*, Springer Verlag, Berlin, 1994.

195. Giesekus, H. *J. Non-Newtonian Fluid Mech.* 1982, 11, 69.

196. Giesekus, H. *J. Non-Newtonian Fluid Mech.* 1984, 14, 47.

197. Giesekus, H. *J. Non-Newtonian Fluid Mech.* 1985, 17, 349.

198. Giesekus, H. *Rheol.* 1995, 34, 2.

199. Giesekus, H. *Rheol.* 2003, 29, 500.

200. Giesekus, H. *Rheol.* 1966, 5, 29.

201. Holz, T., Fischer, P., Rehage, H. *J. Non-Newtonian Fluid Mech.* 1999, 88, 133.

202. Fischer, P., Rehage, H. *Prog. Colloid Polym. Sci.* 1995, 98, 94.

203. Pflaumbaum, M., Rehage, H. *Phys. Chem. Chem. Phys.* 2003, 4, 705.

204. Pflaumbaum, M., Rehage, H., Talmon, Y., Müller, F., Peggau, J. *Tenside Surf. Det.* 2003, 39, 212.

205. Brown, E.F., Burghardt, W.R., Venerus, D.C. *Langmuir* 1997, 13, 3902.

206. Butler, P. *Curr. Opin. Colloid Interface Sci.* 1999, 4, 214.

207. Berghausen, J., Zipfel, J., Diat, O., Narayanan, T., Richtering, W. *Phys. Chem. Chem. Phys.* 2000, 2, 3623.

208. Imai, M., Nakaya, K., Kato, T., Takahashi, Y., Kanaya, T. *J. Phys. Chem. Solids,* 1999, 60, 1313.

209. Myska, J., Stern, P. *Colloid. Polym. Sci.* 1994, 272, 542.

210. Nettesheim, F., Zipfel, J., Lindner, P., Richtering, W. *Colloids Surf. A* 2001, 183, 563.

211. Schmitt, V., Lequeux, F., Pousse, A., Roux, D.C. *Langmuir* 1994, 10, 955.

212. Wheeler, E.K., Fischer, P., Fuller, G.G. *J. Non-Newtonian Fluid Mech.* 1998, 75, 193.

213. Zheng, Y., Lin, Z., Zakin, J.L., Talmon, Y., Davis, H.T., Scriven, L.E. *J. Phys. Chem. B* 2000, 104, 5263.

214. Shchipunov, Y.A., Hoffmann, H. *Rheol.* 2000, 39, 542.

215. Hofmann, S., Hoffmann, H. *J. Phys. Chem. B* 1998, 102, 5614.

216. Liu, C.H., Pine, D.J. *Phys. Rev. Letters* 1996, 77, 2121.

217. Bergins, C., Nowak, M., Urban, M. *Exp. Fluids* 2001, 30, 410.

218. Nowak, M. *Rheol.* 1998, 37, 336.

219. Hu, Y., Wang, S.Q., Jamieson, A.M. *J. Phys. Chem.* 1994, 98, 8555.

220. Hu, Y., Rajaram, C.V., Wang, S.Q., Jamieson, A.M. *Langmuir* 1994, 10, 80.

221. Wang, S.Q., Hu, Y., Jamieson, A.M. In *Structure and Flow in Surfactant Solutions,* Herb, C.A., Prud homme, R.K. Eds., Washington: ACS Symp. Ser. 578, 1994, 578, 278.

222. Keller, S.L., Boltenhagen, P., Pine, D.J., Zasadzinski, J.A. *Phys. Rev. Letters* 1998, 80, 2725.

223. Oda, R., Weber, V., Lindner, P., Pine, D.J., Mendes, E., Schosseler, F. *Langmuir* 2000, 16, 4859.

224. Oda, R., Huc, I., Homo, J.C., Heinrich, B., Schmutz, M., Candau, S. *Langmuir* 1999, 15, 2384.

225. Barentin, C., Liu, A.J. *Europhys. Lett.* 2001, 55, 432.

226. Turner, M.S., Cates, M.E. *J. Phys.:Condens. Matter* 1992, 4, 3719.

227. Bruinsma, R., Gelbart, W.M.B. *J. Chem. Phys.* 1992, 96, 7710.

228. Koch, S. *Rheol.* 1997, 36, 639.

229. Wang, H., Zhang, G. *Riyong Huaxue Gongye* 2001, 31, 1.

230. Wang, S.Q. *J. Chem. Phys.* 1990, 94, 2219.

231. Olmsted, P.D., Lu, C.Y.D. *Phys. Rev. E* 1997, 56, R55.

232. Olmsted, P.D. *Europhys. Lett.* 1999, 48, 339.

233. Bautista, F., Soltero, J.F.A., Perez-Lopez, J.H., Puig, J.E., Manero, O. *J. Non-Newtonian Fluid Mech.* 2000, 94, 57.

234. Berret, J.F., Roux, D.C., Porte, G. *J. Phys. II* 1994, 4, 1261.

235. Britton, M.M., Callaghan, P.T. *Eur. Phys. J. B* 1999, 7, 237.

236. Cappelaere, E., Cressely, R., Decruppe, J.P. *Colloids Surf. A* 1995, 104, 353.

237. Cappelaere, E., Cressely, R. *Colloid. Polym. Sci.* 1997, 275, 407.

238. Decruppe, J.P., Lerouge, S., Berret, J.F. *Phys. Rev. E* 2001, 63, 022501.

239. Decruppe, J.P., Cressely, R., Makhloufi, R., Cappelaere, E. *Colloid Polym. Sci.* 1995, 273, 346.

240. Fischer, E., Callaghan, P.T. *Phys. Rev. E* 2001, 64, 011501.

241. Fischer, P. *Rheol.* 2000, 39, 234.

242. Hartmann, V., Cressely, R. *Colloids Surf. A* 1997, 121, 151.

243. Lee, J.Y., Magda, J.J., Larson, R.G. In *Proceedings of the International Congress on Rheology*, 13th. Binding D.M. Ed., Glasgow: British Society of Rheology, 2000, 208 p.

244. Lu, C.Y.D., Olmsted, P.D., Ball, R.C. *Phys. Rev. Letters* 2000, 84, 642.

245. Mair, R.W., Callaghan, P.T. *Europhys. Lett.* 1996, 36, 719.

246. Pujolle-Robic, C., Olmsted, P.D., Noirez, L. *Europhys. Lett.* 2002, 59, 364.

247. Radulescu, O., Olmsted, P.D., Lu, C.Y.D. *Rheol.* 1999, 38, 606.

248. Walker, L.M. *Curr. Opin. Colloid Interface Sci.* 2001, 6, 451.

249. Yuan, X.F. *Europhys. Lett.* 1999, 46, 542.

250. Berret, J.F., Gamez-Corrales, R., Lerouge, S., Decruppe, J.P. *Eur. Phys. J. E* 2000, 2, 343.

10

Reaction Processes in Self-Assembly Systems

BRIAN H. ROBINSON AND
MADELEINE ROGERSON

CONTENTS

I.   INTRODUCTION

Self-assembly systems, from the viewpoint of this chapter, comprise predominantly micellar and microemulsion systems and vesicle assemblies. Most reactions have, however, been studied in micellar systems, and until recently, these were predominantly organic and inorganic processes, although recently there has been considerable work involving enzyme-catalyzed processes, especially in microemulsion and gel media. Because this topic is now a mature field of study, this chapter will necessarily be selective, and will be mainly focused on the personal interests of the authors.

II.   THEORETICAL CONSIDERATIONS OF REACTIONS IN MICELLAR SYSTEMS

An early and very useful theoretical approach for the interpretation of kinetic data for ground-state reactions in micellar systems was developed by the group of Berezin, working at Moscow State University in the early 1970s.[1-3] Prior to this, there had been a number of experimental studies of organic reactions in micellar and reverse-micellar systems, and one of the pioneers here was Janos Fendler, whose contributions were contained in a significant book published in 1975.[4] However, most emphasis before the 1970s was concerned with characterization and structure of micelles and related systems (by means of an increasingly wide range of physical methods) rather than with reactions in self-assembly systems.

In Reference 1, Berezin and coworkers considered the effects of both partitioning of reactants and products between micelles and the external aqueous medium and their reaction. In their scheme, the overall reaction was considered to take place in one of two possible domains of the reaction medium: either in the aqueous solution or in the micellar pseudo-phase, as shown in Figure 10.1. A and B are two reactants involved in a bimolecular reaction to form the product P. Partitioning

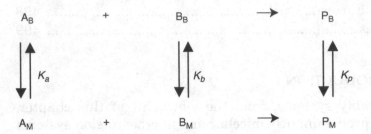

Figure 10.1 Reaction scheme according to Berezin et al.[1] The subscripts M and B refer to micelle and bulk phase.

equilibria for A, B, and P between the bulk and micelle phases are associated with equilibrium constants K_a, K_b, and K_p, respectively.

Of particular interest is the situation where both reactants partition strongly to the micelle environment for surfactant concentrations in excess of the critical micelle concentration (cmc) of the surfactant. Then there is an increase (effectively a step change) in the observed rate of the reaction between A and B in the vicinity of the cmc. For partitioning of both reactants to the micelle, the apparent rate constant for the association reaction between A and B can be effectively increased by factors of up to 10^3 to 10^4 in the cmc region, which might represent only a factor of 2 or so in the micellar concentration. On the other hand, if one reactant partitions to the micelle and the other is predominantly located in the water pseudo-phase, then the rate of the bimolecular reaction will decrease above the cmc, as the two reactants are effectively locating into different microenvironments in the micellar solution. The rate in water (or at surfactant concentration $C < $ cmc) is then the maximum measured rate. The measured equilibrium constants determined at different concentrations of surfactant above the cmc will also reflect these changes in the apparent bimolecular rate constant for reaction between A and B, in the absence of operation of any special factors. The most likely of these is a significant change in the rate constant for the reverse reaction from P to reform A and B.

For the situation where the reaction is accelerated, it might be thought that the micelle is functioning as a catalyst by lowering the energy barrier for the forward reaction, but in fact the rate enhancement is most likely[5] to be due to the reactants A and B locating on the micelle simply as the result of a local concentration enhancement effect. A typical system that demonstrates this effect is for reactions involving divalent or trivalent metal ions in complex formation with neutral hydrophobic ligands, in the presence of micelles having a surface charge opposite to that of the metal ion. Then, for strong partitioning of both the metal ion and ligand to the micelle, as the surfactant concentration is increased above the cmc, the rate enhancement effect and the change (increase) in the apparent equilibrium constant for the reaction become less pronounced.

It is interesting to speculate on the differences between this process and true catalysis, with particular reference to metal ion complex formation. The following points are relevant:

- The micelle concentration is not changed during the reaction. This is a general characteristic of both micelle and true catalysis.
- It is generally found that, comparing the reaction in water with that in the presence of micelles, the activation energy (enthalpy of activation) for the association reaction is not significantly lowered. This is in contrast to the situation in the more conventional type of catalysis, where the energy barrier is lowered for both the forward and reverse reactions, as a result of stabilization of the transition state (reduction of the energy difference between the ground and transition states). The reactants (and products) become associated specifically with the catalyst, which, in the case of simple enzymes, is an "active site."

These points have been discussed in some detail in a recent article by Romsted, Bunton, and Yao.[5] Despite the fact that micellar catalysis is strictly a misnomer, the term continues to be widely used by workers in the field.

Part of the problem with "catalysis" by micelles is that the micelles have a dynamic structure, which is associated with the two relaxation times developed in the relaxation theory of micelle kinetics by Aniansson and Wall[6,7] and tested experimentally by a number of groups, most notably those of Hoffmann and Zana (see Chapter 3, Section III).[8] Both surfactant exchange between the micelle and water and micelle breakdown/formation are important, the former taking place for simple surfactants in the nanosecond-to-microsecond time range, and the latter in the millisecond-to-second time range. Micelles do not usually have in their structure any specific binding sites, which show a resemblance (as found in an enzyme) to an "active site," where the reactants can be specifically adsorbed, and where the reaction between A and B can be truly catalyzed. In the absence of such sites, the reactants can locate to any position on/in the micelle. We will be particularly interested in reactions where both reactants locate near the micelle surface. It is not helpful to give excessive consideration to this process at this stage, since the Berezin theory provides an adequate approach for discussion based on a single "micellar" location.

Clearly a quantitative interpretation of the "hydrophobic effect" would be useful in consideration of the detailed reaction environment for reactions in the presence of micelles, and indeed other self-assembly systems. A detailed review of the hydrophobic effect, directed at chemists, has been reported by Blokzijl and Engberts,[9] who showed that there are a number of difficulties concerned with an interpretation of the hydrophobic effect based on more traditional ideas, e.g., the notion of frozen (low entropy) water around alkyl chains and nonpolar groups, which is transferred to a higher entropic state when the apolar part of a molecule is incorporated in the micelle and the water forms bulk-like water. These authors suggested that neutron diffraction and computational studies offered the best hope for progress in understanding the detailed nature of the interactions involved. Work in both these two areas has accelerated in recent years, but there are still fundamental questions to be answered. However, a detailed discussion of these topics is outside the scope of this chapter.

The diffusional motion of (reactant) molecules when associated with micelles is not generally discussed, but it is reasonable to assume that, when A and B are metal ions and hydrophobic ligands, both will be relatively mobile in the (restricted) micelle environment. Metal ions will be expected to remain hydrated at the micelle surface in that they will not form inner-sphere complexes with the micelle surface group, and will likely diffuse relatively easily over the micelle surface before they eventually desorb from the micelle and re-enter the aqueous pseudo-phase. Because of the generally small size of the micelle, it is likely that the preferred location of both reactant molecules (again for the situation of positive metal ions and anionic micelles) will be close to the micelle surface, with more hydrophobic molecules favoring a less polar micellar environment. In principle, ligand partitioning could be further characterized by defining an additional partition coefficient for fast migration of the ligand between micelle surface and micelle core locations.

It is implicit in the Berezin treatment that partitioning of both reactants between the micelles and water is rapid on the overall time scale of the reaction. The micelle-water partition coefficients then essentially represent a pre-equilibrium situation before the main reaction. Test reactions involving inorganic reactants can cover a very wide time range, from nanoseconds to minutes, but most organic reactions are taking place on time scales of the order of seconds or longer.

In this chapter, first metal-ion/ligand complex formation kinetics will be discussed together with applications in metal extraction, and then organic reactions will be briefly described. The theoretical treatment will then be extended to cover inorganic reactions in microemulsion media.

III. METAL COMPLEX FORMATION

Some of the earliest experimental work on the kinetics of reactions of metal ions with hydrophobic ligands in the presence of anionic micelles was carried out by James and Robinson, who studied in detail 1:1 complex formation between

Ni^{2+}(aq) and hydrophobic dye ligands such as pyridine-2-azo-p-dimethylaniline (PADA).[10] PADA has limited solubility in water ($< 10^{-3}$ mol dm^{-3}), but the solubility is considerably enhanced in micellar solutions, e.g., in the presence of micelles of anionic surfactants such as sodium dodecylsulfate (SDS). This is as a result of a favorable interaction of PADA with the micelle. For surfactant concentrations $>$ cmc, the metal ion will preferentially adsorb to the dodecylsulfate micelle surface and the ligand will also locate to the micelle surface rather than the micelle core, since it is partly polar. (If PADA would locate into the micelle core, with the metal ion confined to the micelle surface, the reaction might not be so effectively promoted, or maybe would even be inhibited.) James and Robinson analyzed their kinetic data in the presence of SDS in terms of a surface reaction, and showed that the kinetics could be explained solely on the basis of a local concentration enhancement effect of the reactants on the micelle surface, with no significant specific effect of the micelle on the reaction. In particular, it was observed that the activation energy for the micelle reaction was not significantly changed when compared with the reaction in water. As discussed for the general case, when the micelle concentration is increased further above the cmc of the surfactant, the reactants are diluted over the micelle surface and the rate enhancement effect becomes less pronounced, as shown in Figure 10.2. The kinetic effect on the reverse reaction — that of dissociation of the metal-ligand complex on the micelle surface to form metal ion and the free ligand — is much less dramatic, but for the Ni^{2+}/PADA system, the rate of reformation on the SDS surface of the discrete metal and ligand species can be increased by a factor of close to 10, essentially independent of the micelle concentration.[11] Work on other metal ions[12] showed that Zn^{2+}(aq) also complexed with PADA in the same way as Ni^{2+}(aq), even though the kinetics with Zn^{2+}(aq) is some orders of magnitude faster than for Ni^{2+}(aq), requiring the use of the temperature-jump method to follow the kinetics. This would suggest that surfactant monomer exchange between the micelle and water is not a significant factor. There was also evidence found for rate enhancement of the forward reaction at concentrations somewhat below the cmc of the surfactant. Reinsborough and

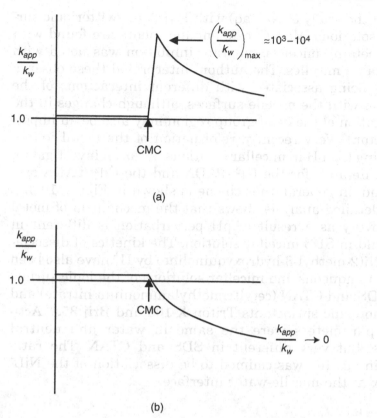

Figure 10.2 Promotion and inhibition of metal ligand complex formation reactions by micelle. (a) Metal ions attracted to micelle, hydrophobic ligand; (b) metal ions attracted to micelle, hydrophilic ligand.

Robinson compared the behavior of a number of partitioning bi-dentate ligands in terms of reaction with $Ni^{2+}(aq)$ in the presence of anionic micelles.[13] More recently, Reinsborough and co-workers also studied in more detail the sub-cmc rate enhancement effects that are evident in surfactant solutions at concentrations near the cmc.[14] It was concluded that, for the Ni^{2+}/PADA reaction, nickel ion-surfactant aggregates could form below the normal cmc, and data from surface tension, dye solubility, and fluorescence probe measurements supported this interpretation. A more recent study has involved the study of Ni^{2+}(aq) with PADA in zwitterionic surfactant solutions. Modest rate enhancements are found with carboxybetaine micelles, but rate inhibition was noted with sulfobetaine micelles. The authors interpreted these observations as being associated with different interactions of the metal ion with the micelle surfaces, although changes in the conformation of the head-group region may also be an important factor.[15] Very recently, reformation of the metal ion on decreasing the pH in micellar solutions has been investigated, in some detail,[11] for the Ni^{2+}/PADA and the Cd^{2+}/PADA systems, and, in general, the scheme is shown in Figure 10.3.

A detailed analysis shows that the mechanism of metal ion recovery as a result of pH perturbation is different in water and in SDS micellar solution. The kinetics of dissociation of Ni(2-methyl-8-hydroxyquinoline) by H^+ have also been studied in aqueous and micellar solutions of the ionic surfactants SDS and CTAN (cetyltrimethylammonium nitrate) and of the nonionic surfactants Triton X-100 and Brij 35.[16] Activation parameters were the same in water and neutral micelles, but very different in SDS and CTAN. The rate-determining step was claimed to be dissociation of the NiL^+ complex at the micelle-water interface.

$$MeL + H^+ \longleftrightarrow MeLH^+ \longrightarrow Me + LH^+$$

Figure 10.3 Reformation of the metal ion on decreasing the pH of the solution. Me is the metal ion, and L and LH^+ represent the ligand and the protonated ligand.

IV. METAL EXTRACTION

Metal extraction continues to be an important industrial process, and a number of reviews have appeared over recent years.[17-19] The bulk of research publications related to this topic have been concerned with extraction of metal ions, particularly Cu^{2+}, into a coexisting organic solvent containing the extractant species. Szymanowski and Tondre reviewed the field[18] in 1994, and it was clear that traditional methods based on emulsion systems have continued to be the preferred approach to extraction. Tondre[19] has recently restated the case for the use of micellar extraction systems, but there still does not appear, so far, to have been much enthusiasm for this approach. Albery and coworkers studied the complexation of Cu^{2+} with an extractant based on an alkylated salicylaldoxime derivative, and they monitored both the complexation and stripping reactions by means of a rotating diffusion cell assembly.[20,21] Other groups have also studied similar systems for copper extraction, but using a range of methods. It is surprising that there still remains much confusion as to the mechanism(s) involved. Stevens, Perera, and Grieser[22] have recently reviewed the field, and suggested a classification of extractants into three groups as follows:

1. Acidic extractants, containing ionizable protons
2. Basic or ion exchangers
3. Solvating extractants

The advantages of using a micellar assembly over a water/organic solvent two-phase system include the following:

• The micellar system is essentially a single homogeneous phase, with a very low organic component. No stirring of the micelle system is required. The two-phase water-organic solvent mixture has a much greater organic component, which is nowadays increasingly environmentally unacceptable.
• A micellar solution may be at its most efficient in terms of extraction just above the cmc of the surfactant, so it is possible to work with very dilute micellar solutions (<1% v/v).
• There is no consumption of ligand or micelles associated with the overall complexation/extraction process.

As already indicated, as well as extracting the metal ion, it is also necessary to recover the metal ion, preferably at a higher metal ion concentration. This is normally achieved by reducing the pH, and this is straightforward for metal ions, which complex with ligands like PADA on the micelle surface. As the pH is reduced, the ligand will become protonated, and in that state the metal ion will complex with the ligand with a much reduced stability constant. As a result, the complex will dissociate. Adjusting the pH back to that used for extraction of the metal ion restores the ligand/micelle system to its original state.

The metal ion can conveniently be recovered from the acid micellar solution by ultrafiltration. Ismael and Tondre[23] demonstrated the use of ultrafiltration in the back extraction of metal ions from micellar solutions containing Ni^{2+} and Cu^{2+} using the extractants 8-hydroxyquinoline (HQ) and an alkylated derivative (C_{11}-HQ) in the presence of cationic micelles of cetyltrimethylammonium bromide and 1-butanol. Decreasing the pH to pH = 3 led to effective extraction of Ni^{2+} in the presence of Cu^{2+} but, for Cu^{2+} extraction, it was necessary to reduce the pH further to less than 1. It is interesting to note that for these systems cationic surfactant systems were used rather than the anionic systems described previously.

Recent work with surfactant assemblies has been reported by Scamehorn et al.[24,25] and especially by the group of Tondre.[26-30] There is a useful review by Tondre, who discusses the coupling of extraction and ultrafiltration in the extraction of metal ions.[19] The Stevens/Grieser/White groups in Melbourne have studied the kinetics of Cu^{2+} extraction by Kelex-100 (7-(4-ethyl-1-methyloctyl)8-hydroxyquinoline) at the heptane-water interface and in a $C_{12}EO_8$ micellar system, and the reaction was modeled using a one-dimensional diffusion model and a diffusion-limited interaction between Kelex-100 and Cu^{2+}.[31] Work has also been done on the extraction of Rh^{3+} complexes into an organic phase of Kelex-100.[32] Although the extraction of both copper and nickel has been well studied, there are very little published data on toxic metals such as cadmium, but a recent study was carried out by Lee and Choi.[33]

There is also active interest in the extraction of precious metals such as Pt and Pd and the use of dioctyl sulphide and Kelex-100 as extractants has been reported by Freiser and Al-Bazi[34] in the extraction of $PdCl_3^-$ into an organic phase. The Secco group in Pisa has studied in some detail the reaction of Pd in the form of $PdCl_3^-$ with PADA in the presence of cationic micelles of dodecyltrimethylammonium chloride.[35] The reaction works well, but to strip the metal ion from the complex it is necessary to reduce the pH to less than 1. Mentasti and co-workers[36] have studied the reaction of gallium and indium with 4-2–pyridylazoresorcinol (PAR) in water and mixed solvents, both in acidic solutions where the pH is less than 3. The dissociation reaction of the metal/PAR complex with hydrogen ions was also investigated, and a quadratic dependence of the rate constant on the hydrogen ion concentration was observed. Interestingly, with these two metals, the ligand should contain a hydroxy binding site, so PADA cannot be used. Recently Secco and coworkers[35] have also studied this reaction in SDS micellar solutions. Catalysis is indeed observed, and again the metal ion (in the case of In^{3+}) can be recovered in strongly acid solution.

At the commercial level, it is essential to use a cheap extractant in order to minimize operating costs. In addition it is probably advisable to work at reasonably high concentrations; however, for toxic (or valuable) metals, it is likely that metal concentrations will be present at only the micromolar (or submicromolar) levels.

V. CATALYSIS OF ORGANIC REACTIONS BY MICELLES

A significant number of studies have been carried out by the groups of Romsted and Bunton,[37–39] who have progressively developed the pseudo-phase ion-exchange (PSIE) model to interpret their data, although some very early work in this field was done by the Menger group.[40,41] They investigated ester hydrolysis in anionic micelles, where inhibition is observed as the reactants are separated, and in cationic micelles where the reaction is promoted. Many studies have been carried out using OH⁻ and cationic micelles, since both OH⁻ and the organic substrate will be preferentially located in the micelle environment.

The main assumptions of the PSIE model are as follows (adapted from V.C. Reinsborough, Reference 42):

- The micelles are a separate phase uniformly distributed through the solution and invariant in composition with micelle concentration.
- The degree of counterion ionization is constant and independent of ion type and surfactant concentration C.
- The micelle surface region can be treated as analogous to an ion-exchange resin.

The model has been tested over many years, and the approach has been reasonably successful. The basic scheme is shown in Figure 10.4. The overall first-order rate constant (adapted from Reference 41) is given by

$$k_{obs} = \frac{k_w + k_m K_S \left(C_T - cmc \right)}{1 + K_S \left(C_T - cmc \right)} \quad (10.1)$$

This equation can be rearranged to give an equation from which k_m and k_s can be evaluated, provided k_w is known:

$$\left(k_{obs} - k_w \right)^{-1} = \left(k_m - k_w \right)^{-1} + \left(k_m - k_w \right) K_S \left(C_T - cmc \right)^{-1} \quad (10.2)$$

This area is not being reviewed in detail as there has recently been an excellent review of this area.[43]

A comparison of the Berezin and PSIE approaches has recently been carried out.[44] However, there are a number of cases where the PSIE model is inappropriately applied.[45]

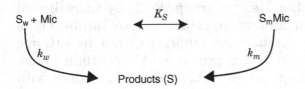

Figure 10.4 Basic scheme for micellar catalysis.

VI. REACTIONS IN MICROEMULSIONS

A. Reminder on Microemulsions

Reactions have been studied in microemulsion media since the
1970s. Some of the earliest work involved organic reactions in
reverse micelles (surfactant-in-hydrocarbon systems in the
absence of water), and Fendler and coworkers observed dramatic rate enhancements for some organic reactions in comparison with the rate in the surfactant–free system (see, for
instance, References 46 and 47). However, microemulsions contain some water, and the most used model system is that of
sodium bis(2-ethylhexyl)sulfosuccinate (Na-AOT)/H_2O/Alkane.
A thermodynamically stable system is readily formed on mixing the three components and shaking for a few seconds. The
microemulsion is a discrete dispersion of nm-sized droplets in
which a water core, containing the counterion to the sulfosuc-
cinate head group, is coated by a curved monolayer of surfactant. The size of the droplets is essentially monodisperse, and
the water droplet core radius is linearly related to the water
to surfactant mole ratio w_0. Core radii are in the region 0–10
nm for w_0 values from 0 to 50. Droplet sizes can be structurally
characterized using a wide range of methods, including small-
angle neutron scattering, photon correlation spectroscopy, time-
resolved fluorescence spectroscopy, and fluorescence correlation
spectroscopy.[48–55] It is possible to substitute water by glycerol
and structural studies of these systems by both SANS and PCS
have been carried out.[56,57] There is again a good linear correlation between the glycerol core radius and the value of the
molar concentration ratio [glycerol]/[surf]. The mobility of AOT
in glycerol and water droplets was studied by quasi-elastic
neutron scattering, and it was shown that changing the core
from water to glycerol only reduced the mobility of AOT by a
factor of 2. Addition of the enzyme alpha chymotrypsin suggested that 250 or so water molecules associated with the
enzyme had a dramatically reduced mobility.[58]

B. Metal-Ligand Complex Formation

Some of the earliest work on metal ligand complex formation
reactions involved the reaction of Ni^{2+}(aq) with the water-

soluble complexing agent ammonium purpurate (murexide) in water droplets with a $w_0 < 20$.[59,60] It was shown that the kinetics in the water droplet system proceeded at a comparable rate to that in bulk water; only a small kinetic effect on the rate constant for loss of water from the metal ion was observed, which is rate-controlling in the overall complex formation process. The initial exchange of reactants between separate water droplets was found to be a much more rapid process, taking place on the millisecond-microsecond time scale. (This is similar to that for surfactant exchange between micelles in aqueous systems). Later work on Na-AOT systems[61] showed that exchange of solubilizates (reactants) between droplets could be expressed by a second-order rate constant, with values in the range 10^6–10^8 $dm^3 mol^{-1} s^{-1}$ being determined. Similar results were found by Pileni et al. based on a stopped-flow and small-angle scattering study.[62] Systematic studies of the exchange of fluorescent probes and of fluorescence quenchers between droplets have been performed.[63–70] The values of the exchange rate constants were also found to be of the order of 10^6–10^8 $dm^3 mol^{-1} s^{-1}$ and to depend on temperature, the water/surfactant molar ratio w_0, nature and length of the oil (alkanes, aromatic solvents), and the nature and chain length of the surfactant and cosurfactant (see Chapter 5, Section VI.F). This clearly demonstrates that the exchange is an intrinsic dynamic property of the droplet interfacial layer.

A number of reactions that are close to diffusion-controlled in the water pools were investigated, including proton transfer, electron transfer, and complexation between $Zn^{2+}(aq)$ and murexide. The rate-limiting solubilizate exchange process was associated with a considerable enthalpy of activation and also a large positive entropy of activation. Data are shown in Table 10.1.

The effect of added cholesterol and benzyl alcohol to Na-AOT was also investigated.[61] Addition of 2×10^{-2} mol dm^{-3} cholesterol decreased the exchange rate constant by a factor of 3, whereas addition of 10^{-1} mol dm^{-3} benzyl alcohol increased the exchange rate constant by the same factor.

TABLE 10.1 $\Delta H^{\ddagger}$ (kJ mol^{-1}) and $\Delta S^{\ddagger}$ (JK^{-1} mol^{-1}) for the Solubilizate Exchange Process in Na-AOT/Water/n-heptane Microemulsions

w_0	Proton Transfer		Electron Transfer	
	$\Delta H^{\ddagger}$	$\Delta S^{\ddagger}$	$\Delta H^{\ddagger}$	$\Delta S^{\ddagger}$
10	70	93	67	116
15	83	140		
20	95	180	87	180
30			108	250

Data from Reference 61.

TABLE 10.2 Kinetic Parameters for the Ni^{2+}/PAP Reaction at 25°C

Medium	k_f (s^{-1})	k_b (s^{-1})	$\Delta H^{\ddagger}_f$ (kJ mol^{-1})	$\Delta H^{\ddagger}_b$ (kJ mol^{-1})
Water/SDS micelles	2000	1.0	48	91
Heptane/water/Na-AOT	3600	1.2	50	93

Data taken from Reference 71.

The kinetics and mechanism of a range of slow metal-ligand substitution processes have been investigated, and a generalized theory was proposed that was very similar to the Berezin model for the interpretation of micelle kinetics in aqueous solutions.[71,72] Ligands of different hydrophobicity were studied in terms of their reaction with Ni^{2+}(aq), and it was found that for pyridine-2-azo-p-phenol (PAP), the rate constants k_f for reaction in SDS micellar solution and water/Na-AOT/heptane systems were comparable, as shown in Table 10.2. (k_f is expressed as a first-order rate constant in the microemulsion.)

For the more hydrophobic ligand PADA, there will be partitioning of the ligand into the bulk heptane phase. This can be expressed in terms of an equilibrium constant

$$K_S^{\text{PADA}} = \frac{[\text{PADA}]_S}{[\text{PADA}]_B [\text{AOT}]} \tag{10.3}$$

When this partitioning is allowed for, there is again excellent agreement between the Ni/PAP and Ni/PADA reactions in Na-AOT microemulsions. Other systems were studied, including reactions of Co^{2+}(aq) and Zn^{2+}(aq), and all the results were consistent with the general model when ligand partitioning was incorporated in the model. Similar reactions have been studied in glycerol-in-heptane microemulsions.[73]

C. Preparation of Nanoparticles

A related field of interest has been the preparation of nanoparticles in microemulsions. A pioneer in this field has been Marie-Paule Pileni, based at Université P et M Curie, Paris. She has studied the synthesis of metals, e.g., Au and Cu, semiconductors (CdS, CdTe) and salts (Ag_2S). She has also made microemulsion dispersions in which Na^+ is replaced by, e.g., Cd^{2+} or Cu^{2+}, and shown that a range of interesting structures can be formed.[74–80] She has recently reviewed her work.[81] Of particular interest is the preparation of nano-sized particles of semiconductors. Early characterization work was carried out by Henglein[82] and Brus.[83] Other significant work on CdS nanoparticles was carried out by the group of Kerry Thomas at Notre Dame University, Indiana.[84] For CdS particle formation in Na-AOT microemulsions, a good correlation was found between the size of the particles and the dimensions of the parent microemulsion.[85] It was shown that the kinetics of particle growth of CdS nanoparticles were consistent with nucleation and growth, with interdroplet exchange of Cd and sulfide solubilizates being rate limiting. This interpretation is similar to the model originally proposed by Smoluchowski for diffusion-controlled reactivity in a homogeneous medium.[86] Hirai at Osaka University has variously studied the formation of titanium dioxide nanoparticles,[87] lead sulfide,[88] and CdS/ZnS composites[89] in AOT reversed micelle systems. Watzke and Fendler have also reported the formation of nano-sized CdS particles in dioctadecyldimethylammonium chloride vesicles in water.[90]

For formation of platinum nanoparticles, the correlation between particle size and droplet size is much weaker.[91–93]

Silver bromide nanoparticles,[94] magnetic colloids,[95,96] and superconducting particles[97] have all been synthesized. Magnetic nanoparticles have also been synthesized in a catanionic vesicle system.[98] The Mann group, originally based in Bath, U.K., also investigated the formation of iron oxide nanoparticles within unilamellar phosphatidylcholine vesicles. The particles were spherical or disk-shaped and had dimensions in the range 1.5–12 nm. Such studies are relevant to biological mineralization.[99] Another system of interest has been the synthesis of nanoparticles of salts such as $BaSO_4$, $BaCO_3$, and $CaSO_4$ in AOT-stabilized microemulsions.[100–102] $BaSO_4$ particle synthesis in cyclohexane microemulsions stabilized by the nonionic dispersant $C_{12}E_4$ produced monodisperse nanospheres of 4–5 nm in radius, whereas $BaSO_4$ prepared in Na-AOT-stabilized microemulsions led to the formation of slightly irregular aggregates 8–50 nm in diameter. $CaSO_4$ formation in microemulsions stabilized by Na-AOT in dodecane produced only 25 nm diameter nano-spheres, whose size was essentially independent of microemulsion composition and reaction conditions. A wide range of particle morphologies are observed in different microemulsions, but the mechanisms involved in their formation are still not understood.

D. Separation Processes and Extraction

Another important area of research is concerned with separation processes and extraction in microemulsions, and a number of systems are discussed by Tondre in this book (see Chapter 5, Section VIII). Of particular interest is the use of Winsor I and Winsor II systems as liquid membranes in extraction processes. The partitioning behavior of simple solutes dimidium ion, p-nitroaniline, and murexide between water and a water-in-oil microemulsion has been studied,[103] and in a follow-up paper, the partitioning of α-chymotrypsin, pepsin, and lysozyme was investigated.[104] High values of the partition coefficient were found when positively charged proteins interacted electrostatically with the negatively charged surfactant head group.

VII. THE MATRIX EFFECT

This concept has been developed by Luisi and coworkers in a series of papers over the last few years, which have been particularly concerned with autopoiesis, as originally defined by Varela and Maturana.[105] The proposition is that vesicles can be considered as possible precursors of protocells.[106,107] Luisi has a continuing interest in molecular evolution, autopoiesis, and the origin of life, and the matrix effect contributes to that debate. An autopoietic unit is a structure capable of self-maintenance, and this includes the option of self-replication.[108] The work has been developed in a series of recent papers.[109–112]

A general discussion of the role of lipid vesicles as key intermediates in the origin of life has also been reported,[113] and there have been a few reports of enzyme–catalyzed reactions in vesicles,[114] although few new insights have emerged and this situation has not significantly changed since the article was written. Previous studies on self-replicating reverse micelles have also been carried out.[115,116] Vesicles made by phospholipids or fatty acids can be extruded to give close to a monodisperse vesicle assembly (e.g., at 50 or 100 nm diameter). When further fatty acid is added to this extruded assembly, the vesicle size of the combined system is very close to that of the original system, but the number of vesicles is increased. The rate at which additional vesicles are formed is much faster than the control system in which fatty acid is simply added to a buffer solution at the same pH, corresponding to a micelle-vesicle transition. In the control process, the vesicles formed are very polydisperse, with sizes ranging from 20 to 1000 nm. Thus the matrix effect is concerned with both a dramatic rate enhancement effect, and essentially a replication of the original vesicle structure. In fact, it is possible to add the additional fatty acid in several stages and still observe the matrix effect. The role of fatty acids is of particular interest since they have been identified in the Murchison meteorite[117] and so may be considered as prebiotic species. Further work carried out recently[118] has shown that the matrix effect is still in evidence even at fatty acid to phospholipid composition ratios of up to 100:1. There is now a need for a better developed theoretical basis to interpret these experimental data.

VIII. POLYMERIZATION IN VESICLE SYSTEMS

It is interesting to consider the possibility of modifying the bilayer vesicle structure to improve its stability/penetration properties. Traditionally, one way to achieve this is to attempt to polymerize the bilayer or to add a reactive monomer which is then polymerized. Early ideas were discussed in the book by Janos Fendler,[119] and the work was developed further by the groups of Ringsdorf[120,121] and O'Brien.[122,123] Pioneering polymerization experiments were carried out by J.K. Thomas[124] and more recently by Hotz and Meier.[125–127] In a useful review, Hotz and Meier[128] have covered both polymerization of reactive lipids and polymerization by adding reactive monomers. In reactive lipids the polymerizable group can be located in the head group adjacent to the surface, or at a defined location within the tail, including the tail end. The authors conclude that the polymerization approach is promising for making nano-capsules for drug delivery and gene therapy as well as for making nano-reactors for the controlled crystallization of inorganic nanoparticles and inorganic-organic composite materials. In an extension of this approach, Pouligny and coworkers[129] have manipulated giant vesicles and latex spheres, studying adhesion of spheres to the vesicle, dragging vesicles by manipulation of the spheres, and inserting and expelling the latex spheres in and out of the vesicles. Polymerization over the bilayer has recently been claimed.[130–132] Kaler and coworkers[130,131] used two monomer species, styrene and divinylbenzene, in a catanionic vesicle system, and this approach looks to be promising for the future. However, during polymerization there is another process that can occur. Phase separation of the polymer within the vesicle can take place and so-called parachute structures are formed on each vesicle.[133,134] The kinetics of the polymerization process has recently been studied in some detail for d-styrene polymerization in didocyldimethylammonium bromide (DODAB) vesicles using online small-angle neutron scattering

(SANS) in real time.[135,136] The process was complete in about 1 h and the final product was an oblate polymer ellipsoid of 20–30 nm in each vesicle.

Marie-Pierre Krafft at the Institut Charles Sadron, Strasbourg, has studied the polymerization of hydrophobic monomers such as isodecylacrylate (ISODAC) in small unilamellar vesicles of perfluoroalkylated phosphatidylcholine (F-PC). The process was compared with that for polymerization of ISODAC in egg PC vesicles. In the case of F-PC vesicles the formed polymer was homogeneously distributed throughout the vesicles whereas for egg-PC, parachute polymerization was found to be operative.[137] Ferro and Krafft[138] have also studied the effect of addition of semi-fluorinated alkanes ($C_mH_{2m+1}C_nF_{2n+1}$, H_mF_n) such as $H_{10}F_6$ on Ca^{2+}-induced fusion and the rate of release of 5,6-carboxyfluorescein referred to as (CF) from the water cores of the vesicles. For $H_{10}F_6$, the rate of fusion was reduced by an order of magnitude, and there was a 40-fold decrease in the rate of release of CF, as compared with phosphatidylserine vesicles as a reference.[138] It will be interesting to see what further changes can be induced in these two processes following polymerization of the bilayer.

IX. ENZYME PROCESSES

"Enzymes can function in organic media." This claim was made by Alexander Klibanov in the 1970s, and he and others demonstrated that this was possible for a number of enzymes in hydrocarbon solvents.[139–144] This work provided the foundation for research by a number of groups who were interested in catalyzing reactions in the opposite direction to that in aqueous solution, so, e.g., condensation would be favored over hydrolysis. Reactions of hydrolases and lipases were thus of particular interest. There was also an interest in regio- and stereo-selective synthesis. Extending the work from organic solvents to a water-in-oil microemulsion system was a natural progression, and indeed this proved to be an interesting new area of research for chemists. However, it was not always clear as to where precisely the enzyme was located. For water-soluble enzymes like α-chymotrypsin (CT), a location in the water core of the droplet was preferred, whereas lipases might be expected to partition to the interface region where the surfactant is located. However, there have been uncertainties for many years over the precise enzyme location. The recent use of fluorescence correlation spectroscopy has suggested that the discrepancies can be resolved when the enzyme is fluorescently tagged and droplets containing tagged α-chymotrypsin were consistently larger than the parent microemulsion for a given water/surfactant ratio.[55] Useful reviews have been published by Luisi.[145,146] A summary of some of the early kinetic data is given in Table 10.3, taken from the review by Luisi, Giomini, Pileni, and Robinson.[147]

In addition to studies in microemulsions, it is of interest to consider the possibility of operating with a more rigid gel system. The preparation of gels based on incorporation of gelatin into a Na-AOT stabilized water-in-oil microemulsion was first reported in 1986.[148,149] Later work[150,151] suggested that an adequate structural description of the system can be based on a rigid network of gelatin-water rods stabilized by Na-AOT in the oil, which coexist with small microemulsion droplets dispersed through the structure. However, the exact type of structure is quite sensitive to the composition of the system. A later paper[152] provided further information on the phase stability of the gels, and it was shown that the frequency dependence of the elastic moduli of the organo-gels was similar to that of aqueous gelatin gels, and characteristic of a viscoelastic solid.

Reverse-enzyme synthesis in microemulsion-based organo-gels has been reported.[153,154] Chromobacterium Viscosum (CV) lipase was used as the catalyst, and ester synthesis was reported for a wide range of primary and secondary alcohols with n-decanoic acid. Some typical data are given in Table 10.4. With primary alcohols the reaction was quite fast and the conversion to the ester was close to 100%. For the secondary alcohol octan-2-ol reaction proceeded to 45%, but some days were needed to reach equilibrium. Selective esterification of the (-) enantiomer was found to occur with enantiomeric excess values of > 90%. No reaction was observed with the tertiary

TABLE 10.3 Kinetic Parameters for Enzymes in AOT Hydrocarbon Reverse Micelles

Enzyme	Aqueous solution		Reverse micelles	
	k_{cat} (s^{-1})	K_m (mM)	k_{cat} (s^{-1})	$K_m$ov (mM)
α-Chymotripsin	0.83	0.6	5.06	0.4
Trypsin	8.7	4.10^{-3}	11.0	$3.5.10^{-3}$
Lipoxygenase	9	$2.7.10^{-2}$	4.50	20.10^{-2}
Lysozyme [a]	3.10^{-4}	4.10^{-2}	4.10^{-4}	5.10^{-2}
Lysozyme [b]	0.14	10.10^{-3}	0.130	2.10^{-3}
LADH	114	0.37	80	0.3
Ribonuclease	0.165	—	1.230	—
Peroxidase	1.100	—	19.100	—

The kinetic values of the micellar systems are compared under optimal pH conditions for reaction, i.e., generally at the same pH values for an aqueous solution.
[a] Using 2,4-dinitrophenol-NAG-3 as substrate (NAG-3 is *N*-acetyl glucosamine).
[b] Using chitin as substrate.
Source: Reference 147

alcohol. Regio-selectivity was also demonstrated, and the gel was found to be effective in sectioned form at temperatures as low as –20°C. The enzyme-containing gel was stable over many months, insofar as the rate of the esterification reaction between 1'-octanol and decanoic acid was unaffected. Reviews of the field have been published[155,156] in which stereo-selective synthesis involving resolution of chiral alcohols was achieved. In a later paper,[157] Na-AOT stabilized water-in-oil microemulsions containing CV lipase were dehydrated by addition of either 4 Å molecular sieve powder or by vapor phase drying with a saturated solution of LiCl. When the water is removed ($w_o < 1$), the enzyme is in a dormant state in the microemulsion. On rehydration after one week the full activity of the enzyme containing microemulsion is immediately restored in terms of a model esterification. After one month a small 20% retardation of the esterification rate is observed, but the extent of conversion is not affected.

TABLE 10.4   CV Lipase-Catalyzed Synthesis from a Variety of Different Alcohols and Decanoic Acid

Alcohol	Product	Incubation time (days)	Recovered Yield (%)
CH_3CH_2OH	$CH_3CH_2O\text{-}\overset{\displaystyle O}{\overset{\|}{C}}\text{-}C_9H_{19}$	1	93
$CH_3(CH_2)_6CH_2OH$	$CH_3(CH_2)_6CH_2O\text{-}\overset{\displaystyle O}{\overset{\|}{C}}\text{-}C_9H_{19}$	1	92
$CH_3(CH_2)_{12}CH_2OH$	$CH_3(CH_2)_{12}CH_2O\text{-}\overset{\displaystyle O}{\overset{\|}{C}}\text{-}C_9H_{19}$	1	93
[structure with OH]	[structure with $O\text{-}\overset{\|}{C}\text{-}C_9H_{19}$, O]	5	45
[structure with OH]	[structure with $O\text{-}\overset{\|}{C}\text{-}C_9H_{19}$, O]	5	0

[a] Batch-type syntheses using 10 mL of sectioned microemulsion-based gel in 30 mL of *n*-heptane containing 0.10 mol each of acid and alcohol. [CV lipase] = 250 μg mL^{-1} of gel. Incubation temperature, 25°C.
Source: Reference 153

X.   THE FUTURE

Reaction kinetics has been studied in micelles and microemulsions for the last 20–30 years and we might ask what has been achieved in terms of applications. Some of these are indicated in the review by Reinsborough,[42] but generally these systems have not to date been exploited as effectively as they should have been by industry. Metal extraction by aqueous micelles and in microemulsions has not been used to a significant extent; industry has preferred to use tried and tested methods that are often based on work carried out 50 years ago. The same is true of reactions in enzyme systems. The authors are not aware of significant take-up of new approaches in reversed enzyme synthesis (except for lipase catalysis in polymer synthesis), despite the large amount of data that is now available. This is not good news; let us hope there will be a more profitable interaction between industry and academia in the future.

REFERENCES

1. Berezin, I.V., Martinek, K., Yatsimirski, A.K. *Russ. Chem. Rev. (Eng Trans)* 1973, 42, 787.

2. Martinek, K. *Dokl Akad Nauk USSR* 1970, 194, 840.

3. Martinek, K., Yatsimirski, A.K., Levashov, A.V., Berezin, I.V. In *Micellization, Solubilization and Microemulsions*, Mittal, K.L., Ed., Plenum Press: New York, 1977, vol. 2, p. 489.

4. Fendler, J.H., Fendler, E.J. *Catalysis in Micellar and Macromolecular Systems*, Academic Press, New York, 1975.

5. Romsted, L.S., Bunton, C.A., Yao, Y. *Curr. Opin. Colloid Interface Sci.* 1997, 2, 622.

6. Aniansson, E.A.G. In *Aggregation Processes in Solution*, Wyn-Jones, E., Gormally, J., Eds., Elsevier, Amsterdam, 1983, p. 70.

7. Aniansson, E.A.G., Wall, S.N. *J. Phys. Chem.* 1975, 79, 857.

8. Aniansson, E.A.G., Wall, S.N., Almgren, M., Hoffmann, H., Kielman, I., Ulbricht, W., Zana, R., Lang, J., Tondre, C. *J. Phys. Chem.* 1976, 80, 905.

9. Blokzijl, W., Engberts, J.B.F.N. *Angew. Chem. Int. Ed.* 1993, 32, 1545.

10. James, A.D., Robinson, B.H. *J. Chem. Soc. Faraday Trans.* 1978, 74, 10.

11. Monteleone, G., Morroni, L., Robinson, B.H., Tine, M.R., Venturini, M., Secco, F. *Colloids Surf. A*, 2004, 243, 23.

12. Holzwarth, J., Knoche, W., Robinson, B.H. *Ber. Bunsenges. Phys. Chem.* 1978, 82, 1001.

13. Reinsborough, V.C., Robinson, B.H. *J. Chem. Soc. Faraday Trans. 1* 1979, 75, 2395.

14. Drennan, C.E., Hughes, R.J., Reinsborough, V.C., Sorijan, O.O. *Can. J. Chem.* 1998, 76, 152.

15. Berberich, K.A., Reinsborough, V.C., Shaw, C.N. *J. Solution Chem.* 2000, 29, 1017.

16. Cai, R., Freiser, H., Muralidharan, S. *Langmuir*, 1995, 11, 2926.

17. Ritcey, G.M., Ashbrook, A.W. *Solvent Extraction: Principles and Applications to Process Metallurgy*, Elsevier, Amsterdam, 1984.

18. Szymanowski, J., Tondre, C. In *Solvent Extraction and Ion Exchange*, Ed., M. Dekker Inc., New York, 1994, p. 873.

19. Tondre, C. In *Surfactant-Based Separations*, ACS Symposium Series 740, American Chemical Society, Washington DC, 2000, p. 139.

20. Albery, W.J., Choudhery, R.A., Fisk, P.R. *Faraday Disc. Chem. Soc.* 1984, 77, 53.

21. Albery, W.J., Fisk, P.R. *Hydrometallurgy* 1981, 5, 1.

22. Stevens, G.W., Perera, J.M., Grieser, F. *Curr. Opin. Colloid Interface Sci.* 1997, 2, 629.

23. Ismael, M., Tondre, C. *J. Colloid Interface Sci.* 1993, 160, 252.

24. Scamehorn, J.F., Christian, S.D., Ellington, R.T. In *Surfactant-Based Separation Processes*, Scamehorn, J.F., Harwell, J.H., Eds., Marcel Dekker, New York, 1989, p. 33.

25. Dunn, R.O., Scamehorn, J.F., Christian, S.D. *Colloids Surf.* 1989, 35, 49.

26. Boumezioud, M., Kim H.S., Tondre, C. *Colloids Surf.* 1989, 41, 255.

27. Son, S.G., Hebrant. M., Tecilla, P., Scrimin, P., Tondre, C. *J. Phys. Chem.* 1992, 96, 11072.

28. Tondre, C., Claude-Montigny, B., Ismael, M., Scrimin, P., Tecilla, P. *Polyhedron* 1991, 10, 1791.

29. Son, S.G., Hebrant. M., Tecilla, P., Scrimin, P., Tondre, C. *J. Phys. Chem.* 1992, 96, 11079.

30. Richmond, W., Tondre, C., Krzyzanowska, E., Szymanowski, J. *J. Chem. Soc., Faraday Trans. 1* 1995, 91, 657.

31. McCulloch, J.K., Perera, J.M., Kelly, E.D., White, L.R., Stevens, G.W., Grieser, F. *J. Colloid Interface Sci.* 1996, 184, 406.

32. Ashrafizadeh, S.N., Demopoulos, G.P. *J. Colloid Interface Sci.* 1995, 173, 448.

33. Lee, S.K., Choi, H.S. *Bull. Korean Chem. Soc.* 2001, 22, 463.

34. Al-Bazi, S.J., Freiser, H. *Inorg. Chem.* 1987, 28, 417.

35. Biver, T., Ghezzi, L., Monteleone, G., Robinson, B.H.,Secco, F., Tine, M.R., Venturini, M., to be submitted to *Langmuir.*

36. Mentasti, E., Baiocchi, C., Kirschenbaum, L. J. *J. Chem. Soc. Dalton Trans.* 1985, 2615.

37. Bunton, C.A., Nome, F., Romsted, L. S., Quina, F.H. *Acc. Chem. Res.* 1991, 24, 357.

38. Bunton, C.A. *J. Mol. Liq.* 1997, 72, 231.

39. Bunton, C.A., Savelli, G. *Adv. Phys. Org. Chem.* 1986, 22, 213.

40. Menger, F.M., Portnoy, C.E. *J. Am. Chem. Soc.* 1967, 89, 4698.

41. Menger, F.M. *Pure Appl. Chem.* 1979, 51, 999.

42. Reinsborough, V.C. In *Interfacial Catalysis*, Volkov, A.G., Ed., Marcel Dekker, New York, 2002.

43. Savelli, G., Germani, R., Brinchi, L. In *Reactions and Synthesis in Surfactant Systems*, Texter, J., Ed., M. Dekker, New York, 2001.

44. Bunton, C.A., Yatsimirsky, A.K. *Langmuir* 2000, 16, 5921.

45. Bunton, C.A., Frankson, J., Romsted, L.S. *J. Phys. Chem.* 1980, 84, 2607.

46. Fendler, J.H., Fendler, E.J., Medary, R.T., Woods, V.A. *J. Am. Chem. Soc.* 1972, 95, 3273.

47. O'Connor, C.J., Fendler, J.H., Fendler, E.J. *J. Am. Chem. Soc.* 1974, 96, 370.

48. Cabos, P.C., Delord, P. *J. Appl. Cryst.* 1979, 12, 502.

49. Robinson, B.H, Toprakciglu, C., Dore, J.C. *J. Chem. Soc. Faraday Trans.* 1984, 8, 13.

50. Kotlarchyk, M., Chen, S.H., Huang, J. S., Kim, M.W. *Phys. Rev. A* 1984, 29, 2054.

51. Zulauf, M., Eicke, H.-F. *J. Phys. Chem.* 1979, 83, 480.

52. Lang, J., Jada, A., Malliaris, A. *J. Phys. Chem.* 1988, 92, 1946.

53. Johannson, R., PhD thesis, Univiversity Uppsala, 1993.

54. Zana, R. In *Surfactant Solutions: New Methods of Investigation*, Zana, R., Ed., Marcel Dekker, New York, 1987, p. 241.

55. Burnett, G.R., Rees, G.D., Steytler, D.C., Robinson, B.H. *Colloids Surf. A*, in press.

56. Fletcher, P.D.I., Galal, M.F., Robinson, B.H. *J. Chem. Soc. Faraday Trans.* 1984, 180, 3307.

57. Fletcher, P.D.I., Galal, M.F., Robinson, B.H. *J. Chem. Soc. Faraday Trans.* 1985, 181, 2053.

58. Fletcher, P.D.I., Robinson, B.H., Tabony, J. *J. Chem. Soc. Faraday Trans.* 1986, 182, 2311.

59. Robinson, B.H., James, A.D., Steytler, D.C. In *Protons and Ions Involved in Fast Dynamic Phenomena*, Laszlo, P., Ed., Elsevier, Amsterdam, 1978, p. 287.

60. Robinson, B.H, Steytler, D.C, Tack, R.D. *J. Chem. Soc. Faraday Trans.* 1979, 175, 481.

61. Fletcher, P.D.I., Howe, A.M., Robinson, B.H. *J. Chem. Soc. Faraday Trans.* 1987, 183, 985.

62. Thain, T.K., Cassin, G., Badiali, J.P., Pileni, M.P. *Langmuir* 1996, 12, 2408.

63. Jada, A., Lang, J., Zana, R. *J. Phys. Chem* 1989, 93, 10.

64. Jada, A., Lang, J., Candau, S.J., Zana, R. *Colloids Surf.* 1989, 38, 251.

65. Jada, A., Lang, J., Zana, R. *J. Phys. Chem.* 1990, 94, 381.

66. Jada, A., Lang, J., Zana, R., Makhloufi, R., Hirsch, E., Candau, S.J. *J. Phys. Chem.* 1990, 94, 387.

67. Lang, J., Mascolo, G., Zana, R., Luisi, P.L. *J. Phys. Chem.* 1990, 94, 3069.

68. Zana, R., Lang, J., Canet, D. *J. Phys. Chem.* 1991, 95, 3364.

69. Lang, J., Lalem, N., Zana, R. *J. Phys. Chem.* 1992, 96, 4667.

70. Lang, J., Lalem, N., Zana, R. *Colloids Surf.* 1992, 68, 199.

71. Fletcher, P.D.I., Robinson, B.H. *J. Chem. Soc. Faraday Trans.* 1984, 180, 2417.

72. Fletcher, P.D.I., Howe, A.M., Perrins, N.M., Robinson, B.H., Dore, J.C. In *Surfactants in Solution*, Mittal, K.L., Lindman, B., Eds., Plenum Press, New York, 1984, vol. 3, p. 1745.

73. Atay, N.Z., Robinson, B.H. *Langmuir* 1999, 15, 5056.

74. Pileni, M.P. *J. Phys. Chem.* 1993, 97, 9661.

75. Pileni, M.P., Lisiecki, I. *Colloids Surf. A* 1993, 80, 63.

76. Pileni, M.P., Veillet, P., Lisiecki, I., Petit, C., Duxin, N., Tanori, J. *Colloid Polym. Sci.* 1995, 273, 886.

77. Tanon, J., Pileni, M.P. *Adv. Mat.* 1995, 7, 862.

78. Lisieki, I., Billoudet, F., Pileni, M.P. *J. Phys. Chem.* 1996, 100, 4160.

79. Pileni, M.P. *Langmuir* 1997, 13, 3266.

80. Pileni, M.P. *Langmuir* 2001, 17, 7476.

81. Pileni, M.P. In *Self Assembly*, Robinson, B.H., Ed., IOS, Amsterdam, 2003, p. 25.

82. Henglein, A. *Ber. Bunsenges. Phys. Chem.* 1982, 86, 301.

83. Steigerwald, M., Alivisatos, A., Gibson, J., Harris, T., Kortan, R., Muller, A., Duncan, T., Douglas, D., Brus, L.E. *J. Amer. Chem. Soc.* 1988, 110, 3046.

84. Lianos, P., Thomas, J.K. *Chem. Phys. Lett.* 1986, 125, 299.

85. Towey, T.F., Khan-Lodhi, A.N., Robinson, B.H. *J. Chem. Soc. Faraday Trans.* 1990, 186, 3757.

86. Smoluchowski, M. *Z. Phys. Chem.* 1918, 92, 129.

87. Hirai, T., Sato, H., Komasawa, I. *IEC Res.* 1993, 32, 3014.

88. Hirai, T., Tsubaki, H., Sato, H., Komasawa, I. *J. Chem. Eng. Jap.* 1995, 28, 468.

89. Sato, H., Hirai, T., Komasawa, I. *IEC Res.* 1995, 34, 2493.

90. Watzke, H.J., Fendler, J.H. *J. Phys. Chem.* 1987, 91, 854–861.

91. Boutonnet, M., Kizling, J., Stenius, P., Maire, G. *Colloids Surf.* 1982, 5, 209.

92. Boutonnet, M., Kizling, J., Stenius, P., Maire, G. *J. Appl. Catal.* 1986, 20, 163.

93. Clint, J.H., Collins, I.R., Williams, J.A., Robinson, B.H., Towey, T.F., Cajean, P., Khan-Lodhi, A. *Faraday Disc.* 1993, 95, 219.

94. Chew, C.H., Gan, L.M., Shah, D.O. *J. Disp. Sci. Technol.* 1990, 11, 593.

95. Rivas, J., Lopez-Quintela, M.A. *J. Colloid Interface Sci.* 1993, 158, 446.

96. Lopez-Quintella, M.A., Rivas, J. *Curr. Opin. Colloid Interface Sci.* 1996, 1, 806.

97. Kumar, P., Pillai, V., Bates, S.R., Shah, D.O. *Mat. Lett.* 1993, 16, 68.

98. Jacobi, L.J., Nunes, A.C., Bose, A. *J. Colloid Interface Sci.* 1995, 171, 73.

99. Mann, S., Hannington, J.P., Williams, R.J.P. *Nature* 1986, 324, 565.

100. Hopwood, J.D., Mann, S. *Chem. Mat.* 1997, 8, 1819.

101. Qi, L., Ma, J., Cheng, H., Zhao, Z. *J. Phys. Chem.* 1997, 101, 3460.

102. Rees, G.D., Evans-Gowing, R., Hammond, S.J., Robinson, B.H. *Langmuir* 1999, 15, 1993.

103. Fletcher, P.D.I. *J. Chem. Soc. Faraday Trans.* 1986, 182, 2651.

104. Fletcher, P.D.I., Parrott, D, *J. Chem. Soc., Faraday Trans.1*, 1988, 84, 1131.

105. Varela, F., Maturana, H., Uribe, R. *Biosystems* 1974, 5, 187.

106. Morowitz, H.J. *Beginnings of Cellular Life*, Yale University Press, New Haven, CT, 1992.

107. Deamer, D.W. *Origins of Life* 1986, 17, 3.

108. Mavelli, F., Luisi, P.L. *J. Phys. Chem.* 1996, 100, 16600.

109. Berclaz, N., Blochlinger, E., Muller, M., Luisi, P.L. *J. Phys. Chem. B* 2001, 105, 1065.

110. Berclaz, N., Muller, M., Walde, P., Luisi, P.L. *J. Phys. Chem. B* 2001, 105, 1056.

111. Lonchin, S., Luisi, P.L., Walde, P., Robinson, B.H. *J. Phys. Chem. B* 1999, 103, 10910.

112. Blochliger, E., Blocher, M., Walde, P., Luisi, P.L. *J. Phys. Chem. B.* 1998, 102, 10383.

113. Luisi, P.L., Walde, P., Oberholzer, T. *Curr. Opin. Colloid Interface Sci.* 1999, 4, 33.

114. Walde, P. *Curr. Opin. Colloid Interface Sci.* 1996, 1, 638.

115. Bachmann, P.A., Walde, P., Luisi, P.L., Lang, J. *J. Am. Chem. Soc. 1990,* 112, 8200.

116. Bachmann, P.A., Luisi, P.L. Lang, J. *Nature* 1992, 357, 57.

117. Deamer, D.W., Pashley, R.M. *Origin Life Evol. Biosphere* 1989, 19, 21.

118. Rogerson, M.L. *Interaction of Phospholipids with Fatty Acids.* PhD diss., 2003, University of East Anglia, Norwich, U.K.

119. Fendler, J.H. *Membrane Mimetic Chemistry*, Wiley, New York, 1983.

120. Ringsdorf, H., Scwarb, B., Venzmer, J. *Angew. Chem.* 1988, 100, 117.

121. Bader, H., Edorn, K., Hupfer, B., Ringsdorf, H. *Adv. Polym. Sci.* 1985, 64, 1.

122. O'Brien, D.F. *Trends Polym. Sci.* 1994, 2, 183.

123. Armitage, B.A., Bennett, D.E., Lamparski, D.F., O'Brien, D.F. *Adv. Polym. Sci.* 1996, 126, 53.

124. Murtagh, J., Thomas, J.K. *Faraday Disc. Chem. Soc.* 1986, 81, 127.

125. Hotz, J., Meier, W. *Adv. Mat.* 1998, 10, 1387.

126. Hotz. J., Meier, W. *Langmuir* 1998, 14, 1031.

127. Meier, W. *Curr. Opin. Colloid Interface Sci.* 1999, 4, 6.

128. Hotz, J., Meier, W. In *Reactions and Synthesis in Surfactant Systems*, Texter, J., Ed., M. Dekker, New York, 2001, p. 501.

129. Dietrich, C., Angelova, M., Pouligny, B. *J. Phys. II France* 1997, 7, 1903.

130. Morgan, J.D., Johnson, C.A., Kaler, E.W. *Langmuir* 1997, 13, 6447.

131. McKelvey, C.A., Kaler, E.W., Zasadzinski, J.A., Coldren, B., Jung, H.T. *Langmuir* 2000, 16, 8285.

132. Poulain, N., Nakache, E., Pina, A., Levesque, G. *J. Polym. Sci. A Polym. Chem.* 1996, 34, 729.

133. Jung, M., Hubert, D.H.W., Bomans, P.H.H. Frederik, P.M., Meuldijk, J., German, A.L. *Langmuir* 1999, 15, 8849.

134. Jung, M., Hubert, D.H.W., van Veldhoven, E., Frederik, P.M., van Herk, A.M., Fischer, H., German A.L. *Langmuir* 2000, 16, 3165.

135. Jung, M., Robinson, B.H., Steytler, D.C., German, A.L., Heenan, R.K. *Langmuir* 2002, 18, 2873.

136. Robinson, B.H., Steytler, D.C., Heenan, R.K., Jung, M. In *Self-Assembly*, Robinson, B.H., Ed., IOS Press, Amsterdam, 2003, p. 196.

137. Krafft, M.P., Schieldknecht, L., Marie, P., Giulieri, F., Schmutz, M., Poulain, N., Nakache, E. *Langmuir* 2001, 17, 2872.

138. Ferro, Y., Krafft, M.P. *Biochim. Biophys. Acta* 2002, 1581, 11.

139. Klibanov, A.M. *Science* 1983, 219, 722.

140. Zaks, A., Klibanov, A.M. *Science* 1984, 224, 1249.

141. Zaks, A., Klibanov, A.M. *J. Am. Chem. Soc.* 1986, 108, 2767.

142. Whitesides, G.M., Wong, C-H. *Angew. Chem. Int. Ed.* 1985, 24, 617.

143. Wong, C.H. *Science* 1989, 244, 1145.

144. Klibanov, A.M. *Acc. Chem. Res.* 1990, 23, 114.

145. Luisi, P.L. *Angew. Chem. Int. Ed,* 1985, 24, 4339.

146. Luisi, P.L., Magid, L.J. *CRC Critical Rev. Biochem.* 1986, 20, 409.

147. Luisi, P.L., Giomini, M., Pileni, M.P., Robinson, B.H. *Biochim. Biophys. Acta* 1988, 947, 759.

148. Hacring, G., Luisi, P.L. *J. Phys. Chem.* 1986, 90, 5892.

149. Quellet, C., Eicke H.-F. *Chimia* 1986, 40, 233.

150. Howe, A.V., Katsikides, A., Robinson, B.H., Chadwick, A.V., Al-Mudaris, A. *Prog. Colloid Polym. Sci.* 1988, 266, 2115.

151. Atkinson, P.J., Grimson, M.J., Heenan, R.K., Howe, A.M., Mackie, A.R., Robinson, B.H. *Chem. Phys. Lett.* 1988, 151, 494.

152. Atkinson, P.J., Robinson, B.H., Howe, A.M., Heenan, R.K. *J. Chem. Soc. Faraday Trans.* 1991, 87, 3389.

153. Rees, G.D., da Graca Nascimento, M., Jenta, T.R., Robinson, B.H. *Biochim. Biophys. Acta* 1991, 1073, 493.

154. Rees, G.D., Jenta, T.R., da Graca Nascimento, M., Catauro, M., Robinson, B.H., Stephenson, G.R., Olphert, R.G.D. *Indian J. Chem.* 1993, 32B, 30.

155. Rees, G.D., Robinson, B.H. *New Scientist* 1991, 43.

156. Rees, G.D., Robinson, B.H. *Adv. Mat.* 1993, 9, 608.

157. Carlile, K., Rees, G.D., Robinson, B.H., Steer, T.D., Svensson, M. *J. Chem. Soc. Faraday Trans.* 1996, 92, 4701.

Index

Printed in the United States
by Baker & Taylor Publisher Services